Global Issues in Water Policy

Volume 14

More information about this series at http://www.springer.com/series/8877

Manuel Lago • Jaroslav Mysiak
Carlos M. Gómez • Gonzalo Delacámara
Alexandros Maziotis
Editors

Use of Economic Instruments in Water Policy

Insights from International Experience

Springer

Editors
Manuel Lago
Ecologic Institut gemeinnützige GmbH
Berlin, Germany

Jaroslav Mysiak
Fondazione Eni Enrico Mattei (FEEM)
Venice, Italy

Carlos M. Gómez
Departamento de Economía
Universidad de Alcalá and IMDEA-Water Foundation
Alcalá de Henares (Madrid), Spain

Gonzalo Delacámara
Madrid Institute for Advanced Studies in Water (IMDEA-Water Foundation)
Alcalá de Henares (Madrid), Spain

Alexandros Maziotis
University of Manchester
Manchester, UK

ISSN 2211-0631 ISSN 2211-0658 (electronic)
Global Issues in Water Policy
ISBN 978-3-319-18286-5 ISBN 978-3-319-18287-2 (eBook)
DOI 10.1007/978-3-319-18287-2

Library of Congress Control Number: 2015945335

Springer Cham Heidelberg New York Dordrecht London

Printed on acid-free paper

Springer International Publishing AG Switzerland is part of Springer Science+Business Media (www.springer.com)

Preface and Acknowledgements

Despite growing interests worldwide, little is known about the actual performance of economic policy instruments (EPIs) in achieving water policy objectives. Fostered by a research grant from the European Commission, this book displays a large body of evidence on the different types, design features and outcomes of water-related economic policy instruments in place and the practice guiding their choice and implementation. Compared to other horizontal reviews of environmental EPIs, this book has an exclusive focus on water uses and services, and the breadth and depth of the analysis is unique from the international perspective. The scope of this review is to explore and identify conditions under which the EPIs perform well in practice and for this purpose; a large number of existing instruments are reviewed and assessed against a common set of assessment criteria. A variety of EPIs presented include selected instruments in place in Cyprus, Denmark, France, Germany, Hungary, Italy, Spain, the UK, Australia, Chile, Israel and the USA.

This book does not advocate for the application of any specific EPI, but sets out the basis for the policymaker (and interested reader) to choose a particular form of EPI in specific circumstances. The book follows three fundamental objectives: (1) to learn more about the practical application of EPIs to specifically achieve water policy objectives, (2) to better understand the policy frameworks under which water-related EPIs are or have been designed and implemented and (3) to advocate the use of economic assessment tools and methods to inform available choices in the development of environmental protection policy at large and, more specifically, decisions regarding the management of water resources. These key objectives can be translated into broad research questions that this book aspires to address: (1) What are the purposes and motives that have led some policymakers around the world to promote the design and implementation of these instruments to achieve specific water policy objectives? (2) How do water EPIs interact and perform as part of complex policy mixes? (3) What is the level of information required and what assessment tools can be applied to impart significance regarding their performance?

The research leading to this book has received funding from the European Union's Seventh Framework Programme (FP7/2007–2013) under grant agreement no. 265213 (EPI-WATER – Evaluating Economic Policy Instruments for Sustainable Water Management in Europe). The EPI-WATER project was carried out by a consortium led by Fondazione Eni Enrico Mattei (FEEM), Italy, and 10 other European institutions: ACTeon, France; Ecologic Institute, Germany; Università di Bologna, Italy; Wageningen University, the Netherlands; National Technical University of Athens, Greece; Instituto Madrileño de Estudios Avanzados – Agua, Spain; University of Valencia, Spain; Middlesex University, Flood Hazard Research Centre, UK; Aarhus Universitet – National Environmental Research Institute, Denmark; and Corvinus University of Budapest, Regional Centre for Energy Policy Research, Hungary. The consortium liaised with overseas experts from Resources for the Future, the Australian University of Sydney, the Australian University of Adelaide, the Hebrew University of Jerusalem, the University of California, the University of Colorado, Kieser & Associates, the University of Richmond, Pontifical Catholic University of Chile and Peking University.

The book is composed of contributions presented at the international conference *Water Management: Review of Empirical Evidence, Experiences and Lessons Learned from Europe and Elsewhere*, held in Berlin from January 26–28, 2012. The authors are grateful for the conference's fruitful discussion, which engaged experts and practitioners, representatives from governments and river basin authorities, EU institutions and non-government and international organizations. We would like to acknowledge the constant support to the project by a panel of experts from AgroParis Tech, the Organization for Economic and Co-operation Development, the Seine-Normandie Water Agency, the Swedish Agency for Marine and Water Management, the Spanish Ministry of Environment, the UK Department for Food and Rural Affairs, World Wide Fund for Nature (WWF), the Committee of Professional Agricultural Organisations and General Committee for Agricultural Cooperation in the European Union, the Romanian Waters National Administration, the EC Directorate-General for the Environment (DG Environment) and the European Environmental Agency.

In addition, we would like to acknowledge the support from Henriette Faergemann and Łukasz Latała (DG ENV, European Commission), Xavier Leflaive (OECD), Tania Runge (COPA-COGECA), Martina Mlinarić (EEB), Lucile De Boisson (WWF), Edi Interwies (InterSus), Bernard Barraqué (CNRS), Thomas Dworak (FreshThoughts), Miguel R. Solanes (IMDEA Agua), Hans-Peter Weikard (WUR), David Zetland, Pedro Andrés Garzón Delvaux and Pierre Strosser (ACTeon).

Last but not least, we would like to thank Carlos Mario Gómez, Gonzalo Delacámara and Alexandros Maziotis for their excellent contributions to editing this book and all lead and contributing authors of this book for their hard work and efforts towards making this publication happen. This book would have not been possible without their constant support. Finally, many thanks to the staff at Springer for their help throughout.

Germany — Manuel Lago
Italy — Jaroslav Mysiak

Contents

Contributors

Tiho Ancev School of Economics, University of Sydney, Sydney, NSW, Australia

Mikael Skou Andersen Department of Environmental Science, Aarhus University, Roskilde, Denmark

Gerardo Anzaldua Ecologic Institute, Berlin, Germany

Tom Ash Western Municipal Water District/Inland Empire Utilities Agency, Riverside, CA, USA

M.A. Samad Azad University of Sydney and Tasmanian School of Business & Economics, University of Tasmania, Sydney, NSW, Australia

Margaretha Breil Fondazione Eni Enrico Mattei (FEEM) and Euro-Mediterranean Centre on Climate Change (CMCC), Venice, Italy

Lorenzo Carrera Fondazione Eni Enrico Mattei (FEEM) and Euro-Mediterranean Centre on Climate Change (CMCC), Venice, Italy

Pierre Defrance AequitO, Saint-Denis, France

ACTeon, Colmar, France

Gonzalo Delacámara Madrid Institute for Advanced Studies in Water (IMDEA-Water Foundation), Parque Científico-Tecnológico de la Universidad de Alcalá, Alcalá de Henares (Madrid), Spain

Ariel Dinar School of Public Policy, University of California Riverside, Riverside, CA, USA

Guillermo Donoso Department of Agricultural Economics, Pontificia Universidad Católica de Chile, Santiago, Chile

Fabio Farinosi Fondazione Eni Enrico Mattei (FEEM) and Euro-Mediterranean Centre on Climate Change (CMCC), Venice, Italy

Carlos M. Gómez Departamento de Economía, Universidad de Alcalá and IMDEA-Water Foundation, Alcalá de Henares (Madrid), Spain

Colin Green Flood Hazard Research Centre, Middlesex University, London, UK

Francesc Hernández-Sancho Department of Applied Economics II, Faculty of Economics, University of Valencia, Valencia, Spain

Charles W. (Chuck) Howe Environment & Society Program, Institute of Behavioral Science (IBS), University of Colorado-Boulder, Boulder, CO, USA

Estefanía Ibáñez Madrid Institute for Advanced Studies in Water (IMDEA-Water Foundation), Parque Científico-Tecnológico de la Universidad de Alcalá, Alcalá de Henares (Madrid), Spain

Iddo Kan Department of Environmental Economics and Management and The Center for Agricultural Economics Research; The Robert H. Smith Faculty of Agriculture, Food and Environment, The Hebrew University of Jerusalem, Rehovot, Israel

Mark S. Kieser Kieser & Associates, LLC, Kalamazoo, MI, USA

András Kis Regional Centre for Energy Policy Research, Corvinus University of Budapest, Budapest, Hungary

Yoav Kislev Department of Environmental Economics and Management and The Center for Agricultural Economics Research; The Robert H. Smith Faculty of Agriculture, Food and Environment, The Hebrew University of Jerusalem, Rehovot, Israel

Maggie Kossida Laboratory of Hydrology & Water Resources Management, Department of Water Resources and Environmental Engineering, School of Civil Engineering, National Technical University of Athens, Athens, Greece

SEVEN-Engineering Consultant, Athens, Greece

Carolyn Kousky Resources for the Future, Washington, DC, USA

Manuel Lago Ecologic Institute, Berlin, Germany

Antonio Massaruto Dipartimento di Sscienze Economiche e Statistiche, Università di Udine, Udine, Italy

Verena Mattheiß ACTeon, Colmar, France

Alexandros Maziotis School of Mechanical, Aerospace and Civil Engineering (MACE), University of Manchester, Manchester, UK

Jamie L. McCarthy Kalamazoo River Watershed Council, Kalamazoo, MI, USA

Simon McCarthy Flood Hazard Research Centre, Middlesex University, London, UK

Katriona McGlade Ecologic Institute, Berlin, Germany

Maria A. Mimikou Laboratory of Hydrology & Water Resources Management, Department of Water Resources and Environmental Engineering, School of Civil Engineering, National Technical University of Athens, Athens, Greece

María Molinos-Senante Departamento de Ingeniería Hidráulica y Ambiental, Pontificia Universidad Católica de Chile, Santiago, Chile

Jennifer Möller-Gulland PricewaterhouseCoopers Private Limited, Gurgaon, Haryana, India

Jaroslav Mysiak Fondazione Eni Enrico Mattei (FEEM) and Euro-Mediterranean Centre on Climate Change (CMCC), Venice, Italy

Helle Ørsted Nielsen Department of Environmental Science, Aarhus University, Rønde, Denmark

Joanna Pardoe Flood Hazard Research Centre, Middlesex University, London, UK

Anders Branth Pedersen Department of Environmental Science, Aarhus University, Roskilde, Denmark

C. Dionisio Pérez-Blanco Centro Euro-Mediterraneo sui Cambiamenti Climatici, Fondazione Eni Enrico Mattei (FEEM), Venice, Italy

Meri Raggi Department of Statistical Sciences, Alma Mater Studiorum University of Bologna, Bologna, Italy

Judit Rákosi ÖKO Co. Ltd, Budapest, Hungary

Marta Rodríguez Madrid Institute for Advanced Studies in Water (IMDEA-Water Foundation), Parque Científico-Tecnológico de la Universidad de Alcalá, Alcalá de Henares (Madrid), Spain

Ramón Sala-Garrido Department of Mathematics for Economics, Faculty of Economics, University of Valencia, Valencia, Spain

Laura Sardonini Department of Agricultural Sciences, Alma Mater Studiorum University of Bologna, Bologna, Italy

Anastasia Tekidou Laboratory of Hydrology & Water Resources Management, Department of Water Resources and Environmental Engineering, School of Civil Engineering, National Technical University of Athens, Athens, Greece

Francesca Testella Department of Political Sciences, Communication and International Relations, University of Macerata, Macerata, Italy

Gábor Ungvári Regional Centre for Energy Policy Research, Corvinus University of Budapest, Budapest, Hungary

Davide Viaggi Department of Agricultural Sciences, Alma Mater Studiorum University of Bologna, Bologna, Italy

Christophe Viavattene Flood Hazard Research Centre, Middlesex University, London, UK

Michele Vollaro Department of Agricultural Sciences, Alma Mater Studiorum University of Bologna, Bologna, Italy

Andrew J. Yates Department of Economics and Curriculum for the Environment, University of North Carolina at Chapel Hill, Chapel Hill, NC, USA

Michael D. Young Water Economics and Management, School of Earth and Environmental Sciences, The University of Adelaide, Adelaide, Australia

Chapter 1
Defining and Assessing Economic Policy Instruments for Sustainable Water Management

Manuel Lago, Jaroslav Mysiak, Carlos M. Gómez, Gonzalo Delacámara, and Alexandros Maziotis

Abstract This first chapter sets the scene for the work presented in this book. Based on a review of the literature, the chapter introduces a definition of economic policy instruments (EPIs) and a classification of broad categories of EPIs relevant for water policy that will be used to present the following parts of the book (prices, trading and other instruments) and following chapters/case studies under each part. A literature review is presented to justify the relevance on the selection of the three broad categories of instruments selected. Further, this chapter introduces the state of the art in the application of water EPIs and their ex-post evaluation, which is followed by the presentation of the criteria that is used for the evaluation of economic policy instruments that has been applied to all the case studies in the book. In this context, criteria are grouped into three outcome criteria and three

M. Lago (✉)
Ecologic Institute, Pfalzburger Strasse. 43/44, 10717 Berlin, Germany
e-mail: Manuel.lago@ecologic.eu

J. Mysiak
Fondazione Eni Enrico Mattei (FEEM), Isola di San Giorgio Maggiore, 30124 Venice, Italy
e-mail: jaroslav.mysiak@feem.it

C.M. Gómez
Departamento de Economía, Universidad de Alcalá and IMDEA-Water Foundation, c/ Plaza de la Victoria s/n, 28803 Alcalá de Henares (Madrid), Spain
e-mail: mario.gomez@uah.es

G. Delacámara
Madrid Institute for Advanced Studies in Water (IMDEA-Water Foundation), Punto Net 4, floor 1, Edificio ZYE, Parque Científico-Tecnológico de la Universidad de Alcalá, 28805 Alcalá de Henares (Madrid), Spain
e-mail: gonzalo.delacamara@imdea.org

A. Maziotis
School of Mechanical, Aerospace and Civil Engineering (MACE), University of Manchester, Manchester M13 9PL, UK
e-mail: alexandros.maziotis@gmail.com

M. Lago et al. (eds.), *Use of Economic Instruments in Water Policy*, Global Issues in Water Policy 14, DOI 10.1007/978-3-319-18287-2_1

process criteria. Outcome-oriented criteria describe how the EPIs perform. They include intended and unintended economic and environmental outcomes and the distribution of benefits and costs among the affected parties. These steps consider the application of cost effectiveness and cost benefits analysis for example to assess ex-post performance of the EPI. Process criteria describe the institutional conditions (legislative, political, cultural, etc.) affecting the formation and operation of the studied EPI (particularly relevant if we are assessing the possible impacts from the use of economic instruments), the transaction costs from implementing and enforcing the instruments and the process of implementation.

Keywords Economic policy instruments • Water policy • Definition and categories • Ex-post assessment • Outcome-oriented and process-oriented evaluation criteria

1.1 Background

Economic Policy Instruments (EPIs) are incentives designed and implemented with the purpose of adapting individual decisions to collectively agreed goals. They include incentive pricing, trading schemes, cooperation (e.g. payments for environmental services), and risk management schemes. EPIs can significantly improve an existing policy framework by incentivising, rather than commanding, behavioural changes that may lead to environmental improvement. They can have a number of additional benefits, such as creating a permanent incentive for technological innovation, stimulating the efficient allocation of water resources, raising revenues to maintain and improve the provision of water services, promoting water use efficiency, etc.

EPIs have received widespread attention over the last three decades, and have increasingly been implemented not just to raise revenue but also, most importantly, to achieve environmental policy objectives. However, whereas EPIs have been successfully applied in some policy domains (such as climate, energy and air quality), their application to tackle environmental issues such as droughts/water scarcity, floods and water quality control are beset by many practical difficulties. In recent years, however, an increasing number of local, national and international EPI experiences in water management have appeared, and key legislative and policy documents, including the EU Water Framework Directive 2000[1] (WFD) and the recent EU communication Blueprint to Safeguard Europe's Waters[2] (2012), now support their wider use.

Following prior policy oriented references (NCEE 2001; Stavins 2001; Kraemer et al. 2003; UNEP 2004; PRI 2005; ONEMA 2009; OECD 2011; EEA 2013), EPIs for sustainable water management are consequently designed and implemented

[1] http://eur-lex.europa.eu/resource.html?uri=cellar:5c835afb-2ec6-4577-bdf8-756d3d694eeb.0004.02/DOC_1&format=PDF

[2] http://eur-lex.europa.eu/legal-content/EN/TXT/PDF/?uri=CELEX:52012DC0673&from=EN

both to induce some desired changes in the behaviour of all water users in the economy (being individuals, firms or collective stakeholders) and to make a real contribution to water policy objectives, in particular reaching the broad environmental objectives of water policy (e.g. EU Water Framework Directive or US Clean Water Act[3]), at least cost for society.

Three ideas are crucial when thinking of EPIs: incentives, motivation, and voluntary choice. Rather than prescribing a particular type of behaviour that the user should comply with, EPIs create or harness economic incentives to encourage or discourage certain behaviour, but finally leave it to the user to devise his/her way of dealing with those incentives based on individual motivations. An EPI must result in voluntary changes (i.e. of practices, technology, etc.) that contribute to improving the status of ecosystems and meeting relevant environmental objectives. In saying so, not all economic instruments may induce changes that contribute to meeting environmental objectives. For instance, an increase in water tariffs to recover the cost of drinking water supply might not necessarily result in reducing water use. To be environmentally effective, tariffs should be designed by taking into account how users may respond to the price signal.

Four main forms of EPIs can be broadly distinguished: pricing, trading, cooperation, and risk management schemes:

- In pricing mechanisms, incentives are usually introduced via tariffs, charges or fees, taxes or subsidies;
- Trading relies on the exchange of rights or entitlements for abstracting or using water, or polluting the water environment;
- Cooperative mechanisms are based on the voluntary adoption of new practices leading to reduced pressure on the water environment. They can either be self-motivated – without monetary incentives – or accompanied with some form of payments (e.g. subsidies);
- Risk-based mechanisms rely on the influence of differential insurance premiums and compensation levels.

Table 1.1 presents in more detail the main characteristics of the four main types of EPIs and introduces the opportunities they can bring in for water policy.

Besides influencing the behaviour of water users to reach environmental objectives, Each type of EPI can have a number of additional benefits (OECD 2001, 2010, 2012), notably by:

- Increasing the economic efficiency of governmental action. EPIs allow water users to meet environmental targets by adopting practices and/or technologies at least cost. Water users with lower marginal abatement costs will find an incentive to reduce pollution first, so the overall aggregate costs of meeting environmental targets are lower than if all water users are targeted indiscriminately. Finally, EPIs may maximise overall benefits by allocating water resources to most valuable uses;

[3] http://www.epw.senate.gov/water.pdf

Table 1.1 Revised typology of EPIs relevant to water management

Type of instrument		Definition	What can the EPI deliver for water policy?
Pricing	Tariffs	Price to be paid for a given quantity of water or sanitation service, either by households, irrigators, retailers, industries, or other users	Encouraging technological improvements or changes in behaviour leading to a reduction in water consumption or in the discharge of pollutants. In addition, they generate revenues for water services or infrastructures
	Taxes	Compulsory payment to the fiscal authority for a behaviour that leads to the degradation of the water environment	Encouraging alternative behaviour to the one targeted by the tax, for example the use of less-polluting techniques and products
	Charges (or fees)	Compulsory payment to the competent body (environmental or water services regulator) for a service directly or indirectly associated with the degradation of the water environment	Discouraging the use of a service. For example, using charges in a licensing scheme may discourage users to apply for a permit
	Subsidies on products	Payments from government bodies to producers with the objective of influencing their levels of production, their prices or other factors	Leading to a reduction in the price of more water-friendly products, resulting in a competitive advantage with comparable products
	Subsidies on practices	Payments from government bodies to producers to encourage the adoption of specific production processes	Leading to the adoption of production methods that limit negative impacts, or produce positive impacts, on the water environment
Trading	Trading of permits for using water	The exchange of rights or entitlements to consume, abstract and discharge water	Encouraging the adoption of more water efficient technologies
			May improve the allocation of water amongst water users
	Trading of permits for polluting water	The exchange of rights or entitlements to pollute the water environment through the discharge of pollutants or wastewater	Encouraging the adoption of less water polluting technologies
			Improve the allocation of abatement costs amongst water users.
Cooperation		Negotiated voluntary arrangement between parties to adopt agreed practices often linked to subsidies or offset schemes	Encouraging the adoption of more water-friendly practices

(continued)

Table 1.1 (continued)

Type of instrument		Definition	What can the EPI deliver for water policy?
Risk management schemes	Insurance	Payment of a premium in order to be protected in the event of a loss	Water users' aversion to risk and willingness to pay for income stabilisation. When properly designed, insurance premiums signal risk and discourage behaviours that increase risk or exposure
	Liability	Offsetting schemes where liability for environmental degradation leads to payments of compensation for environmental damage	Liability as a means to incentivise long-term investments in water efficient devices

Source: Delacámara et al. 2013

- Generating financial resources to maintain and improve the delivery of water services. EPIs may help recover capital and operational costs, as well as so-called environmental and resource costs (as required by the EU WFD);
- Creating permanent incentives for continued technological innovation, as opposed to regulatory instruments that may only provide incentives to innovate until compliance is achieved;
- Flexibility and the capacity to adjust to shifting conditions with minimal transaction costs (e.g. option value that informs infrastructure design and investment).

1.2 Review of Application

The use of EPIs in water management clearly faces several challenges, notably due to lack of information and misconceptions on their "real" costs and benefits, and limited interest or, in some cases, political resistance. While the theoretical literature argues that EPIs are more "adaptable" and easier to reform than other instruments, adjusting EPIs can in reality face similar rent-seeking practices and constraints to other policy instruments. As with any other policy instruments, the choice, design and implementation of EPIs must be complemented by a careful analysis of the environmental, social and economic context, and embedded in critical debate on their relevance, limitations, and their potential synergies and conflicts with other forms of governmental action.

In practice, a wide range of EPIs have been applied at different spatial scales (e.g. national, regional, river basin, etc.) and on in different sectors (e.g. water utilities, industry, agriculture, tourism, hydropower generation, etc.). Tariffs, taxes and charges are by far the most recurrent EPIs, followed by subsidies and cooperative schemes. While trading schemes on water quantity have been limited to a few cases in Europe (e.g. Spain, England and Wales), they have been more popular elsewhere, notably in Australia, the semiarid Western states of the USA, or Chile.

However, the actual use of economic instrument differs among countries and among policy areas. Notwithstanding well-established theoretical foundation, the implementation of EPI lacks follow-through. Whereas positive experience abound in other areas of environmental management (notably air quality and recently climate governance), the application of EPI in the context of water (particularly demand) management is relatively recent (PRI 2005; Cantin et al. 2005).

In the context of Europe, a survey by the European Commission on the use of economic instruments in the WFD first river basin management plans shows that a minority of actions have been taken by individual Member States to comply with the requirements of Article 9 on cost recovery for environmental and resource costs through water pricing of the WFD. Further, the details of the actions often referred to water pricing, were unclear and did not provide any details on what was effectively proposed to adapt existing water pricing policies. Where economic instruments are mentioned, mostly it referred to subsidies for eco-system services (where the sources of funding mostly come from the EU Rural development program) and water and waste water charges or taxes.

With the programmes of measures for the achievement of the objectives of the EC WFD being developed and then finalised, Member States in Europe have shown increasing interest in economic instruments. The very high costs of the proposed programmes of measures have raised the issues of (cost-)effectiveness of proposed measures and of financing and revenue raising. In practice and policy terms, although the application of economic instruments are often justified on economic efficiency grounds, attention is mostly given to the financing dimension of economic instruments, i.e. how they contribute to collecting new revenue that feeds into the central government budget or can support “good practice” in water use and management.

The examples in the interest in the application of EPIs to tackle water management issues abound in Europe; Sweden has started to investigate new pollution permit-fee schemes that include the potential for water pollution permit trading in the medium term; with Denmark and Norway showing similar interest in the application of the same EPI. And there are signs of renewed interest in France for water markets, following the publication of a report that concluded that water markets established in Australia and California could be considered as applicable in France (Barthélémy et al. 2008). In the Netherlands, a review of existing economic instruments applied to water management in Europe (Mattheiß et al. 2009) was launched with the objective of identifying new opportunities for economic instruments that would support the implementation of the WFD and in particular measures dealing with hydromorphology, ecology and biodiversity. Most experiences and policy discussions on tradable permits and water markets in Europe are from Spain. See for example: Calatrava and Garrido 2005; Gómez-Limón and Martínez 2006.

Very interestingly, the review for the Dutch Government has stressed the very wide range of economic instruments already implemented in individual Member States such as: innovative water tariffs structure to limit water demand; electricity premium to hydropower for good hydromorphological practices/restoration; tradable permits for both quantity and quality; subsidies for the construction of green roofs aimed at improving rainwater management/reducing excess water; voluntary agreements for restoring flood plains and shifting practices to good environmental practices (both in urban areas and for agriculture), etc. The review also emphasised:

- The importance of extending the policy focus of economic instruments to be investigated and proposed, from economic instruments separated between "water quantity" and "water quality" to economic instruments targeting: (i) water scarcity and drought, (ii) excess water (floods), (iii) pollution management and (iv) ecology/biodiversity.
- The limited knowledge available on (i) the functioning/implementation and (ii) the performance of these economic instruments, stressing the need for more rigorous assessments of the innovative approaches developed by individual Member States.

There are several key reasons why EPI are not more widely used in water management, or why implementation in Europe has been focused mainly on water tariffs, environmental charges and taxes and dedicated subsidies (mainly agriculture-related):

- Uncertainty – Not enough is known about the effectiveness of many instruments in contributing to the achievement of environmental goals, that is whether economic instruments will spur the change needed in the given time frame and without unintended drawbacks. This applies to economic instruments that require the development of "new markets" (such as tradable permits or payments for environmental services). It also applies to many innovative instruments already in place in selected countries for which no knowledge is available. It also applies to "traditional" water tariffs and environmental charges for which expected changes in water demand or pollution discharged is rarely translated into environmental and ecological status of aquatic ecosystems. The same holds true for the actual implementation/transaction costs and their distribution. When uncertainties abound about what can be delivered by the EPI and whether predetermined policy objectives will be met, the policy makers are inclined to make use of prescriptive regulatory instruments (such as environmental standards and best available technologies).
- Path dependency – EU countries already have a set of fairly sophisticated regulations for the management of water quality and water quantity issues. Changing these systems to incorporate EPIs might offer (uncertain) efficiency gains in the longer term, but will inevitably require additional efforts (and hence costs) by regulators and regulatees during the adaptation process. Hence, we are more likely to see EPI applied in fields that were hitherto unregulated, or in areas where a significant regulatory reform is necessary anyway (for instance, where competencies are re-organised within a federal governance structure).
- Transaction costs – It is often assumed that the supposedly superior efficiency of economic instruments stands against the higher transaction costs associated with EPI. For instance, tradable permit systems require a regular allocation of permits, ongoing monitoring, reporting and verification, and of course the trade

itself. All of these activities impose additional efforts onto the regulatee, which need to be balanced against the expected efficiency gains. In most cases, however, there is no information on transaction costs that such new instruments would imply, the transaction cost argument being used on a rather emotional basis. Furthermore, command and control mechanisms have also their own transaction costs that are rarely analysed nor quantified.

- Heterogeneity of impacts – the efficiency of EPI is maximised if the unit to which they are applied is completely homogeneous across space and time, i.e. if 1 kg of nitrogen released or 1 l of groundwater abstracted has the same marginal impact anytime, anywhere. While this condition is satisfied e.g. for greenhouse gas emissions, it is typically not the case for water management issues. There are options to account for this heterogeneity of marginal impacts, but they will necessarily drive up transaction costs for regulatees and regulator alike.

Although arguments in favour of using EPIs to make water decisions more flexible and adaptable have been put forward, it is expected that such arguments in favour or against an extended adoption of EPIs have to be based on proven facts and testable empirical evidence. At this moment, there is a gap in the literature about the evaluation of performance of water EPIs that this book aims to fill in. In this context, this book sets to shed light into assessing the effectiveness and the efficiency of implemented EPIs in achieving water policy goals, and to identify the preconditions under which they complement or perform better than alternative (e.g. regulatory) policy instruments or together with them as part of complex policy mixes. Case studies from Cyprus, Denmark, France, Germany, Hungary, Italy, Spain, and the UK (European Union), as well as from Australia, Chile, Israel, and the USA, are included in this book. The development of a consolidated assessment criteria helps clarify (and where possible, quantify) the effectiveness of each EPI and helps with the establishment of relevant cross-reference between the different analysed EPIs.

1.3 EPIs Performance Evaluation

Policy assessment is a necessary tool for the design of new policies and improvement of existing ones. These tools are these days part of good governance approaches and used to justify increased transparency in policy making. Often policies are designed with assumptions, guesses and expectations as to how they will affect outcomes, and ex ante impact assessments to inform policy choices are only required in a handful of countries (see Thaler et al. 2014). The lack of ex-ante forecasts, combined with even more-frequent lack of ex-post evaluation, often impedes the evaluation of performance of implemented policies or the design of future policies.

An ex-post assessment of any given EPI in order to understand and explain its success or failure must explain relevant aspects in relation with the EPI contribution towards the achievement of its stated objectives and provide clear explanation of the specific surrounding settings of its implementation. All the EPIs evaluated in this book have been assessed in relation with two types of broad criteria divided in terms of those that are output oriented and those that help understanding the EPI specific context relevant for its design and implementation.

An analysis of the so-called output oriented criteria of the EPI include an understanding of its: (i) environmental outcomes, (ii) economic costs and benefits and (iii) distributional or social equity impacts.

An analysis of the so-called context criteria of the EPI Water assessment framework and it is intended to deal with: (i) the institutional set up in place and the one required for the EPI to deliver its full potential; (ii) the transaction costs associated to the EPI implementation and how the institutional set-up and the design have dealt with this; (iii) the design and implementation of the EPI and why it has succeeded or failed in the situation analyzed.

Table 1.2 provides clear definitions of each of the assessment criteria used to understand the selected EPIs.

1.4 Objectives, Scope and Structure of the Book

We aim to present in this book most of the case studies that were reviewed *ex post* in the EPI-WATER (FP7-265213) project.[4] The highest added value of the work done in this project is the breath of the information that came out from the review process of specific EPIs. This basically includes the review of application of EPIs in different countries, institutional contexts and situations but performed through the lens of relevant assessment criteria that allow drawing some comparability conclusions.

This book is designed to increase knowledge about the application of economic policy instruments to tackle water management challenges relevant for the implementation of water policy (e.g. restoration of water ecosystems, tackling pollution, etc.). It also sheds light on key concepts and definitions, and conveys the benefits, limitations, transaction costs, and opportunities of using EPIs in water policy. It illustrates real challenges associated with the use of EPIs with ad-hoc examples and case studies based on a wide set of implemented EPIs within and outside the EU.

[4]The EU-funded research project **EPI-WATER** (standing for: *Evaluating Economic Policy Instruments for Sustainable Water Management in Europe*) was launched in January 2011 for a 3-year period. Its main aim was to assess the effectiveness and the efficiency of Economic Policy Instruments (EPIs) in achieving water policy goals. In a first *ex-post* assessment, the project studied 30 EPIs in Europe and around the world. The second phase of the project carried out in-depth *ex-ante* assessments of the viability and the expected outcome of EPIs in five EU areas facing different water management challenges (flood risk and waterlogging in Hungary, water scarcity and drought risk in Spain, biodiversity and ecosystem service provision in France, water scarcity in Greece and water quality in Denmark). For more information on the EU-funded EPI-WATER research project: http://www.feem-project.net/epiwater/

Table 1.2 Proposed assessment criteria for the evaluation of EPIs performance

Output oriented assessment criteria	
Environmental outcomes	Environmental outcomes are assessed by comparing actual outcomes with alternatives (no action or regulation, for example) and evaluating positive and negative side effects. This criterion connects behaviours that have direct or indirect impacts on water (e.g. irrigation, use of pesticide) to the status of ecosystems and the value of ecosystem services to humans. Environmental characteristics are embodied in measures of water pollution, water abstractions, and so on
Economic costs and benefits	The economic criterion evaluates EPI efficiency according to cost-benefit analysis, cost-minimization or other methods. Economic efficiency is often evaluated with proxy variables such as the income generated from the use of the EPI, financial costs related to the implementation of the EPI and/or the cost of water delivery
Distributional or social equity impacts	The distribution of goods and burdens across different stakeholder groups affects social equity and acceptability of EPIs. This criterion focuses primarily on assessing the nature of the distribution, highlighting inequalities in the allocation of goods and burdens as a result of the implementation of EPI (i.e. material living standards, health, education, personal activities including work, political voice and governance, social connections and relationships, environment and insecurity)
Context related assessment criteria	
Institutions	Institutions are the formal rules and informal norms that define choices. Most institutions are difficult to describe, highly adapted to local conditions, and effective in balancing many competing interests. Institutional constraints vary in strength, according to their permanence (from culture and religion to constitutions to laws to rules and regulations). Institutions often determine the difference between success and failure of an EPI, due to the way that they can strengthen or weaken the EPI's mechanism, i.e., they are either reliable and robust or unstable and rigid. We separate institutions and transaction costs (TCs) by associating institutions with exogenous impacts on EPIs and TCs with the endogenous fixed costs of implementing an EPI and variable costs of using it. A water market, for example, is established with fixed TCs and operated with variable TCs, but both are affected (positively and negatively) by institutions
Transaction costs	Transaction costs (TCs) represent friction, i.e., the time and money cost of moving from idea to action to conclusion, or the costs of implementing and using EPIs. Ex-ante TCs (from, e.g., negotiating new property rights) are equivalent to fixed costs; ex-post TCs (e.g., from monitoring) are equivalent to variable costs. TCs are identified by examining the steps from design and implementation (ex-ante) to monitoring and enforcement (ex-post)
Design and implementation	Policy implementation reflects the cost and challenge of moving from a theoretical idea to practical application of an EPI. This criterion considers the adaptability of the EPI, public involvement, institutional factors, and external factors (e.g., EU sectorial policies)

Source: Zetland et al. 2013

The book has a practical remit and is aimed to anyone interested in finding out more about the use of economic instruments in water. It is expected that the book will be relevant for academic researchers, consultants and practitioners working in the water management/economics field. More specifically this book aims to:

- Support national decision-makers and experts in the development and implementation of EPIs in water management; and
- Raise awareness of EPIs, so that interested parties can engage effectively with decision-makers and experts on the development and implementation of EPIs.
- Help to increase understanding through the use of practical examples about the ex-post evaluation of public policy interventions.

The structure of the book is organized in three main parts in terms of the broad categories of economic instruments covered through case studies: PART I (pricing and taxes), PART II (trading) and PART III (other types of incentives, such as cooperation and risk management schemes). Each part includes a short introductory chapter highlighting cross-cutting problems, challenges, design and implementation issues of the broad instrument category. Each introductory chapter also highlights some conclusions in terms of cross-cutting issues for that specific broad category of instrument. The consecutive chapters in each part present specific case studies in the application of those EPIs. Case study chapters aim to follow a similar presentational structure mindful of the application of the proposed assessment framework. Each chapter aims to discuss the review of application of the EPI in question in terms of each of the assessment criteria towards which the economic instruments are assessed, including environmental outcomes, economic efficiency, financial revenues, transaction costs for regulator and regulated entities, social impact and equity issues and policy implementability. Mediating factors such as institutional set-up are also explored.

The overall structure of the book is as follows: Chap. 2 illustrates a short introduction to Part I of the book on water pricing and taxes and Chaps. 3, 4, 5, 6, 7, 8, 9, 10, 11, 12, and 13 present the related case studies in this topic. Chapter 14 illustrates a short introduction to Part II of the book on water trading and Chaps. 15, 16, 17, 18, 19, 20, and 21 present the related case studies on the review of practical application of these EPIs. Chapter 22 illustrates a short introduction to Part III of the book on other relevant economic instruments and Chaps. 23, 24, 25, 26, and 27 present the related case studies. Chapter 28 provides a concluding chapter relevant for the three parts. Conclusions will be outcome oriented per type of challenge that the EPIs can address: (i) Water quality, (ii) Water scarcity, (iii) Flood risk and (iv) Ecosystem conservation.

1.5 Book Chapter Outline

This book has been divided into the following chapters:

Chapter 2: Water Pricing and Taxes: An Introduction
Chapter 3: Effluent Tax in Germany
Chapter 4: The Water Load Fee of Hungary

Chapter 5: Water Abstraction Charges and Compensation Payments in Baden-Württemberg (Germany)
Chapter 6: The Danish Pesticide Tax
Chapter 7: Subsidies for Drinking Water Conservation in Cyprus
Chapter 8: Residential Water Pricing in Italy
Chapter 9: Water Tariffs in Agriculture: Emilia Romagna Case Study
Chapter 10: Corporatization and Price Setting in the Urban Water Sector Under Statewide Central Administration: The Israeli Experience
Chapter 11: Water Budget Rate Structure: Experiences from Several Urban Utilities in Southern California
Chapter 12: Green Energy Certificates and Compliance Market
Chapter 13: Subsidies for Ecologically Friendly Hydropower Plants Through Favourable Electricity Remuneration in Germany
Chapter 14: Water Trading: An Introduction
Chapter 15: Water Quality Trading in Ohio
Chapter 16: Nitrogen Reduction in North Carolina
Chapter 17: Evaluation of Salinity Offset Programs in Australia
Chapter 18: Water Trading in the Tagus River Basin (Spain)
Chapter 19: Chilean Water Rights Markets as a Water Allocation Mechanism
Chapter 20: Unbundling Water Rights as a Means to Improve Water Markets in Australia's Southern Connected Murray-Darling Basin
Chapter 21: The Development of an Efficient Water Market in Northern Colorado, USA
Chapter 22: Other Types of Incentives in Water Policy: An Introduction
Chapter 23: Cooperative Agreements Between Water Supply Companies and Farmers in Dorset (E)
Chapter 24: Financial Compensation for Environmental Services: The Case of the Evian Natural Mineral Water (France)
Chapter 25: New York City's Watershed Agricultural Program
Chapter 26: Voluntary Agreement for River Regime Restoration Services in the Ebro River Basin (Spain)
Chapter 27: Voluntary Agreements to Promote the Use of Reclaimed Water at Tordera River Basin
Chapter 28: Key Conclusions and Methodological Lessons From Application of EPIs in Addressing Water Policy Challenges

References

Barthélémy, N., Verdier, L., & CGDD. (2008). Les marchés de quotas dans la gestion de l'eau: les exemples de l'Australie et de la Californie, Evaluation Letter, November 2008.

Calatrava, J., & Garrido, A. (2005). Modelling water markets under uncertain water supply. *European Review of Agricultural Economics, 32*(2), 119–142.

Cantin, B., Shrubsole, D., & Ait-Ouyahia, M. (2005). Using economic instruments for water demand management: Introduction. *Canadian Water Resources Journal, 30*(1), 1–10.

Delacámara, G., Dworak, T., Gómez, C. M., Lago, M., Maziotis, A., Rouillard, J., & Strosser, P. (2013). EPI-water deliverable 5.3: Guidance on the design and development of economic policy instruments in European water policy. EPI-water – evaluating economic policy instruments for sustainable water management in Europe.

European Environmental Agency (EEA). (2013). Assessment of cost recovery through water pricing. Technical report, No 16. European Environment Agency, Copenhagen.

Gómez-Limón, J. A., & Martínez, Y. (2006). Multi-criteria modelling of irrigation water market at basin level: A Spanish case study. *European Journal of Operation Research, 173*(1), 313–336.

Kraemer, R. A., Castro, Z. G., da Motta, R. S., & Russell, C. (2003). Economic instruments for water management: Experiences from Europe and implications for Latin America and the Caribbean. In: Regional Policy Dialogue Study Series: Inter-American Development Bank.

Mattheiß, V., Le Mat, O., & Strosser, P. (2009). Which role for economic instruments in the management of water resources in Europe? In search for innovative ideas for application in the Netherlands. Report for the Dutch Ministry of Transport, Public Works and Water Management.

National Center for Environmental Economics (NCEE). (2001). *The United States experience with economic incentives for protecting the environment*. Washington, DC: U.S. Environmental Protection Agency. EPA-240-R-01-001.

OECD. (2001). *Environmentally related taxes in OECD countries 15*. Paris: Organisation for Economic Co-operation and Development.

OECD. (2010). *Pricing water resources and water and sanitation services*. Paris: Organisation for Economic Co-operation and Development.

OECD. (2011). Economic instruments for water management. Environment Directorate, Environment Policy Committee. Paris, 27–28 Oct 2011. ENV/EPOC/WPBWE(2011)13.

OECD. (2012). *A framework for financing water resources management*. Paris: Organisation for Economic Co-operation and Development.

ONEMA. (2009). Economic instruments to support water policy in Europe – Paving the way for research and future development. Workshop Synthesis report, Paris, 9 & 10 December 2009. Published by ONEMA and ACTeon. Available online at http://www.onema.fr/IMG/EV/meetings/instruments-economiques-UK-BD.pdf

Policy Research Initiative (PRI). (2005). *Economic instruments for water demand management in an integrated water resources management framework*. Synthesis report. Policy Research Initiative Project Sustainable Development. PH4-18/2005E-PDF. ISBN 0-662-39330-9. Government of Canada Publications.

Stavins, R. N. (2001). Experience with market-based environmental policy instruments. Resources for the future, discussion paper 01–58.

Thaler, T., Boteler, B., Dworak, T., Holen, S., & Lago, M. (2014). Investigating the use of environmental benefits in the policy decision process: A qualitative study focusing on the EU water policy. *Journal of Environmental Planning and Management, 57*(10), 1515–1530.

UNEP. (2004). Economic instruments in biodiversity-related multilateral environment agreements (UNEP/ETB/2003/10). ISBN: 92-807-2390-1.

Zetland, D., Dworak, T., Lago, M., Maziotis, A., & Viavattene, C. (2013). EPI-WATER discussion paper 02: An orientation to the assessment framework for evaluating economic policy instruments for managing water. EPI-WATER – Evaluating economic policy instruments for sustainable water management in Europe.

Chapter 2
Water Pricing and Taxes: An Introduction

Jaroslav Mysiak and Carlos M. Gómez

Abstract Water pricing embraces a range of distinct policy instruments that affect the scale and/or the pattern of production and *resource*-exploitation costs. Ideally, water prices should reflect *financial costs* of service delivering water infrastructure, *environmental costs* arising from harm induced to ecosystems and ecosystem services, and *resource costs* attendant to social welfare losses from not using the water for the most socially beneficial purpose. What is straightforward and unchallenged in economic theory may not translate into clear and uncontested principles to be followed in practice. The information asymmetries, pre-existing water permits or entitlements adhering to different legal doctrines, and hostile reception of water policy reform may antagonise introduction of pricing policy instruments. This chapter provides an overview of the empirical studies from different European countries, supplemented by studies from California and Israel, comprised in the first book section. Although the collection is not meant to be exhaustive or thorough, it offers insightful overview of design principles and choices made to put in place a variety of instruments designed to cope with water pollution, water stress, and hydrological and morphological modifications of water bodies. The majority of the chapters in this section addresses residential and industrial water supply provision and wastewater discharge. The remaining chapters examine the application of EPIs in agriculture, for cost recovery of irrigation services and pollution control; and in hydroelectricity generation, for curbing the environmental impact of water impoundments. The common structure of all showcased studies is a result of meticulous efforts to highlight the scope of the analysed instruments, the embedding legislative and regulatory environment, and the evidence collected so as to substantiate the performance assessment.

Keywords Water pricing • Cost recovery • Water Framework Directive (2000/60/EC) • Environmental taxes • Subsidies

---

J. Mysiak (✉)
Fondazione Eni Enrico Mattei (FEEM), Isola di San Giorgio Maggiore, 30124 Venice, Italy
e-mail: jaroslav.mysiak@feem.it

C.M. Gómez
Departamento de Economía, Universidad de Alcalá and IMDEA-Water Foundation,
c/ Plaza de la Victoria s/n, 28803 Alcalá de Henares (Madrid), Spain
e-mail: mario.gomez@uah.es

M. Lago et al. (eds.), *Use of Economic Instruments in Water Policy*,
Global Issues in Water Policy 14, DOI 10.1007/978-3-319-18287-2_2

2.1 The Role of Water Prices and Taxes in Water Policy

Water pricing embraces a range of distinct policy instruments that affect the scale and/or the pattern of production and *resource*-exploitation costs. Staged by means of *incentives* (i.e. subsidies) or *disincentives* (i.e. taxes or charges), these instruments eventually affect the price paid for goods or services that either make use of water resources or otherwise affect natural water bodies. Characteristically, pricing instruments are put to use to rectify *market failures* that arise when social costs or benefits of production and consumption are not reflected through prices determined by *free* markets.

Water is notoriously known as both, an economic and social good; essential for life, economic development, social cohesion, and the environment. The multitude of the at least to some extent incompatible uses of water and their impacts on natural water bodies makes public water policy choices both value-laden and *intractable*. What is more, availability of water is unevenly distributed over time and space, implying that there is not enough water to permanently or temporarily satisfy all demands. As a result, economic costs of water and water services, that should ideally be reflected in the price users pay for them, is a combination of *financial costs* of service delivering water infrastructure, *environmental costs* arising from harm induced to ecosystems and ecosystem services, and *resource costs* attendant to social welfare losses from not using the water for the most socially beneficial purpose. With other words, designing pricing instruments for a sustainable water management is as challenging as are the public choices themselves about what is the appropriate and sustainable way of managing water resources.

To qualify as *economic policy instruments* (EPIs, see also Chap. 1), price interventions ought to deliver discernible *environmental* improvements in regard to the predetermined water policy objectives. This is only the case if the demand for water or water services is elastic, that is when the quantity demanded of a good or service responses to a change of its price. Notably, *price elasticity* depends on a host of factors, including the income and availability of substitutes. It has been demonstrated in numerous instances (Mansur and Olmstead 2012; Olmstead et al. 2007; Olmstead and Stavins 2009; Olmstead 2010), including the studies featured in this book, that although demand is *relatively inelastic*, it is nevertheless different from zero. This implies that sizeable changes in demand require considerable price adjustment. If the demand was *entirely* inelastic, demand for water and water services would not respond to price intervention and pricing instruments would merely serve *financial purposes*, i.e. generating revenues. But even in that case, if the revenues were earmarked for implementing measures helping to safeguard the environmental health of water bodies, pricing can contribute to accomplishing public water policy goals.

What is straightforward and unchallenged in economic theory may not translate into clear and uncontested principles to be followed in practice. The information asymmetries, pre-existing water permits or entitlements adhering to different legal doctrines, and hostile reception of water policy reform may antagonise introduction of pricing policy instruments. As a consequence and despite the sound theoretical foundation, the experiences reported in this book still mark rather early stages of managing water as an economic resource. Accordingly, the 2012 EU Water Policy Review[1] lamented a limited application of '*incentive and transparent water pricing*', concluding that '*not putting a price on a scarce resource like water can be regarded as an environmentally-harmful subsidy*' (EC 2012, p. 10). Noting the practical difficulties and the necessary mind-set change, we argue that the policy analysis should not be centred only on how much water and water services should be priced in principle, but rather how water prices should be designed so as to best respond to the challenge of managing water resources effectively. This shifts the emphasis away from the determining the optimal price levels alone to choosing the pricing schemes and combination of instruments that are tailor-made for the specific policy contexts, taking due account of the existing institutions and competing policy objectives.

This *book section* features a compilation of empirical studies, organized in separate chapters that examine applications of water pricing instruments in different European countries, member states of the European Union (EU), which are supplemented by noteworthy studies from California and Israel.

Although the collection is not meant to be exhaustive, it offers insightful overview of design principles and choices made to put in place a variety of instruments designed to cope with water pollution, water stress, and hydrological and morphological modifications of water bodies. More than that, all analysed instruments are explored in the same way, making sense of all available evidence in support of assessing the instruments' environmental, economic and social outcomes. The majority of the chapters in this section addresses residential and industrial water supply provision and wastewater discharge. The remaining chapters examine the application of EPIs in agriculture, for cost recovery of irrigation services and pollution control; and in hydroelectricity generation, for curbing the environmental impact of water impoundments. The common structure of all showcased studies is a result of meticulous efforts to highlight the scope of the analysed instruments, the embedding legislative and regulatory environment, and the evidence collected so as to substantiate the performance assessment driven by the framework outlined in the Chap. 1.

The *Polluter Pays Principle* (PPP), already featured in the First *European Environment Action Programme* (1973–1976), made its way into the EC Treaty in the 1987[2] and successively in the secondary European legislation (e.g. Water Framework Directive 2000/60/EC, the Directive on Industrial Emissions 010/75/EU). The effluent tax in Germany (Chap. 3), introduced in 1976, was among the first applications of environmental taxes in Europe implementing the PPP. The tax that is still applied to the authorized discharges is calculated in terms of damaging units,

---

[1] Communication from the Commission to the European Parliament, the Council, the European Economic and Social Committee and the Committee of the Regions *A Blueprint to Safeguard Europe's Water Resources* COM (2012) 673 final.

[2] Article 130r of the *Single European Act* (SEA). In the currently in force Lisbon Treaty the PPP is covered by the Article 191(2) of TFEU.

estimated as the equivalents of ten contaminants. The water load tax in Hungary (Chap. 4), introduced incrementally shortly before Hungary joined the EU, operates in a similar way. The tax is determined by nine contaminants contained in the discharged wastewater, but unlike the German tax it takes into account the environmental sensitivity of the receiving environment and the way the sludge is eventually disposed. In both cases the municipal wastewater disposal is the most affected sector and the tax is eventually paid by households as the final consumers. The taxes contributed to an earlier implementation of the *Urban Waste Water Directive* (91/271/EEC) among others by allowing that the polluters' investments into better wastewater treatment was deducted from the amount of tax due. While in Germany the tax revenues are earmarked for pollution control executed by the state authorities, in Hungary they contribute to consolidating public finances.

The Danish pesticides tax (Chap. 6) was designed to protect the surface and groundwater bodies, the latter being source of drinking water provision usually without treatment, and to contribute to fulfil the objectives of the Danish pesticide policy. It replaced the previous general tax levied on pesticides wholesale prices that proved unable to curb the use of pesticides. Implemented as a product tax, levied on the sales prices, the instrument differentiates the categories of use, rather than the toxicity levels. Designed in revenue-neutral way, the collected tax revenues are reimbursed to farmers through lower land taxes and subsidies for organic and environmentally friendly farming. In doing so, the design of the tax is amenable to the principles of environmental tax reform.

The design of water tariffs for residential water uses is particularly intrigued as it is often called to conciliate solidarity principle of affordability of water service provision for economically disadvantages households (ability-to-pay principle) with principles of full economic cost recovery and efficient use of resources. The studies of water tariffs analysed in this book complementary to some extent. In all cases the tariffs are designed so as to recover financial costs of the service provision, and discourage *disproportionate* (beyond what is understood as reasonable) use of water resources.

Chapter 8 shows how this reconciliation was accomplished in the residential water pricing scheme in the *Emilia Romagna* administrative region (Italy). As a natural monopoly frequently managed through *concessive model* exemplifying the *public-private* partnerships, the organisation of residential water supply and sanitation services (WSS) and the water tariff setting are narrowly regulated. Amidst the institutional reform implemented since the 1990s, the administrative region of *Emilia Romagna* waged a modification of tariff method in a way that rewards a better service and environmental performance of water utilities, and in contrary, penalises utilities whose performance is judged substandard. The rewards and penalties aimed at utilities and could not be passed on to the final consumers. The modified tariff system also privilege economically vulnerable households by cross-subsidising their water consumption by higher price levels in the upper tiers of the increasing block tariffs.

The application of increasing block-rate (IBR) water budgets in three water districts in southern California, covered in the Chap. 11, applies similar tiered price structure but pioneers tailor-made block sizes specific for households characteristics and environmental conditions. Prompted by equity issues and financial viability of water utilities, the reform of water tariffs involves specification of a reasonable use of water in the first (indoor) and second (outdoor) block, the consumption beyond which is deemed inefficient (third block) or even excessive (fourth block). The *reasonable use* of water is determined by state regulation (e.g. around 200 l per day and household member), empirical evidence (e.g. real time monitoring of evapotranspiration), and individual household/property information (e.g. irrigated area). Whereas the revenues collected from the first two block rates and the fixed component of the tariff are design to recover the financial costs of the service provision, the penalising tariffs for the water use beyond what is considered reasonable is destined for exploitation of additional or alternative water sources.

Volumetric water tariffs may play perhaps even more important role in agriculture, especially in temporarily or permanently water stress countries in the Southern Europe. Chapter 9 brings this to the point by analysing empirical evidence from the *Tarabina* irrigation district in the Emilia Romagna administrative region (Northern Italy). The irrigation districts relies on water supplied by the *Canale Emiliano-Romagnolo* (CER), which is one of the largest water transfer projects in Italy, from the Po river. Although Po river (basin) is usually water abundant, recent prolonged drought spells (2003, 2006–2007) have induced water shortages that prompted water restrictions throughout the river basin. The volumetric water tariff was introduced both as a mean to foster both, water re-allocation to higher value uses during periods of restricted water supply, and a more equitable distribution of irrigation-related costs among the farmers within the irrigation board. The volumetric tariff resulted in a demonstrable reduction of about 50 % of water demand on average, and a sizeable reduction of costs for farmers who irrigate less or do without.

The subsidies-related EPIs in this book are represented in this book by Chaps. 5, 7, and 13. These studies address different policy goals. In Cyprus study (Chap. 7), the subsidies were meant to restrain domestic demand for potable water by encouraging greater use of alternative water sources, from aquifer or recycled wastewater. The assessment of these subsidies yielded mixed results. Although a limit was imposed on groundwater abstraction for newly installed borehols, the weak monitoring of the actually abstracted water might have increased the pressure of the aquifers. Hence although the subsidies contributed to restructure outdoor water demand, especially during the extreme 2007–2008 drought, it is not obvious to what extent they contributed to greater water conservation. On opposite side, the subsidies did not succeed to stimulate larger interest in wastewater recycling that would have generate long-lasting reduction of water withdrawal.

The compensation payments for less intense agricultural practices in vulnerable areas are discussed in Chap. 5 as a part of a bundle of policy instruments addressing nitrate water pollution and untenable water abstraction. First pursued as a partial compensation for production losses prompted by strict regulation in the water protection areas, the subsidies were later extended, under different design, to other areas in which nitrate pollution persist. The water abstraction charge complements the policy mix, especially after the revision in 2010 that reinforced the incentives to conserve and protect water resources and incentivised investments by large water users.

Yet another subsidy scheme from Germany, presented in Chap. 13, revisit the economic incentives of hydropower producers to reduce the environmental impacts of water impoundments through higher remuneration for electricity produced. Introduced in 2004, the scheme bears a resemblance to feed-in tariff, further explored in the next chapter on example of Italy. The schemes guarantees an incentive price for hydropower supplied from plants with better environmental performance, specified by considering plant's design criteria (storage capacity, biological passability) and management practice.

The Chap. 12 wraps up the collection of pricing related instruments, by reviewing a mix of EPIs designed separately but all acting together in a way hydropower potential was exploited in Italy. Feed-in tariffs (FIT) and especially tradable green energy certificates (GEC) had been introduced to build supply-side competition among the RES and to curtail the costs of renewables. The actionable concession award or operating large hydropower plants are an opportunity to coerce environmental improvement. The chapter goes on to discuss the roles of water abstraction fees and charges that can be designed in a way that is sensible to the environmental impacts, and at the same time limit the development of hydropower in less or not suitable places.

References

EC. (2012). *A blueprint to safeguard Europe's water resources*. Communication from the Commission to the European Parliament, the Council, the European Economic and Social Committee and the Committee of the Regions. COM(2012) 673 final. European Commission, Brussels.

Mansur, E. T., & Olmstead, S. M. (2012). The value of scarce water: Measuring the inefficiency of municipal regulations. *Journal of Urban Economics, 71*, 332–346. doi:10.1016/j.jue.2011.11.003.

Olmstead, S. M. (2010). The economics of managing scarce water resources. *Review of Environmental Economics and Policy, 4*, 179–198. doi:10.1093/reep/req004.

Olmstead, S. M., & Stavins, R. N. (2009). Comparing price and nonprice approaches to urban water conservation. *Water Resources Research, 45*, 1–10. doi:10.1029/2008wr007227.

Olmstead, S. M., Hanemann, W. M., & Stavins, R. N. (2007). Water demand under alternative price structures. *Journal of Environmental Economics and Management, 54*, 181–198. doi:10.1016/j.jeem.2007.03.002.

Chapter 3
Effluent Tax in Germany

Jennifer Möller-Gulland, Manuel Lago, Katriona McGlade, and Gerardo Anzaldua

Abstract The exceptionally high growth in pollution-intensive sectors (such as energy, chemicals, and construction) in the post-war period in Germany caused serious environmental problems as the construction of wastewater treatment facilities did not keep pace and posed a serious threat to future water supply. This chapter analyses the policy mix of economic and regulatory instruments, which was introduced in the Federal Republic of Germany to address this threat. The policy mix consists of discharge permits (Federal Water Act, 1957), discharge limits and technical standards (Waste Water Ordinance, 1997) and the effluent tax (Effluent Tax Act, 1976). The effluent charge, the focus of the chapter, was introduced in 1976 as a reaction to the insufficient implementation of direct regulation (Federal Water Act, Waste Water Ordinance) of effluent discharges by the water management administrations of the Federal States of Germany and the resultant non-compliance with prescribed discharge standards in the private and municipal sectors.

While the policy mix and the environment in which it acts makes it difficult to single out the impact of the effluent tax, it was found that the overall quantity and harmfulness of discharged effluents was decreased substantially since the introduction of the policy mix. Wastewater plants were upgraded to state of the art technologies, with 92.6 % of effluents receiving tertiary treatment today. As a result, the quality of water bodies increased substantially, with 85 % of all surface water bodies achieving a water quality II chemical status.

In this chapter it is illustrated that a policy mix consisting of regulatory and economic instruments can be very powerful in implementing and enforcing policies to address direct effluent emissions. However, it also shows the importance of setting the right incentive structure and discusses the factors preventing this from happening in the case of the German effluent tax. Further, enabling and disabling factor

J. Möller-Gulland (✉)
PricewaterhouseCoopers Private Limited, 17th Floor, Building 10 Tower C, DLF Cyber City, Gurgaon 122002, Haryana, India
e-mail: jennifer.m.gulland@in.pwc.com

M. Lago • K. McGlade • G. Anzaldua
Ecologic Institute, Pfalzburger Strasse. 43/44, 10717 Berlin, Germany
e-mail: Manuel.lago@ecologic.eu; Katriona.McGlade@ecologic.eu; gerardo.anzaldua@ecologic.eu

M. Lago et al. (eds.), *Use of Economic Instruments in Water Policy*, Global Issues in Water Policy 14, DOI 10.1007/978-3-319-18287-2_3

related to the implementation of the EPI are discussed, as well as the EPI's economic, social and distributional effects on the German society. Germany's role as pioneer in the field of environmental taxation as well as the implications of extending the policy mix to the former German Democratic Republic after Germany's reunification in 1990 provide additional interesting angles of analysis.

Keywords Effluent tax • Discharge permits and standards • Germany • Policy mix

3.1 Introduction

Following the broad typologies of Economic Policy Instruments presented in Chap. 1, a pollution (effluent) charge is a fee or tax to be paid on discharges into the environment, based on the quantity and/or quality of discharged pollutants (UN 1997). Pollution charges are commonly linked to different characteristics of the polluter (e.g., sector, processes), the effluents (volume or pollutant concentration) or the recipient type of water body (e.g., surface or groundwater). Unitary rates can differentiate between quantities of pollutants emitted and the level of the economic activity that causes the pollution. Regarding their practical application, a recent review throughout Europe on the applications of the polluter pays and cost recovery principles according to the EC WFD by the European Environment Agency found out that effluent charges are set in most European countries in a way that clearly is aimed at recovering the costs of running the regulatory functions of the responsible authorities (EEA 2013). Although pollution charges remain as the most applied policy tool employed in most European countries to control point source emissions to water, little information is available about the understanding of their interaction with other regulatory or economic instruments that complement the application of charges as part of a policy mix.

This chapter analyses the policy mix of economic and regulatory instruments introduced in Germany to reduce point source pollution.

The policy mix consists of the following instruments[1]:

- Discharge Permits (Federal Water Act, implemented in 1957)
- Effluent Tax (Effluent Tax Act; implemented in 1976)
- Discharge limits and technological standards (Waste Water Ordinance; implemented in 1997)

While all of the above mentioned instruments are considered in the analysis, the focus lies on the effluent tax.

[1] Federal Water Act (Wasserhaushaltsgesetz); Effluent Tax Act (Abwasserabgabengesetz); Waste Water Ordinance (Abwasserverordnung).

3.1.1 Definition of the Analysed EPI and It's Purpose

In Germany all discharges of effluent require a permit. This permit is issued only if the effluent to be discharged is kept as low as possible for the required process and with the best available technology. In 2004, the emission-related requirements, such as pollutants limits and technical standards, were further specified for 57 areas of origin and production sectors by enforcing the Waste Water Ordinance. Permits can be granted temporarily or permanently and can be withdrawn if concerns regarding water protection and management arise (Kraemer 1995).

The effluent tax should implement the "polluter pays principle", i.e. lead to the internalisation of external costs. In conjunction with direct regulations on the discharge of effluents, the effluent tax shall provide an economic incentive to avoid or reduce harmful effluent discharges. The explicit objectives of the effluent charge include (1) mitigating and avoiding the discharge of pollutants into waterways, soil, and drainage systems; (2) maintaining clean water bodies; (3) keeping water treatment plants consistent with the state of the art; (4) developing production processes with less or no wastewater development; (5) and appropriately distributing the costs to mitigate, eliminate, and balance damage to water bodies (Sächsisches Staatsministerium für Umwelt und Landwirtschaft 2011). In addition to its incentive function, the effluent tax should help solve the "implementation deficit" of the states' administrations because part of the revenue can be used for capacity building activities (Kraemer, op cit., p. 8). With the "polluter pays principle" being anchored in the EC Treaty only in 1987, Germany can be said to be among the pioneers in the field of environmental taxation.

3.1.2 Design of the Effluent Tax

The effluent tax (Abwasserabgabe) is based on the aforementioned permits, rather than on actual measurements. The tax rate is based on damage units, which are calculated as the equivalents of pollutants in the discharged effluent. Measured pollutants include phosphorous, nitrogen, organic halogen, mercury, cadmium, chromate, nickel, lead, copper, and indicators on the chemical oxygen demand and the toxicity for fish eggs. It was decided to increase the effluent tax per damage unit stepwise between 1981 (EUR 6.1) and 1986 (EUR 20.5).

Charges can be reduced by 50 % (75 % before 1998) if abatement measures are introduced or sewage treatment plants are constructed or improved. Furthermore, dischargers have the option to "offset the costs of investments in pollution control equipment against their charges," which in the case of municipalities can take the shape of 3-year exemption from the tax (OECD 1997: 41, Smith and Vos 1997: 41). During the first decade of the tax, a hardship clause "allowed for a reduction or even annulment of the tax" (ECOTEC et al. 2001b). This provision was removed in 1989.

Table 3.1 Increase in effluent tax per damage unit, 1981–1997

Year (January)	Effluent charge per damage unit (annual)	
1981	12 DM	EUR 6.1
1982	18 DM	EUR 9.2
1983	24 DM	EUR 12.3
1984	30 DM	EUR 15.3
1985	36 DM	EUR 18.4
1986	40 DM	EUR 20.5
1991	50 DM	EUR 25.6
1993	60 DM	EUR 30.8
1997	70 DM	EUR 35.8

Source: BMU 2005

If the permitted discharge is exceeded in quantity or concentration, disproportionately rising charges apply (BMF 2003: 26). Should this occur more than once, the water authorities of the Länder (the Federal States of Germany) impose additional fees (ECOTEC et al. 2001b: 84). Fines for non-compliance are regulated via the standard fiscal code.

Given the federal nature of Germany, a distinction needs to be made between laws passed at federal level and those passed at Länder level. In Germany, two federal laws determine essential elements of water management: the Federal Water Act (WHG) of 1957 and the Effluent Tax Act (AbwAG) of 1976. These laws are obligatory for the Länder.

The Federal Water Act and Federal Effluent Taxes Act[2] were passed as framework laws, which had to be transposed into the federal state legislation before coming into force.[3] Most Länder introduced the effluent tax in 1981, with others following in 1982–1983. After the reunification of the FRG and the GDR in 1990, the five new federal states adopted the tax as of 1991.

The Federal Effluent Tax Act has been amended several times, leading to substantial revisions with respect to the calculation of damage unit rates (Table 3.1), inclusion of pollutants, and regulations designed to promote investments in water pollution abatement (Kraemer 1995).[4] Despite these amendments, the character of the effluent charge has not fundamentally changed over the years (ECOTEC et al. 2001a: 84).

[2]Current EU legislation has been transposed into the national legislation. As such, the Water Framework Directive has been transposed via the Federal Water Act: the Urban Wastewater Directive via the Federal Effluent Tax Act and the IPPC via both, the Federal Water Act and the Federal Effluent Tax Act.

[3]As part of the Federalism Reform in 2006 the framework law of the Federal Water Act was amended and is now partially replaced by full regulations controlled by the federal government (concurrent legislation).

[4]Amendments were made in 1984, 1986, 1990, 1994, 1996, 1997, 1998, 2001, 2004, 2009, and 2010. For more information, please consult Kraemer (1995): 12–20, Bundesministerium der Justiz (2005).

While indirect discharges into municipal treatment systems are not covered, the sewage charges imposed by municipalities and other (public) operators of sewerage systems allow that effluent charges are passed through to indirect emitters (Gawel and Ewringmann 1994).

The revenue of the effluent tax is earmarked for investments in water quality programs by the Länder, such as the construction of municipal sewage treatment facilities and the administration of water quality programmes (Article 13, AbwAG). The earmarking is intended to complement the tax's incentive effect in improving water quality.

The monitoring and enforcement of effluent charges is the responsibility of the water management authorities. Besides the legal requirement of the operators of water pollution abatement facilities to monitor themselves (Eigenkontrolle) – an activity which can be contracted out to accredited institutions – the water management authorities "monitor the self-monitoring"(Kraemer 1995).

3.2 Setting the Scene: Background on the Introduction of the EPI

Since 1949, 4 years after the end of the Second World War, Germany was divided into the western Federal Republic of Germany (FRG), which had a multi-party democratic system and a social market economy, and the eastern German Democratic Republic (GDR), which was ruled by the communist party and adhered to a planned economy. The exceptionally high growth in pollution-intensive sectors (such as energy, chemicals, and construction) in the post-war period caused serious environmental problems as the construction of wastewater treatment facilities did not keep pace. In addition, Germany did not have the option to dispose wastewater from its industrial areas directly to the sea, which led to highly polluted river systems. Under invariable conditions, a future acceptable water supply as well as other water uses would have been under a serious threat (SRU 1974).

Following re unification in 1990, 75 % and 94 % of the population from the former GDR Länder and FRG Länder respectively, were connected to the public sewage system. By 2007, 96 % of the total population was connected (Destatis 2009).

Between 1975 and 2001, wastewater discharges from public sources increased smoothly by 74 %. Private wastewater discharges, however, reached their peak in 1987 following an increase of 70 % between 1975 and 1987. Between 1987 and 2001, private wastewater discharges decreased by 18 %. Total effluent discharges decreased by 4 % between 1983 and 2001 (see Fig. 3.1).

Regarding industrial wastewater, only 14.3 % is discharged indirectly, i.e., into municipal wastewater treatment plants. The remaining 85.6 % is discharged directly into water bodies. The main industrial sectors directly discharging wastewater into water bodies include the chemical industry (49 %), mining of coal and lignite (22 %),

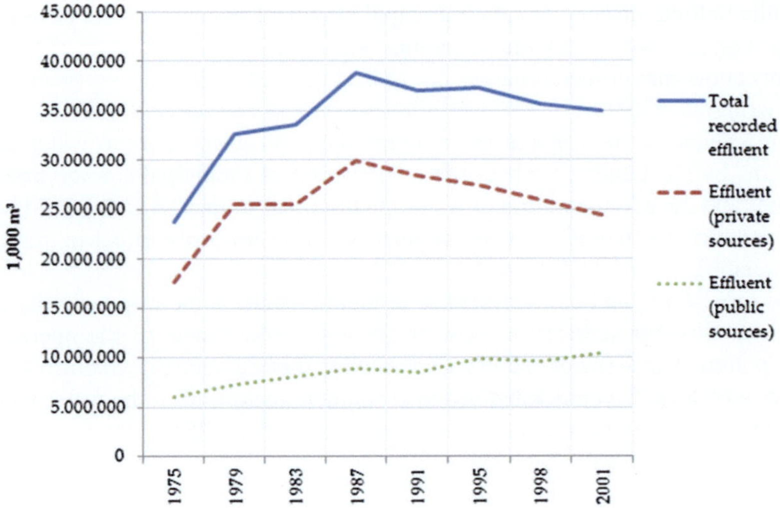

Fig. 3.1 Wastewater discharges by private and public dischargers in Germany, 1975–2001 (Source: UBA (1975–2001))

quarrying earth and other mining (7 %), and the paper industry (6 %). The sectoral breakdown remained stable between 1991 and 2007 (Destatis 2011a).

The effluent charge was introduced in 1976 as a reaction to the insufficient implementation of direct regulation (Federal Water Act) of effluent discharges by the water management administrations of the Länder and the resultant non-compliance with prescribed discharge standards in the private and municipal sectors (Kraemer 1995).

3.3 The German Effluent Tax in Action

3.3.1 The Effluent Tax and the Policy Mix Contribution

Please note that the effluent tax functions complementary to regulatory instruments, i.e., the Federal Water Act and the Waste Water Ordinance, as described earlier in this chapter. As the individual elements of this policy mix are all designed to achieve the same objectives, the single impact of the effluent tax is difficult to disentangle.

3.3.1.1 Environmental Outcomes

The effluent tax sent a signal to effluent dischargers that the government is determined to achieve the objectives set out in the direct regulation. This along with the announcement of the increasing effluent tax rate led to changes in economic agents' behaviour.

Following the introduction of the effluent tax, polluters had the option to choose between investing in pollution abatement, either through effluent treatment or by changing production processes, or paying the effluent tax (Kraemer 1995).

A survey carried out following the announcement of the tax – but before its implementation – found that three-quarters of private enterprises and two-thirds of municipalities had increased, accelerated, or modified their abatement measures for water pollution in anticipation of the tax (Barde and Smith 1997). Investments in equipment for water pollution abatement increased markedly more than 3 years before the effluent tax was instituted (Erwingmann et al. 1980). Barde and Smith (1997) noted that, in this case, even the announcement of an economic instrument was useful for inducing pollution abatement. It increased awareness of the need and potential for water pollution control (Kraemer 1995).

It is generally accepted that the option to offset the effluent tax with investment expenditures in abatement measures has promoted the construction and extension of effluent treatment installations because industrial direct emitters were incentivized to maintain or reduce their number of permits (e.g. Deutscher Bundestag 1994; Kraemer 1995).

The independent Council of Advisors on the Environment (Sachverständigenrat für Umweltfragen) determined the marginal abatement cost curve (MACC) for wastewater treatment plants to assess the optimal effluent tax rate before its introduction (SRU 1974). The MACC shows that costs to remove pollutants of the equivalent of 33 % (standard mechanical treatment) – 70 % of Biological Oxygen Demand (BOD) remain rather constant. Costs increase exponentially beyond the removal of pollutants of an equivalent of 70 % BOD.

As can be seen in Fig. 3.2, the percentage of effluents undergoing secondary and tertiary treatment has increased substantially over the years, with tertiary treatment first being introduced in 1991. In 2007, 92.6 % of effluents in Germany underwent

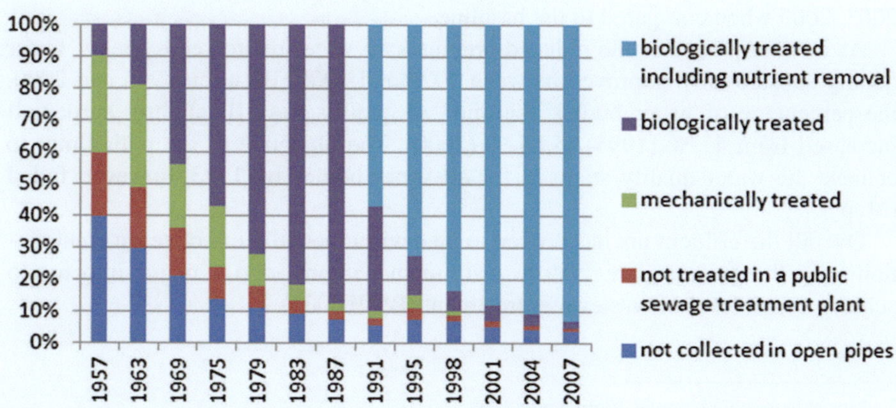

Fig. 3.2 Public effluent disposal per treatment technology in Germany, 1957–2007. Note: Data before German unification in 1990 only includes *Länder* of the Federal Republic of Germany. Values for 1991 represent values for the FRG (*left*), the *Länder* of the former GDR (*centre*) and the average (*left*) (Source: BMU 2011)

tertiary treatment, a percentage which, when compared to other Western European countries, makes Germany a frontrunner.

As the main (direct) effluent discharger, the chemical industry introduced abatement measures that led to significant reductions in discharged pollutants between 1995 and 2006. Reduced pollution discharges included, for example, AOX[5] (−74 %), COD[6] (−55 %), phosphorous (−50 %), and nitrogen (−57 %) (VCI 2006). The paper industry, Germany's fourth largest (direct) effluent discharger, changed production processes to reduce the average waste water volume needed to produce 1 ton of paper from 46 m^3/tonne in 1974 to 11 m^3/tonne in 2002. Clear attribution to the effects of the effluent tax however, remain uncertain.

Investments for effluent treatment by the government, privatised wastewater treatment facilities, and industry totalled EUR 16 billion in the year 2000, exceeding total investments in waste removal, air pollution prevention, and noise abatement (Destatis 2003). Of this figure, around 56 % was used to cover operational expenditures while 44 % covered capital expenditures. A European comparison by the BDEW (2010) revealed that Germany's average investments relating to wastewater (EUR 1.18/m^3) are higher than in the Netherlands (EUR 0.93/m^3), France (EUR 0.97/m^3), and England and Wales (EUR 1.03/m^3). Only Austria showed higher investment levels with EUR 1.44/m^3. Investments for water protection exclusively by enterprises have been decreasing constantly from EUR 914,454,000 in 1992 to EUR 568,005,000 in 2002 (−38 %) (Destatis 2011b). While it cannot be assumed that these investments are exclusively used for effluent abatement measures, it does indicate that abatement measures have achieved the state of the art within the limits of the marginal abatement function of enterprises.

As can be seen in Figs. 3.3 and 3.4 the discharges of mercury and nitrogen to surface water bodies have been reduced significantly from point sources.[7] When compared to the baseline, discharge of mercury could be reduced by 99 % from direct industrial dischargers and by 65 % from municipal treatment plants in 2003–2005. Nitrogen discharges have been reduced by 76 % from point sources in 2003–2005 when compared to the baseline.

As a consequence of the reduced pressures on water-related ecosystems, water quality substantially improved between 1975 and 2000. Between 1995 and 2000, the percentage of water bodies classified as quality class II (slightly burdened) increased from 47 % (1995) to 65 % (2000). The objective of the policy mix to achieve the water quality status II for all water bodies by 1985, however, failed (Map 3.1).

Overall the effluent tax has proven to be environmentally effective. In combination with the enhanced regulatory instruments, it provided a major impetus to achieve a high level of wastewater treatment (BMF 2003).

[5] Adsorbable organic halogen compounds.

[6] Chemical Oxygen Demand.

[7] Mercury and nitrogen are chosen as representatives for heavy metal and nutrient pollutants. Additional pollutant discharges can be requested from the authors.

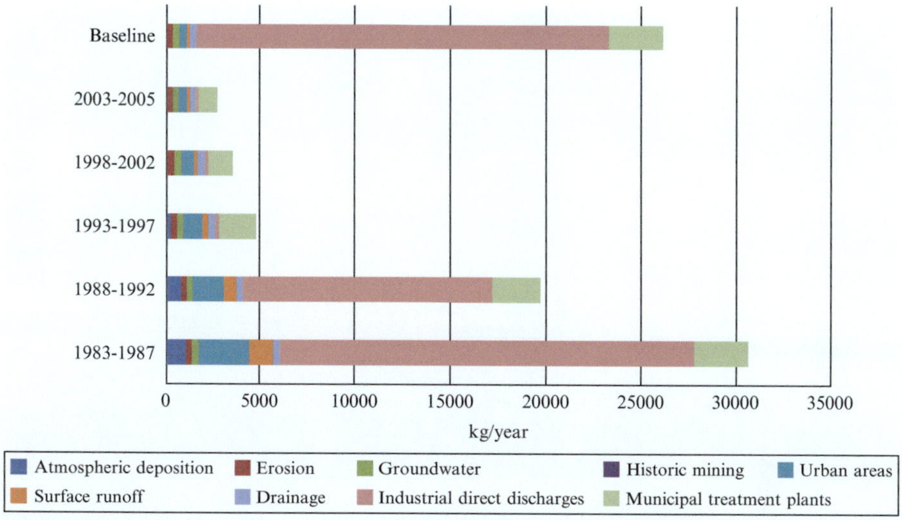

Fig. 3.3 Discharge of mercury from diffuse and point sources into surface water bodies, 1983–2005 including a baseline. Note: The baseline assumes direct discharges from 1983 to 1987 to remain the same, while diffuse source pollution are based on 2003/2005 data (Source: UBA 2010; authors' estimation)

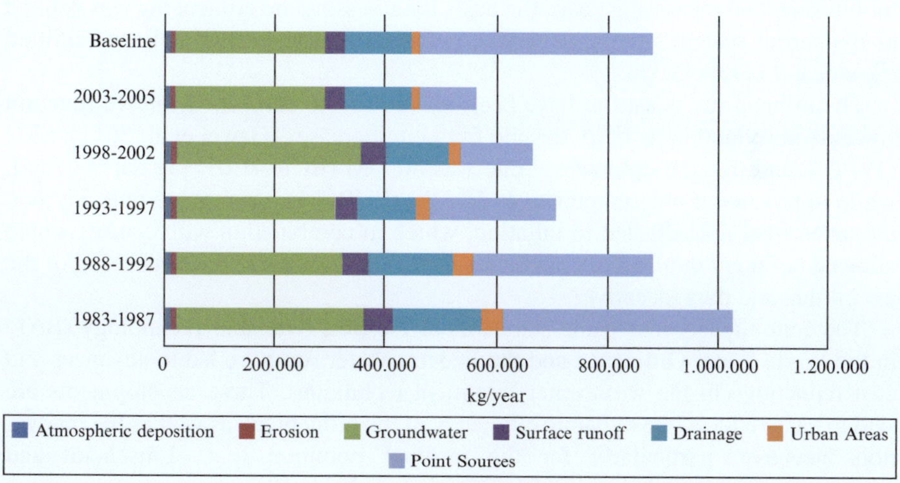

Fig. 3.4 Discharge of nitrogen from diffuse and point sources into surface water bodies, 1983–2005, including a baseline (Source: UBA 2010; authors' estimation)

3.3.1.2 Economic Outcomes

The design of the adopted effluent tax was marked by political compromise and considerations of administrative reality rather than optimising incentive structures. Information asymmetries, e.g. the abatement cost curve of the polluters, further aggravated the design of the EPI along incentives. To reduce administrative,

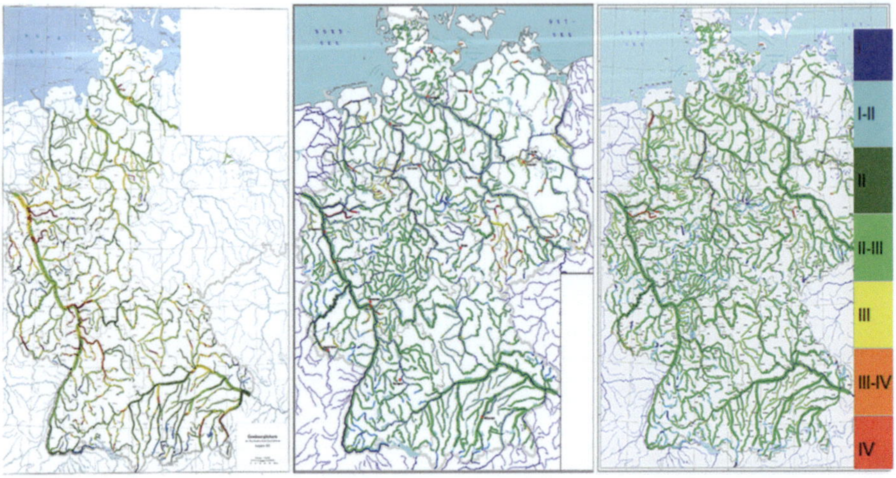

Map 3.1 Water quality classes of German surface water bodies, 1975 (*left*), 1995 (*center*) and 2000 (*right*). Note: The map illustrating the water quality classes in 1975 does not include the water bodies located in the *Länder* of the former GDR. Water quality class I is the best (Source: UBA 2009)

monitoring, and measuring costs, the basis for assessing the effluent tax was defined as the permit system (Art. 4, AbwAG), rather than on the actual effluents emitted (Gawel and Fälsch 2011).

The effluent tax is said to have been set too low to fulfil its incentive function since its introduction in 1976, despite frequent increases (Gawel et al. 2011). SRU (1974) found that the optimal tax rate was 80 DM (EUR 41.03) per damage unit, while in practice it only amounted to 12 DM (EUR 6.1) per damage unit. Further, the taxes were not adjusted to inflation, which in combination with clauses which allowed polluters to offset tax payments over time, led to a real depreciation of the tax burden and thus incentive.

The continuously increasing standards of the Best Available Technology (BAT) in the Waste Water Ordinance and the Federal Water Act have led to advances and cost reductions in the wastewater treatment techniques. These developments are said to have reduced the dynamic efficiency of the effluent tax as well as the innovation incentive, particularly for the residual pollution (e.g., Linscheidt and Ewringmann 1999; Rahmeyer 2001; Gawel et al. 2011: 10).

While differentiation based on regions and water quality levels has been included in the concept of the effluent tax, it is not being applied, leading to common criticism on the incentive alignment (Gawel et al. 2011).

Public sewage treatment plants and industries react differently to the incentives created by the effluent tax (ECOTEC et al. 2001a: 323–234). As public sewage treatment plants do not follow the objective of profit maximisation, they are unlikely to be incentivised to improve compliance beyond technological guidelines (i.e., the regulation). However, they are incentivized not to exceed the thresholds mentioned in the technological guidelines to avoid being penalized (i.e., forego the 50 % reduction of taxes which is granted with compliance and be forced of pay a fine). Industries on the other hand, are profit maximizers and, as such, keen to remain in compliance with the technological guidelines on the one hand, and, in addition, to reduce discharges where the marginal costs of abatement are less than or equal to the effluent tax.

In 1983, the revenues from effluent charges amounted to between EUR 110–205 million in the Länder of the FRG. In 1993 (following reunification), the revenues of the Länder of the FRG, i.e. the "old Länder", steadily increased 3.8-fold. Interestingly, after 1995 the revenue has decreased, despite the latest increase of the effluent tax in 1997. Lower revenue from effluent taxes are expected to indicate the effectiveness of the effluent tax (Kraemer 1995: 34), i.e. as less point source pollution takes place. The revenue from the new Länder, i.e., the Länder formerly belonging to the GDR, does not show any noteworthy trend (Fig. 3.5).

3.3.1.3 Distributional Effects and Social Equity

It could be expected that municipalities carry a disproportionate burden of the effluent tax, as they pay 60 % of the revenues of the effluent tax (RIZA 1995). In Germany, however, fees for water and wastewater are set to recover financial costs fully, thus shifting the potential burden to the consumers. As the effluent tax only makes up 4 % of the annual wastewater taxes to consumers, little or no difference has been noticed by the general public (ATT et al. 2011).

(Potentially) high water polluters, such as the chemical and the paper industry in Germany, can be said to have been disproportionately affected by the introduction of the effluent tax and the increase in regulatory requirements.

However, while no reliable data could be identified as basis for a quantitative analysis, it was stated that the effluent tax had no significant impact on the profits or the competitiveness of the paper industry, as the cost of water, when compared to other cost factors (raw materials, energy, personnel costs) is rather low (PTS Paper

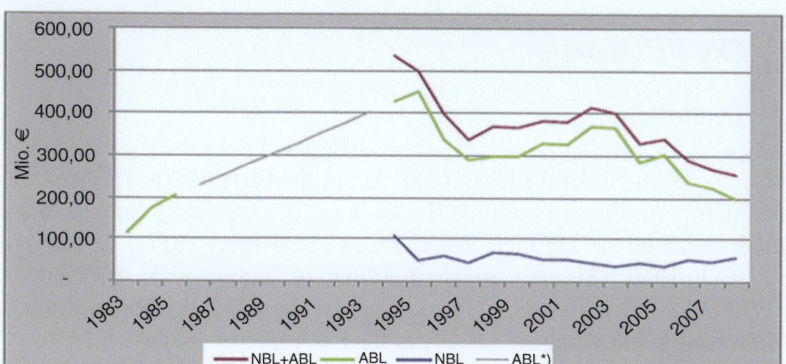

Fig. 3.5 Revenue from effluent taxes, 1983–2007. Note: Data before 1991 only includes the Länder of the FRG. From 1991 onwards, revenue from the new Länder is included. *Red* line: GDR and FRG; *green* line: FRG, i.e. old Länder; *blue* line: Länder of the former GDR, i.e. new Länder; *grey* line: estimation of FRG before 1990 (Source: Gawel et al. 2011: *104*)

2011). On the other hand, the chemical industry, Germany's main effluent discharger, sees the effluent tax as a "pure penalty tax," which only absorbs capital via costly administrative procedures and thus harms the industry's global competiveness (VCI statement n.d.).

The hardship clause, which provides for a reduction or in certain cases annulment of the tax, was removed in 1989. This, together with the stepwise increase of the tax rate, were intended to minimize the negative effects caused to economic agents. While the increasingly stringent regulation may have led to considerable disadvantages in comparison to foreign competitors with lower additional costs for waste water treatment (Rudolph and Block 2001), the effluent tax was found to have only a small effect on competitiveness.

The fact that the effluent tax is based on permits, rather than on actual emissions, can result in an imbalanced burden of the effluent tax. For a more equal distribution of the tax and an improved steering function, the association for local public utilities in Germany (VKU) calls for the reflection of the polluter pays principle (VKU 2011).

Generally it can be said that the constant need to adapt to abatement requirements and the incentive to innovate has brought greater technological development and efficiency improvements to the German industry, in turn strengthening its global competitive advantage in this area (Rudolph and Block 2001).

As the regulatory instruments, i.e., the Waste Water Regulation, are administered by the same authorities who administer the economic instrument, i.e., the effluent tax, the revenue from this is partly used to cover administrative costs and to employ additional staff. The increased information requirements, such as surveying and modeling water bodies, and the documentation of effluent discharges allowed for the development of a solid basis of information with which administrative functions could be improved. The introduction of effluent taxes further led to increased coordination between the water management administration and the water dischargers, improving conflict resolution mechanisms and intervention capacities.

3.3.2 The EPI Set-Up

3.3.2.1 Institutions

The inception of the federal Effluent Tax Act in the early 1970s occurred during a re-orientation period in the political life of the FRG, following the election of the first government of the German Social Democratic Party (SPD) and the Liberal Party (FDP) in 1969. This government identified protection of the environment as a major new policy area and initiated measures to establish the institutional framework for environmental policy, notably at the federal level (Kraemer 1995).

This re-orientation was mainly necessary due to the exceptionally high growth in pollution-intensive sectors (such as energy, chemicals, and construction) in the post-war period, which caused serious environmental problems, as the construction of wastewater treatment facilities did not keep pace (SRU 1974).

Since the introduction of the effluent tax, external factors have influenced its design. As such, the recession following the oil shock in the mid 1970s resulted in a reduction of the planned tax rate and a deferral of its introduction. The third amendment of the effluent tax occurred at the same time as the massive algal blooms in German coastal waters and the consequent widespread decline in seal populations in the North Sea and thus benefited from an intense public interest in water pollution. In this climate, nitrogen and phosphorous were included in the damage units and the effluent tax rate was increased substantially – the last point was revoked in the fourth amendment (1994) in the face of increased investment needs in the Länder of the former GDR[8] following reunification and an economic recession (Kraemer 1995).

3.3.2.2 Transaction Costs

The transaction costs related to administration as a percentage of revenue generated from the effluent tax for the Länder are illustrated in Table 3.2.

The transaction costs were significantly reduced from 47–48 % of revenues in 1982 to 13–21 % of revenues in 2006–2009, showing that administrative procedures need time to adapt and be optimized. Bavaria achieved the highest reduction from 122 % in 1982 to 22 % in 2006–2009.

It should be noted that the percentage of administrative costs varies with the amount of wastewater discharged and with the amount of dischargers offsetting tax obligations with investment expenditures. The high variance between the Länder can be further explained by the heterogeneity of the assessment methodologies of the Länder. No federal guidelines exist for the definition of administrative costs for effluent taxes – as such, some Länder may include further cost factors.

While the public sector faces annual transaction costs of approximately EUR 32.5 million, the private sector is burdened with a charge of around EUR 65 million annually to comply with the information requirements introduced by the effluent tax (Destatis 2008). Interestingly, the most frequent and second most expensive transaction is the proof of eligibility for tax exemption or reduction. This illustrates that despite the high cost of this transaction, offsetting expenditures is still a rational economic decision. The most expensive transaction (questionnaire on effluent quality and quantity if no permit has been received) is similarly a specialised and thus the least occurring one.

---

[8] In the former GDR, sewage systems were in a very poor condition, if they existed at all, and were not widely available. At the time of unification in 1990, the heavily contaminated water bodies in the GDR required substantial and sustainable sanitation measures. As such, more than 2,000 treatment plants were constructed and complete industrial sectors were improved to match Western German standards (BMU 2001).

Table 3.2 Annual administrative costs as percentage of revenue from effluent taxes for the Länder, 1982 and 2006–2009

	1982	2006–2009
Baden-Württemberg	54 %	8–28 %
Bavaria (Bayern)	122 %	22 %
Berlin	n/a	1.5–3 %
Brandenburg	n/a	n/a
Bremen	n/a	25–40 %
Hamburg	14 %	1–3 %
Hesse (Hessen)	27 %	n/a
Mecklenburg Western Pommerania (Mecklenburg-Vorpommern)	n/a	13 %
Lower Saxony (Niedersachsen)	49 %	2.5–11 %
North Rhine-Westpahlia (NRW)	36 %	7–21 %
Rhineland-Palatinate (Rheinland-Pfalz)	47–51 %	12 %
Saarland	n/a	7–45 %
Saxony (Sachsen)	n/a	15 %
Saxony-Anhalt (Sachsen-Anhalt)	n/a	20 %
Schleswig-Holstein	49 %	38–48 %
Thüringen	n/a	7–8 %
Average	47–48 %	13–21 %

Source: Gawel and Fälsch (2011) and Kraemer (1995)
Data for 1982 does not include *Länder* from the former GDR. Values for the years 2006–2009 need to be treated with caution–for some *Länder* static values for a given year were provided; for others a percentage of revenues was provided. Years of the underlying data may differ. To the authors' best knowledge, no more up to date data was available to the time of publication

3.3.2.3 Policy Implementability

The effluent tax is a rather flexible economic instrument and was changed before and after its implementation to account for current circumstances. The original proposal of the effluent tax had to give way to political compromise and administrative realities in order to achieve its implementation (Kraemer 1995).

As the Effluent Act is a framework law and had to be transposed by the Länder into federal state legislation, the Länder had the power to adapt a number of aspects such as treatment of rainwater run-off, schedules or exemptions for small emitters, procedures relating to indirect emitters, and administrative procedures. Thus Länder could influence the level, and thus the economic impact, of effluent taxes (Kraemer 1995).

Additionally, the effluent taxes were amended several times–mainly to adjust the calculation of damage unit rates, inclusion of pollutants, and regulations designed to promote investments in water pollution abatement (Kraemer 1995).

When the effluent tax came into force in the Länder of the former GDR, industries that were not expected to pay taxes previously were subject to the tax first in 1993 instead of in 1991 (ECOTEC et al. 2001a).

It can be assumed that the reduction of the effluent tax rate, its stepwise increase and the hardship clause increased the acceptance of the introduction of the effluent tax.

Before the adoption of the effluent tax, certain Länder were against the introduction of such an instrument, arguing that the administrative costs would be too high–particularly regarding the measurement of pollutants. These Länder appear also to have the highest administrative charges following the enforcement of the Effluent Tax Act (Michaelis 1996).

Several industries were concerned about how these additional costs would harm their competitiveness; in Cologne a survey showed that 10 % of the companies feared that the effluent tax would threaten their future existence (ECOTEC et al. 2001a: 86). It can be said that the participation of the dominant players, i.e., the Länder municipalities and industry, led to the rejection of the design of the effluent tax that would have led to maximal economic efficiency and impact (Troja 1998: 81).

3.4 Conclusions and Lessons Learnt

This case study illustrates that a policy mix consisting of regulatory and economic instruments can be very powerful in implementing and enforcing policies to address direct effluent emissions. While the policy mix and the environment in which it functions make it difficult to single out the impact of the effluent tax with certainty, it can be stated that the policy mix as a whole achieved most of its objectives:

- The quantity of overall discharges of pollutants into water ways was reduced by 4 %, while discharges of private emitters were decreased by 18 %. The harmfulness of effluents was decreased substantially. Mercury discharges were reduced by 99 % from industrial dischargers and by 65 % by municipal treatment plants in 2003–2005, when compared to the baseline of 1987. Nitrogen discharges from point sources were reduced by 76 % in 2003–2005 when compared to the baseline of 1987.
- The quality of water bodies increased substantially, with 65 % of all surface water bodies achieving a water quality II status. The concrete objective, however, of improving all water bodies to water quality II status by 1985, failed.
- Waste water treatment plants were upgraded to the state of the art. In 2012, 92.6 % of effluents in Germany underwent tertiary treatment–a percentage which, when compared to other Western European countries, makes Germany a frontrunner of advanced wastewater treatment standards.
- Industries, such as the paper industry, developed production processes which required less wastewater development. Others, like the chemical industry, invested in effluent abatement measures and considerably reduced their discharge of pollutants.
- The costs to mitigate, eliminate, and balance damage to water bodies were distributed among the polluters, which reflects a successful implementation of the polluter pays principle.

While good results have been achieved in terms of environmental outcomes, the policy mix has been deprived of the effluent tax's essential contribution to achieve its objectives. This is mainly due to the challenge to create the right incentive structure to achieve the targeted objectives. It was found that the effluent tax rate has been set too low since its introduction in 1979, and has not been adjusted to inflation. As the cost of measures for abatement have increased with inflation, and as the standards for BATs in the Waste Water Ordinance have become more stringent, the effluent tax could not develop its full potential for setting innovation incentives to abate residual pollution.

The reasons for the failure to create the correct incentive structure can be found in the policy implementation process and the institutional settings. The participation of dominant players, i.e., the Länder, municipalities, and industries led to the rejection of the effluent tax design that would have had the optimal incentive structure. It can be said that political compromise and administrative realities, such as capacity and budget issues, shaped the effluent tax's current design in order to simplify implementation. Further, external shocks influenced the effluent tax rate. While the economic crisis aggravated the potential to increase tax rates, the algae bloom in German waters, which led to a widespread decline in seal populations, raised public awareness in water pollution and led to a slight increase of the effluent tax rate.

A further finding of this case study shows that the incentives created by the effluent tax may be different for private and public dischargers. Mostly profit-seeking agents (i.e., private industry) changed their behaviour as a reaction to the effluent tax, while municipalities prioritised the (mere) compliance with standards, foregoing further possible reductions in the effluent tax.

Finally, the introduction of the effluent tax led to significant capacity building in the water management administration and a consequent decrease in public administration costs over time.

Given that most point source pollution does not anymore pose a serious issue in Germany and that the incentive to abate residual pollution is weak, the effluent tax should be updated to reflect today's conditions. In addition, an analysis of the potential for a discharge permit trading system would enlighten discussions about Germany's future policy options.

References

ATT (Arbeitsgemeinschaft Trinkwassertalsperren e. V), Bundesverband der Energie- und Wasserwirtschaft e. V. (BDEW), Deutscher Bund der verbandlichen Wasserwirtschaft e. V. (DBVW), Deutscher Verein des Gas- und Wasserfaches e. V. – Technisch-wissenschaftlicher Verein (DVGW), Deutsche Vereinigung für Wasserwirtschaft, Abwasser und Abfall e. V. (DWA), & Verband kommunaler Unternehmen e. V. (VKU). (2011). *Branchenbild der deutschen Wasserwirtschaft*. Bonn: wvgw Wirtschafts- und Verlagsgesellschaft.

Barde, J. P., & Smith, S. (1997). Do economic instruments help the environment?, *The OECD observer,* No. 204. Paris: OECD

BDEW. (2010). *VEWA- Vergleich Europäischer Wasser- und Abwasserpreise*, Berlin: Bundesverband der Energie und Wasserwirtschaft e.

BMF. (2003). *Die Förderung des Umweltschutzes im deutschen Abgabenrecht. Fachblick: Finanz- und Wirtschaftspolitik*. Berlin: Bundesministerium der Finanzen.

BMJ. (2005). *Bundesgesetzblatt Jahrgang,* Teil I Nr. 5. Berlin: Bundesministerium der Justiz.

BMU. (2001). *Der Wassersektor in Deutschland Methoden und Erfahrungen, Gewässerschutz – Kostenfaktor und Entwicklungsmotor für Industrie und Umweltschutz,* Berlin: Bundesumweltministerium.

BMU. (2005). *AbwAG – Abwasserabgabengesetz (Gesetz über Abgaben für das Einleiten von Abwasser in Gewässer) vom 13. September 1976, mehrfach novelliert, Bekanntmachung der aktuellen Fassung des Gesetzes am 18. Januar 2005*. Berlin: Bundesumweltministerium.

BMU. (2011). *Das neue Wasserhaushaltsgesetz: Endlich ein bundeseinheitliches Abwasserrecht, das modernen Erfordernissen Rechnung trägt*. Berlin: Bundesumweltministerium http://www.bmu.de/binnengewaesser/gewaesserschutzrecht/deutschland/doc/47394.php. Accessed 12 Nov 2011.

Deutscher Bundestag. (1994). *Bericht der Bundesregierung an den Deutschen Bundestag über die Auswirkungen der 2. Und 3. Novelle zum Abwasserabgabengesetz auf die Gewässer.* Drucksache 12/8344, 1 Aug 1994.

Ecotec et al. (2001a). ECOTEC in association with CESAM, CLM, University of Gothenburg, UCD and IEEP "Economic and environmental implications of the use of environmental taxes and charges in the European union and its member states." Final report. http://ec.europa.eu/environment/enveco/taxation/environmental_taxes.htm

Ecotec et al. (2001b). Waste water. In *Study on the economic and environmental implications of the use of environmental taxes and charges in the European Union and its Member States*. Final Report. http://ec.europa.eu/environment/enveco/taxation/pdf/ch7_waste_water.pdf.

Erwingmann, D., Kibat, K., & Schafhausen, F. J. (1980). *Die Abwasserabgabe als Investitionsanreiz, Berichte, 8/80*. Berlin: Erich Schmidt Verlag.

European Environment Agency. (2013). *Assessment of cost recovery through pricing of water*. EEA technical report no 16/2013. ISSN 1725-2237.

Gawel, E., & Ewringmann, D. (1994). *Abwasserabgabengesetz und Indirekteinleitung. Finanzwissenschaftliche Forschungsarbeiten, Neue Folge 61*. Berlin: Dunker & Humblot.

Gawel, E., & Fälsch, M. (2011). Zu den Lenkungswirkungen von Wasserentnahmeentgelten. DWA Hennef: *Korrespondenz Wasserwirtschaft*, 4. Jg., forthcoming.

Gawel, et al. (2011). *Weiterentwicklung von Abwasserabgabe und Wasserentnahmeentgelten zu einer umfassenden Wassernutzungsabgabe*. Endbericht.

Kraemer, R. A. (1995). *The effectiveness and efficiency of water effluent charge systems: Case study on Germany*. Paris: OECD.

Linscheidt, B., & Ewringmann, D. (1999). *Innovationseffekte von Umweltabgaben in der Industrie*. In Linscheidt, B. (Ed.), *Umweltinnovationen durch Abgaben*. Berlin: Duncker & Humblot.

Michaelis, P. (1996). *Ökonomische Instrumente in der Umweltpolitik. Eine anwendungsorientierte Einführung*. Heidelberg: Physica-Verlag.

OECD. (1997). *Evaluating economic instruments for environmental policy*, (pp. 40–41). Paris: Organisation for Economic Co-operation and Development. http://www.oecd.org//env/policies/publications.htm#eistrpub

PTS Paper. (2011). Personal communication with Holger Jung. November 11, 2011.

Rahmeyer, F. (2001). *Organisationsformen und Umweltabgaben im kommunalen Gewässerschutz.* Augsburg: Volkswirtschaftliche Diskussionsreihe, Institut für Volkswirtschaftslehre der Universität Augsburg 212

RIZA. (1995). *Waste water charge schemes in the European Union, Part 1–2*. Lelystad: Institute for Inland Water Management and Waste Water Treatment.

Rudolph, K. -U., & Block, Th. (2001). Der Wassersektor in Deutschland – Methoden und Erfahrungen –Wissenschaftliche Bearbeitung der Dokumentation in Kooperation des Bundesministeriums für Umwelt, Naturschutz und Reaktorsicherheit und des Umweltbundesamtes, 1. Aufl., Oktober 2001, ISBN 3-934898-25-4, Hrsg.: Bundesministerium für Umwelt, Naturschutz und Reaktorsicherheit und Umweltbundesamt. Dessau.

Sächsisches Staatsministerium für Umwelt und Landwirtschaft. (2001). *Abwasserabgabe*. http://www.umwelt.sachsen.de/umwelt/wasser/1909.htm. Accessed 01 Sept 2011.

Smith, S., & Vos, H. B. (1997). *Evaluating economic instruments for environmental policy*. Paris: OECD.

SRU. (1974). *Die Abwasserabgabe: Wassergütewirtschaftliche und gesamtökonomische Wirkungen, Sondergutachten des Sachverständigenrates für Umweltfragen*. Wiesbaden: W.Kohlhammer GmbH.

Troja, M. (1998). *Umweltpolitik und moderne Ökonomik: Der Beitrag der Neuen Politischen Ökonomie und der Neuen Institutionenökonomik zur Erklärung umweltpolitischer Entscheidungsprozesse*. Münster: LIT Verlag.

UBA. (2009). *Biologische Gewässergüte*. Oberflächengewässer– Atlantis, Umweltbundesamt. http://gis.uba.de/website/web/atlantis/karten/gewguete.htm. Accessed 10 Nov 2011.

UBA. (2010). *Beherrschung der Gewässereutrophierung,* Umweltbundesamt http://www.umweltbundesamt.de/wasser/themen/trinkwasser/eutrophierung.htm. Accessed 10 Oct 2011.

United Nations. (1997). *Glossary of environment statistics* (Studies in methods, Series F, Vol. 67). New York: United Nations.

VCI. (2006). *Vergleich Europäischer Wasser- und Abwasserpreise Studie, German Chemical Industry, Responsible Care Report, Verband der Chemischen Industrie e.V.* Bonn: WVGW.

VCI. (n.d.). *VCI-Stellungnahme zum Konzept des Bundesministeriums für Umwelt, Naturschutz und Reaktorsicherheit für eine "Ökologische Industriepolitik"*. Verband der Chemischen Industrie e.V.

VKU. (2011). *Weiterentwicklung von Abwasserabgabe und Wasserentnahmeentgelten zu einer umfassenden Wassernutzungsabgabe*. Berlin: Forschungsbericht Verband kommunaler Unternehmen e.V.. http://www.vku.de/wasser/wirtschaft/abwasserabgabe/abwasserabgabe.html. Accessed 11 Oct 2011.

Data Sources

Destatis. (2003). *Wirtschaft und Statistik*. Wiesbaden: Federal Statistical Office Germany.

Destatis. (2008). *Informationspflichten* (für Gesetz AbwAG). Wiesbaden: Federal Statistical Office Germany. https://www-skm.destatis.de/webskm/online/online;jsessionid=413E9AB15116714D895DB6BEAFAF53D8.tomcat_SKM_1_1?operation=previous&step=0. Accessed 6 Nov 2011.

Destatis. (2009). *Fachserie 19 Umwelt, R. 2.1 Öffentliche Wasserversorgung und Abwasserbeseitigung (1995, 1998, 2001, 2004, 2006)*. Wiesbaden: Statistisches Bundesamt. http://www.bmu.de/binnengewaesser/statistik/doc/2833.php. Accessed 12 Nov 2011.

Destatis. (2011a). Online Database "GENESIS" *Abwasser: Deutschland, Jahre, Abwassereinleitungsarten, Produktionsbereiche*. Wiesbaden: Federal Statistics Office (Destatis). www.destatis.de/genesis. Accessed 12 Nov 2011.

Destatis. (2011b). Online Database "GENESIS" *Investments for environmental protection by enterprises in industry*, (pp. 1992–2009).

Destatis. (2011c). *Volkswirtschaftliche Gesamtrechnungen – Bruttoinlandsprodukt, Bruttonationaleinkommen, Volkseinkommen – Lange Reihen ab 1950*. Wiesbaden: Statistisches Bundesamt.

Chapter 4
The Water Load Fee of Hungary

Judit Rákosi, Gábor Ungvári, and András Kis

Abstract The chapter reviews the operation of and experience with the Water load fee (WLF) introduced in Hungary in 2004. The WLF is an effluent charge imposed on industrial facilities and wastewater utilities that discharge their effluents directly into surface water. This instrument supplements a command and control regulation that sets pollution limits and imposes fines in case of non-compliance. The chapter inspects the interaction of the two instruments, while also assessing their institutional background. The latter is important in understanding how the evolving institutional structure within a transition economy puts limits to developing efficient EPIs, while the conflicting goals and priorities of the stakeholders can further distort the design and operation of the instrument. The allowance provision of the WLF offers an example of a ripple effect generating inefficient allocation of investment resources in the adjoining market of laboratory services. The case provides an example for the different roles an EPI can play in environmental policy as a regulatory instrument to influence behaviour or an instrument to raise revenue for further defined goals based on environmental principles.

Keywords Effluent charge • Economies in transition • Environmental tax • Command and control regulation • Discharge limits

4.1 Introduction

The chapter summarizes the case study of the Water load fee (WLF), an effluent charge introduced in Hungary in 2003. The WLF had been long planned as the cornerstone of environmental regulation, but finally it was not implemented as a

J. Rákosi (✉)
ÖKO Co. Ltd, Attila út 16, 1013 Budapest, Hungary
e-mail: juditrakosi@gmail.com

G. Ungvári • A. Kis (✉)
Regional Centre for Energy Policy Research, Corvinus University of Budapest, Fővám tér 8, 1093 Budapest, Hungary
e-mail: gabor.ungvari@uni-corvinus.hu; ungvarigabor@gmail.com; andras.kis2@uni-corvinus.hu

M. Lago et al. (eds.), *Use of Economic Instruments in Water Policy*, Global Issues in Water Policy 14, DOI 10.1007/978-3-319-18287-2_4

stand-alone solution, only as a supplement to the pre-existing command and control regulation. The latter was introduced in 2001 in order to comply with the EU waste water standards defined by Directive 91/271/EEC on Urban Wastewater Treatment, and the subsequent WLF was not harmonised with it, even though the two instruments are imposed on a comparable set of polluters. The resulting policy mix generated marginal environmental benefits compared to the command and control regime, with moderate changes in polluting behaviour.

After several failed attempts Act No. 89 of 2003 on Environmental Load Fees was passed in 2003 as a result of the Ministry of Finance's promotion of the bill as part of an attempt to improve the revenue source of the public budget. The act introduced three kinds of fees: an air load fee, a water load fee and a soil load fee.

The WLF is imposed on point sources and it is assessed based on the total measured amount of pollutants and the estimated damage assigned by the regulation to each pollutant. Nine contaminants are regulated: COD, phosphorus, inorganic nitrogen, mercury, cadmium, chrome, nickel, lead and copper. All polluters that discharge contaminants into surface water are required to pay the WLF. Water utility companies recover the tax through their wastewater tariffs, thus the final users of wastewater services also pay their share of the fee.

The environmental load fees had originally been envisioned by the Environmental Protection Act (EPA, Act No. 53 of 1995) to reach a complex set of goals: to encourage polluters to reduce their pollution (incentive function); to enforce the polluter-user pays principle as each unit of emission is subject to payment; and to earmark a significant share of fee revenues for the reduction of the environmental burden. As shown within chapter, these goals have been attained with various levels of success.

The case study, especially when compared to the effluent charge system of Germany (described in Chap. 3), helps to illustrate that even a single instrument can be introduced in multiple ways and with various designs, leading to materially different outcomes. Fine-tuning an effluent charge based on the targeted pollution reduction and the existing regulatory environment seems indispensable.

4.2 Setting the Scene: Challenges, Opportunities and EPIs

The transformation of the Hungarian economy in the beginning of the 1990s bankrupted the most out-dated heavy industries of the country and introduced incentives for rational resource use, manifesting itself, among others, in declining water consumption and lower effluent discharge levels. The newly built industrial facilities employed more advanced technologies, lowering the per capita environmental impact of economic growth. The impact of the upgrade of core technologies on pollution abatement, nevertheless, has its limits, and additional efforts, reducing specifically effluent discharges were needed. Moreover, the uptick of economic activities in the early 2000s generated additional pressures on the environment.

As another consequence of economic transition, a growing share of pollution originated from households, since investments into municipal wastewater treatment plants lagged behind industrial pollution abatement efforts. During the late 2000s 85–90 % of all effluent discharge originated from the water and wastewater utilities (NRBMP 2010c), about half of which took place in Budapest which did not have its final wastewater treatment plant completed until 2010. The development of municipal wastewater treatment became the most critical measure to reduce effluent discharges.

The legacy of the economic downturn that accompanied the market transition of the 1990s created strong interests against imposing additional burden on the industry. At the same time the EU accession process and the demand of society for reduced environmental threats advanced in line with the strengthening of environmental and community regulations. These opposing forces resulted in a regulatory structure with insufficient resources and a weak mandate to exercise increased regulatory authority.

Prior to the eventual introduction of the WLF in 2004, a fundamental change had occurred in the regulation of water protection. In order to reduce effluent discharges and to be in compliance with EU requirements (91/271/EEC), the water protection regulation was completely reorganised in 2001. A new system of licensing, discharge limit values, area categories, monitoring, self-monitoring, data submission, transition periods, fines, etc. was created (Government Decree 203/2001, later replaced by Government Decree 220/2004 and its implementation decrees). In accordance with the water protection regulation, the prescribed limit values were to be fulfilled by already existing industrial facilities and wastewater treatment plants by the 31st of December 2010, while newly built facilities were subject to it immediately.

The impact assessment of this regulation envisioned a significant improvement of the environment (ÖKO Co. Ltd 2001). Altogether an approximately 30–40 % decline in the level of damage caused by industrial polluters after the expiration of the initial transition period had been foreseen.[1] The effectiveness of the regulation of discharge limits was aided by a system of fines on excess pollution. Substantial efforts to reduce pollution were already under way when the WLF was introduced.

4.3 The Water Load Fee in Action

4.3.1 The EPI Contribution

4.3.1.1 Environmental Outcomes

Pollutant emissions in 2007 were already significantly lower than their 2002 level, the decline in BOD, nitrogen and phosphorous emissions was 83 %, 50 % and 57 %, respectively (NRBMP 2010a; Ministry of Environment and Water, Government of

[1] Estimated quantities of pollutants were converted into "dangerousness units" based on pre-set rates defined by the regulation in order to create a uniform measure of damage.

Hungary 2005). While it is impossible to quantitatively separate the impact of the regulation on discharge limits and that of the WLF, a larger portion of the abatement is assumed to be associated with emission limit values, while the WLF has delivered an additional, but lower overall impact.

The WLF regulation is only applied to point source pollution and does not cover all pollutants. However, in accordance with the WFD approach, the WLF covers a number of important substances (organic matters, nutrients, and other dangerous substances). The status of the waters is also influenced by contaminants other than pollution from point sources (e.g. diffuse sources) and also other impacts (hydromorphological intervention, abstraction, and other pressures like recreation, transport, excess water diversion, etc.).

A survey on the experience of the first 2 years after the introduction of the water load fee was carried out in 2006 among public utility companies with the participation of 21 water and wastewater utilities (Bereczné et al. 2006). The survey revealed that 24 % of the companies (five water utilities) modified their development/investment plans and the technology of existing wastewater treatment as a consequence of the introduction of the water load fee. Investments to reduce ammonia, OSE, phosphorous, and dichromate oxygen were planned, entitling these companies to a reduction of WLF payment according to the provisions on rebate (for details see the next section on "Economic Outcomes" (Sect. 4.3.1.2)). It was clear, however, that the introduction of the WLF alone would not have been enough to bring about substantial investments, like the construction of a new wastewater treatment plant or a full technological upgrade of an existing one. But in conjunction with the regulation on emission limits, it accelerated pollution abatement measures. It also provided an incentive for the continuous monitoring and improvement of the existing wastewater treatment technologies in order to make them more efficient.

Due to the combined effect of the discussed regulatory changes and subsequent investments, but also other forces (e.g. improved cost recovery) the tariffs charged by water utilities increased. Consumers respond to higher tariffs by lowering their consumption, although the demand elasticity of water utility services is generally low. The average annual per capita water use declined from 39 m^3/year in 2003 to 35.9 m^3/year in 2009 (NRBMP 2010b), and a share of this decline may have been due to the price increasing impact of the WLF.

As an indirect, longer term beneficial effect on the environment, the measurement of the quality of emitted wastewaters improved as a result of the introduction of the WLF, since for the first 7 years after its introduction, the WLF regulation allowed dischargers to retain part of their WLF payment if they spent it on monitoring equipment.

4.3.1.2 Economic Outcomes

Assessing the economic efficiency of the WLF regulation is difficult for two reasons. First, as already described above, separating the impacts of the regulation on discharge limits and the WLF is virtually impossible. Second, no formal regulatory impact assessment has been carried out since the introduction of the WLF. Prior to its adoption, impact assessments had been conducted, but not in conjunction with the regulation on discharge limits (ÖKO CO. Ltd. 2000).

The main reason for the introduction of the environmental load fee was the need to generate revenues in order to fill part of the deficit of the central budget. To shield the polluting entities from a sudden burden, the fees determined by the Act were phased in gradually. In the first few years, during 2004–2007, only an annually rising share of the calculated fees had to be paid, starting from 20 % in 2004 to reaching 100 % by 2008.

In order to promote pollution reducing activities, in certain cases the WLF regulation allowed for significant reductions of fee payments. The rationale for the reduction of the payment was that the burden falling on the organizations carrying out infrastructural investments serving environmental protection goals would be eased and they would thus be encouraged to undertake these investments. The Act on Environmental Load Fees defines circumstances under which given expenditures can be deducted from payments to the central budget as follows:

- Firms that carry out investments that cut effluent discharges directly into surface waters are eligible for a 50 % water load fee reduction during the years of the investment, up to a maximum of 5 years. This rule is still in force today.
- In the year of the purchase, 80 % of the purchase price of measurement instruments of water quality and quantity can be deducted from the WLF advance fee paid by the polluter. There has been only one substantial amendment in the environmental load fee regulation since 2004: from January 2011 this allowance is no longer available.

For 2004 budgetary income of about EUR 55.6 million[2] was planned from the WLF, based on the forecasts of the socio-economic impact assessments. Nevertheless, actual revenues were well below the expected amounts. Between 2004 and 2013 the annual income of the central budget from WLF ranged between EUR 7.5 and 31.8 million, as a combined result of increasing WLF rates, the fluctuating use of the allowances for investments and instrument purchase, and declining effluent discharge. The incoming revenue is not earmarked.

In spite of the previously mentioned incentives it was generally expected that the major wastewater treatment investments would be carried out even in the absence of the WLF regulation, especially as the development of urban wastewater treatment infrastructure was addressed in the framework of the National Wastewater Programme financed with the help of EU grants. This assumption was reinforced during interviews with several water utility service providers and their association, MAVÍZ (Bereczné et al. 2006). The 50 % WLF discount related to pollution abatement investments did not provide much incentive in itself. The low level of motivation is also a consequence of the fact that the total amount of the WLF can be passed to the users, i.e. the actual burden was borne in part by those using the service (the general population, institutions, industry). Meanwhile, due to their local

[2] 2004 current prices, exchanged from HUF on the average annual exchange rate for 2004.

embeddedness a number of water utilities tried to reduce the amount of the WLF paid by consumers, considering the poor economic position of these actors.

The situation of industrial wastewater dischargers was similar inasmuch as those emitting above prescribed discharge limit values were very likely to carry out investments independently of the existence of the WLF. As a result of the stringent water protection regulation (high fines and other sanctions), industrial dischargers are compelled to reduce their emissions.

A significant share of the water utility companies – though to different extents and with different levels of agility – took advantage of the 80 % rebate option offered by the Environmental Load Fee Act for the purchase of measuring devices. The utilities claim that buying measurement instruments was practical and beneficial – nevertheless, these claims are difficult to verify. There are contradictory opinions as well, according to which too many of these instruments were purchased by the water utilities and some of the devices were handed over to others through leasing contracts. The technical level of existing laboratories, nevertheless, significantly improved and this contributed to compliance with self-reporting requirements. The purchase allowance for measurement instruments, however, adversely affected private laboratories. It clearly had a market distorting, anti-competitive effect. Thus, on the whole, this policy resulted in a needlessly expensive and ineffective allocation of resources.

Between 2003 and 2012 the average drinking water tariff in Hungary increased from EUR 0.67 to 1.14/m^3, a 70 % rise. During the same period the average waste water tariff climbed from EUR 0.57 to 1.29/m^3, a 126 % increase (KSH 2014). Wastewater tariffs rose more steeply primarily because of the large scale investments into sewers and municipal wastewater treatment plants, with an additional, but less significant effect of the WLF being included in wastewater tariffs. At present, the WLF makes up around 0.5–11 % of the average sewage tariff with large variations among water utility companies, settlements and service users. There are multiple reasons for this wide range: the level of the WLF burden itself differs, for some utilities it is just a few euro cents per cubic meter, while in some cases it reaches EUR 0.15/m^3. Wastewater tariffs themselves also largely vary. In 2009 the average non-household sewage tariff was 43 % higher than the average household tariff, while a 23-fold difference was observed between the lowest and the highest sewage tariff within the country.

In sum, the WLF was introduced primarily with the goal of revenue generation and it has more or less fulfilled this role, even though environmental load fee revenues did not reach the originally intended level. The WLF provides limited incentives to reduce effluent discharges. The fee level and the structure of incentives provided by the regulation are not sufficient to trigger large scale pollution abatement investments, but they can have a role in optimising technical processes in order to reduce emissions. While no formal assessment of the WLF scheme has been carried out, it is widely assumed that the economic efficiency of this instrument is mediocre at best.

4.3.1.3 Distributional Effects and Social Equity

The main stakeholders and social groups affected by the WFL are the general population (households), wastewater utilities and business entities.

Water utilities are responsible for most of the WFL payment. Altogether, in 2005 municipal wastewater treatment amounted to 90 % of the total WLF revenue. The fee liability amounted to 1.5 % of the revenue and 26 % of the after-tax profit in the public water utility sector, but this was still before the Budapest Central Wastewater Treatment Plant started to operate (NRBMP 2010c).

Theoretically, water utilities are only intermediaries, since they collect the WLF from users and pay it into the central budget. However, most water utilities make steps to decrease their effluent discharge and thus lower the WLF obligation and this way reduce the burden falling on their customers.

As a result of the system of allowances for pollution reducing investments and the purchase of measurement instruments, water utilities, nevertheless, also benefited from the introduction of the regulation.

The majority of WLF payments originate from the consumers using public sewers as they pay their service providers a WLF surcharge within the wastewater bills – most water utilities pass their WLF costs to their customers. The service provider then transfers the collected fees to the central budget. As already mentioned, the WLF component makes up between 0.5 % and 11 % of the wastewater bill, depending on the settlement.

The national river basin management plan contains an analysis on affordability of drinking water and sewage services (NRBMP 2010b). According to this in 2009 water and sewage costs amounted to 3.4 % (water price: 1.8 %, wastewater price: 1.6 %) of the average net household income in Hungary. Naturally, these figures vary significantly from region to region. Despite the level of their drinking water consumption being only 70 % of the national average figure, the average burden of the population in the lowest income decile is 6 % of their income, spending 3.2 % of their income on drinking water and 2.8 % on wastewater.

Medium and high income households are unlikely to be notably affected by the WLF. Low income households in areas where the WLF makes up more than just a trivial portion of the wastewater bill, however, may be adversely affected, occasionally supplying themselves from – often polluted – groundwater sources, instead of relying on the public utility water supply, thus creating health risks.

For industrial facilities the WLF has increased the costs of production and thus influenced the total amount of profit at a rate that depends on the market situation. In 2005 industrial facilities directly discharging into surface water – as opposed to the public sewer – had an 8 % share in total WLF payments. In the same year, when the payment obligation was only 30 % of the total fee, the WLF amounted to 0.005 % of net industry revenue and 0.07 % of profit (NRBMP 2010c).

The sectors were affected differently by the regulation. According to the preliminary social and economic assessment (ÖKO Co. Ltd. 2003), compared to the sector level GDP the following sectors were affected to a higher degree than the average: fisheries, the wood-working industry, food industry, metallurgy, metal-working and the chemical industry.

Industrial facilities discharging into the public sewer or directly into surface water need to be distinguished. The latter can directly control their discharges and therefore the WLF payment, while the former depend on the technology and abatement efficiency of the public wastewater treatment plant.

Finally, the introduction and implementation of the WLF raised awareness in relation to the theme of water pollution and the Polluter Pays Principle (PPP). This principle was accepted by the industry, the water utilities and the public, and the level of environmental awareness has increased in the past 10 years, especially in the early years of the scheme.

4.3.2 The EPI Setting Up

4.3.2.1 Institutions

It is important to briefly review the institutional background of the WLF, as it considerably impacts the efficiency of this policy instrument.

The system of the central environmental and water administration and the regional organizations – directorates, inspectorates (authorities) – has been changing continuously since the transition period in 1989. After each change of government, and often even during government terms, new rounds of radical organizational restructuring (splits and mergers) have taken place. These changes generate uncertainty in the affected organizations, strengthen the dependence of regional entities on the headquarters that are also constantly reorganised, and weaken the enforcement of the regulation.

The regionally competent Inspectorates for Environment, Nature and Water – there are ten inspectorates in Hungary – regularly monitor wastewater emissions according to the applicable rules[3] by means of sampling and on-site control.

At the ministerial level, until 2010 the WLF had been under the direction of the Ministry of Finance. Today, the Ministry of Rural Development is responsible for environmental protection. Taxation duties related to the WLF are carried out by the National Tax and Customs Administration (NTCA).

The inspectorates audit the emission data. In the course of monitoring, if disparities are found in the submitted data, the NTCA is informed. However, practice shows that the NTCA is concerned only about the tax revenues, but it is not really interested in environmental monitoring. In practice, the inspectorates do not seem to be aware that the emission data serves as the basis for calculating the WLF payments. The NTCA's monitoring power only covers payments, the schedule, and, in

[3] MoEW Decree No. 27/2005 (6.12.) on the detailed rules of the control of used and wastewater discharges.

particular, the verification of allowances and exemptions. This indicates that in reality the WLF (along with the air load fee and the soil load fee paid by enterprises) function(s) as a tax.

Most settlements are served by municipally owned water utilities.[4] Under the present scheme of the financing system for infrastructure and development the municipality invests in infrastructure and other assets, and the water utility is responsible for operation and maintenance. In practice, the financing of pollution reducing infrastructure development depends on limited state and/or EU resources, municipalities do not have sufficient own resources for this purpose. Since funds for investments are frequently not available, neither the municipalities nor the water utilities are in a decision-making position when it comes to large scale pollution abatement as a response to the WLF regulation. In case of water utilities therefore the incentives of the WLF are usually limited to low cost amendments of existing wastewater treatment technologies in order to improve their efficiency.

4.3.2.2 Transaction Costs and Design

The costs of introduction were covered partly by the public administration, partly by the wastewater emitters (water utilities and industrial plants). However, the final cost bearers are those using the public wastewater utility: the general population and industry.

The obligations to submit emissions data and carry out self-monitoring are required by the regulation. The polluters are required to report their actual emissions and to fulfil their payment obligation. As a consequence of the obligation of self-monitoring, the cost of the establishment and operation of a laboratory, or alternatively, the cost of hiring an external contractor, needs to be covered by the polluter. The purchase of measuring devices did not fully require the resources of the dischargers, since 80 % of the costs was financed from the WLF allowances. Nevertheless, even if polluters did not have to devote additional resources to measuring instruments, this still counts as a transaction cost from the perspective of the WLF.

The introduction of the WLF-related regulation also led to a minor, operational change for water utilities. It required the modification of the pollution registry and the accounting system and changes in the internal rules of operation. The nature of the task required the co-operation of the technical staff, examination laboratory and the financial department. In general, the data collection and management tasks did not require additional employees and the supplementary cost is not significant.

The National Tax and Customs Administration (NTCA) acquired additional responsibilities: the development and introduction of a WLF declaration form, data processing, monitoring, etc. No information is available on these expenditures.

[4]The rest, about 28 % of the population is served by five large state owned regional water utilities.

In order to determine the supplemental cost of the WLF system's operation, the central question is how many additional measurement and control functions are defined by the WLF regulation as compared to the command and control regulation on discharge limits. With respect to measurement, the scope of the pollutants is much wider in the command and control regulation and covers not only the nine substances affected by the WLF regulation. In accordance with the command and control regulation, the measurements must be carried out annually. In accordance with the WLF regulation, the emitters must determine and transfer an advance WLF payment on a quarterly basis. This quarterly obligation demands additional work, mostly in the form of an increased number of measurements compared to the command and control regulation's requirements.

There is no information on the actual operational cost falling on state administration. According to the preliminary socio-economic impact assessment (ÖKO Co. Ltd. 2003.), a staff of approximately 24 people are required to administer and monitor the WLF at the national level. Specific wage costs can amount to EUR 400–500,000 (2003 average exchange rate) for 24 persons annually as a consequence of the characteristics of the required professions. On top of this, job creation costs amount to approximately EUR 190,000. The total of these sums represent about 2 % of the annual revenue of the central budget from environmental load fees.

Since the inspectorates did not have a substantial enforcement role, no significant transaction costs arose for them.

4.3.2.3 Implementation

The legal basis for applying environmental load fees was established by the Environmental Protection Act (Act No. 53 of 1995) which required that an environmental load fee regulation had to be formulated and passed by the end of 1996. In order to introduce the fees, the Ministry of Environment and Water, responsible for environmental protection at the time, prepared several concepts with different versions of fee rates along with the socio-economic impact assessments and submitted the corresponding bills to inter-ministerial negotiations three times between 1996 and 2000. All of these attempts were rejected by the Ministry of Finance.

The Ministry of Finance agreed with the position of one of the main stakeholders, the Confederation of Hungarian Employers and Industrialists' (BusinessHungary), that the environmental load fee would damage competitiveness and economic profitability and was thus opposed to its introduction. MAVÍZ, the association of water utilities also raised objections, mainly because of the expected rise of wastewater prices.

Resistance within the government against the WLF diminished in 2003, when the introduction of the regulation was initiated by the Ministry of Finance and not by the Ministry of Environment and Water, with the explicit purpose of increasing the income of the central budget. Finally, Act No. 89 of 2003 on environmental load fees was passed.

In spite of the expressed interests and opinions of the stakeholders, the introduced WLF was more unfavourable for them than originally foreseen. The total level of the WLF unit fees almost doubled compared to the original concept from 2000. At the same time, the ratio that could be spent on direct pollution reducing infrastructure development decreased. Only 50 % of the fees could be spent on this purpose, compared to the 92 % figure of the original concept.

According to the initial proposal, the WLF revenue would have gone into the Environmental Fund of the State and could have been used as earmarked revenue for pollution reducing investments. Contrary to the original concept, however, the adopted regulation channelled environmental load fee payments directly to the central budget. Moreover, by 2004 the management and financial system of state environmental protection had been changed and the Water and Environmental Fund was abolished.

4.4 Conclusions

The long process of introducing the WLF provides a fitting example of the conflict between economic and environmental goals in transition economies. Originally the WLF concept was developed by the ministry responsible for the environment in order to create incentives to reduce effluent discharges. Between 1996 and 2002 the proposal failed several times due to stakeholder resistance conveyed by the Ministry of Finance. Finally, in 2003, it was exactly this ministry that embraced and promoted the WLF in order to enhance the income of the state budget. From this point on, however, environmental considerations were of secondary importance.

There was also a conflict between the goals of revenue generation and limiting the burden falling on polluters. To constrain the burden, the full WLF rate was introduced gradually in 5 years, giving time for polluters to make adjustment. This is viewed as a sensible rule. In addition, dischargers could retain part of their payment obligation if they purchased measurement instruments. This rule resulted in the inefficient allocation of resources: an oversupply of such devices coupled with a distorted laboratory market. Half of the payment could also be retained for investments that reduce effluent discharges. Monitoring the adherence to these rules generated significant transaction costs. Overall, the exemptions reduced some of the burden falling on the polluting facilities, while also lowering the WLF revenues of the central budget.

The WLF was introduced after a more stringent command and control regulation developed to meet EU requirements had already been implemented. The two instruments were not harmonised. Since both of them target effluent discharges, the independent effect of the WLF cannot be determined or quantified. In fact, since its introduction in 2004 no impact assessments have been carried out. Field experience suggests, however, that the WLF alone would not have had a major pollution abatement impact, while in conjunction with the command and control regulation it probably accelerated the realization of environmental goals.

Both from an environmental, as well as from an economic, point of view, one fundamental lesson to be drawn is that parallel regulations and double taxation (in this case fines and the WLF) should either be completely avoided, or should be introduced and operated in a harmonized fashion to fulfil well defined adjacent goals.

The sector most affected by the WLF is urban wastewater services. The national wastewater program, under which all of the municipal wastewater treatment plants have been built, was mainly financed out of state and EU sources and only to a limited extent by the municipalities. Wastewater utilities, or the municipalities owning them have not had the resources necessary to execute large scale investments that would substantially reduce effluent discharges. Actual WLF rates did not provide incentives for utilities to reduce their pollution, but even extremely high rates would have stayed ineffective due to the lack of own resources on the part of the utilities. It can be concluded that the national wastewater program and its grants had a much higher impact on effluent discharges than the WLF.

For most settlements the WLF contributed to a minor increase in wastewater prices that had already steeply risen as a result of the wastewater programme. For settlements the wastewater of which was not treated, the WLF increased the sewage tariff, which was usually below average due to the lack of treatment, by several percentage points. By now most of the collected wastewater is treated as a result of EU and government funded investments, so this disparity is not a problem any more.

The failure to harmonize the operation of the regulatory structure is an important observation. While the management of the command and control regulation on discharge limits is under the governance of the Ministry responsible for environmental protection and its regional bodies, the collection and monitoring of the WLF falls under the responsibility of the tax authority. As a result of this institutional, political situation the WLF system has been driven entirely by a fiscal perspective. Important information about the basis for the WLF fees, the amount of pollutants and operational and transaction costs, is not readily available to the competent authorities. This example clearly illustrates the outcome of the difficulties that arise in handling an environmental, emission-based regulation solely from the perspective of revenue generation.

References

Bereczné Eszterhai Valéria, Csuport, László, Istvánné Szabó. (2006). *Determination of the service fee, enforcement of total return*. Study prepared for the Water Utilities Sectoral Dialogue Committee.

Council Directive 91/271/EEC of 21 May 1991 concerning urban waste-water treatment. http://eur-lex.europa.eu/legal-content/EN/TXT/?uri=CELEX:31991L0271

Government Decree 203/2001 (X. 26.) Kormány rendelet a felszíni vizek minősége védelmének egyes szabályairól (Decree of the Hungarian government on certain rules in order to protect the quality of surface waters).

Government Decree 220/2004 (VII. 21.) Korm. rendelet a felszíni vizek minősége védelmének szabályairól (Decree of the Hungarian government on rules to protect the quality of surface waters).

KSH. (2014) – Household expenditures by products. http://www.ksh.hu/docs/hun/xstadat/xstadat_eves/i_zhc004b.html. Accessed 1 Sept 2014.

Ministry of Environment and Water, Government of Hungary. (2005). Country report according to the Directive 2000/60/EC of the European Parliament and of the Council establishing a framework for Community action in the field of water policy on analysis of the characteristics of the Hungarian part of the Danube River Basin District, and review of the environmental impact of human activities and economic analysis of water uses. Reporting deadline: 22 Mar 2005.

NRBMP. (2010a). *National River Basin Management Plan 7. 3. Background material: Economic and water management forecast until 2015*. Water load forecast. http://www.vizeink.hu/files2/100505/7hat.zip

NRBMP. (2010b). *National River Basin Management Plan, Main Text*. http://www.vizeink.hu/files2/100505/Orszagos_VGT0516.pdf

NRBMP. (2010c). *National River Basin Management Plan 7.5 background material: Evaluation of costs and expenditures related to water services, examination of the return level, critical analysis of the situation and elaboration of proposals regarding the development*. http://www.vizeink.hu/files2/100505/7hat.zip

Öko Co. Ltd. (2000). A víz- és talajterhelési díjak bevezetésének várható gazdasági, társadalmi, környezeti hatásai, ÖKO 2000 September (*Probable economic, social and environmental impacts of the introduction of water and soil load fees*. Decision support analysis for the Ministry of Environment and Water Management).

Öko Co. Ltd. (2001). Szennyvízkibocsátásokra vonatkozó, az EU szabályozással konform technológiai határértékrendszer hazai kidolgozása analitika és mérési eljárások, a szabályozás tervezet gazdasági-társadalmi hatásvizsgálata és vizsgálati elemzése, 2001 február (*Elaboration of a Hungarian technological limit value system regarding wastewater emissions conforming with the EU regulation, analytics and measurement methods, socio-economic impact assessment and analysis of the draft regulation*. Background material for the Ministry of Environment and Water Management for the preparation of the 220/2004 Gov. decree).

Öko Co. Ltd. (2003). A környezetterhelési díjakról szóló törvény tervezetéhez szükséges hatásvizsgálat elvégzése. Vízterhelési díj részfeladat 2003. August (*Impact assessment necessary for the draft act on environmental load charges, sub-task: water load charge*. Impact assessment of the environmental load fees for the Ministry of Environment and Water Management).

Chapter 5
Water Abstraction Charges and Compensation Payments in Baden-Württemberg (Germany)

Jennifer Möller-Gulland, Manuel Lago, and Gerardo Anzaldua

Abstract This chapter analyzes the policy mix of economic and regulatory instruments introduced in the German state of Baden-Württemberg in order to address two key water management problems: excessive nitrate concentrations in groundwater and unsustainable water abstraction. Three different policy instruments have been applied: the Regulation on Protected Areas and Compensatory Payments (SchALVO) introduced in 1988 (a regulatory and economic instrument), water abstraction charges, and Market Relief and Cultural Landscape Compensation for farmers (MEKA), a voluntary instrument introduced in 1992.

The analysis of the policy mix shows the MEKA and SchALVO measures have been considerably successful in reducing groundwater nitrate concentrations. However, their success may have been higher if monitoring activities had been expanded and enforcement measures had been imposed. Water abstraction charges allow for the internalization of environmental and resource costs, but the compensation payments from the MEKA and SchALVO programs arguably contradict the "polluter pays principle", going against Article 9 of the Water Framework Directive.

Positive outcomes include the fact that transaction costs can be reduced by introducing joint applications for compensatory measures (e.g., for MEKA and SchALVO) and by harmonizing administrative procedures to already existing economic or regulatory instruments (e.g., the water abstraction charge was linked to existing procedures of the effluent tax).

Keywords Policy mix • Baden-Württemberg • Abstraction charge • Compensation payment • Amendments and exemptions • Negotiation

---

J. Möller-Gulland (✉)
PricewaterhouseCoopers Private Limited, 17th Floor, Building 10 Tower C, DLF Cyber City, Gurgaon 122002, Haryana, India
e-mail: jennifer.m.gulland@in.pwc.com

M. Lago • G. Anzaldua
Ecologic Institute, Pfalzburger Strasse. 43/44, 10717 Berlin, Germany
e-mail: Manuel.lago@ecologic.eu; gerardo.anzaldua@ecologic.eu

M. Lago et al. (eds.), *Use of Economic Instruments in Water Policy*, Global Issues in Water Policy 14, DOI 10.1007/978-3-319-18287-2_5

5.1 Introduction

In Baden-Württemberg, a *Land* (German Federal State) located in south-western Germany, problems relating to groundwater quality, especially high nitrate levels, have been known of since the 1970s. Since 2000, the overall objective has been to achieve "good ecological status" for all water bodies – following the goals of the EC Water Framework Directive (WFD) – and to reduce nitrate values at all measuring stations to below the threshold stated in the Drinking Water Directive, i.e. 50 mg/l by 2015. Prior to 2000, but still relevant, a long-term goal is that all water protection areas should be categorized as "low-risk zones" according to the EC Nitrates Directive (Landtag Baden-Württemberg 2008). Further, the *Länder* need to set provisions for compliance with Article 9 of the WFD on full cost recovery of water services.

This chapter introduces and evaluates the performance of the policy mix of economic and regulatory instruments introduced in Baden-Württemberg to address water management problems, such as high nitrate levels in groundwater. The policy mix consists of the following instruments[1]:

- Regulation on Protected Areas and Compensatory Payments (SchALVO)
- Market Relief and Cultural Landscape Compensation (MEKA)
- Water Abstraction Charges

5.1.1 Introducing the Instruments' Objectives

The objective of the SchALVO is to protect the ground and surface waters in water protection areas from agricultural runoff, particularly nitrates, pesticides and microbial pollutants. In addition, previously polluted water shall be rehabilitated (LTZ 2010). However, no quantitative targets were set with the introduction of the instrument. In addition to the SchALVO measures, the MEKA program was introduced in 1992 to cover ground and surface water bodies outside of water protection areas, and since 2001, those in low risk areas, which do not receive SchALVO compensations. Its objectives include the maintenance of the cultural landscape, support for the agricultural market, and the introduction of environmentally-friendly and extensive farming practices. As the environmental impact of measures covered in the MEKA programs are sufficiently documented, the targets of these programs are based partially on area-wide coverage and levels of acceptance, rather than on quantitative environmental goals (see Table 5.1).

While considerations to introduce the water abstraction charge started with the decision to introduce and need to finance compensation payments to farmers, such

[1] Regulation on Protected Areas and Compensatory Payments (Schutzgebiets- und Ausgleichs-Verordnung – SchALVO); Market Relief and Cultural Landscape Compensation (Marktentlastungs- und Kulturlandschaftsausgleich – MEKA); Water Abstraction Charges (Wasserentnahmeentgelten).

Table 5.1 Goals of the MEKA III programme

	MEKA III (2007–2013)		
	Plan 2013	2007–2009	(%)
# of farms participating	35,000	33,515	96
Area covered by MEKA (ha)[a]	1,520,000	1,548,430	102
Physical area covered by MEKA (ha)	900,000	864,616	96
Area covered by MEKA measures to improve water quality	500,000	2,962	59
EUR spent	657.1 million	295.7 million	45

Source: IFLS (2010)
[a]The area covered by MEKA measures exceeds the physical area of agriculturally used land, as one physical area may be supported by multiple MEKA measures

as SchALVO (Bergmann and Werry 1989: 2–4), the policy objectives of the water abstraction charge itself were focused on the following[2]:

- Despite the current water abundance in Baden-Württemberg, water shall be seen as a valuable resource by its users, as its current availability may be reduced in the future by competing uses and climate change-related impacts on hydrology (*awareness raising and precautionary principle*);
- As such, the water abstraction charge shall incentivize water-saving behaviour by its users (*incentive function*);
- Furthermore, the water abstraction charge shall reduce the economic advantage (*Sondervorteilsabschöpfung*) of agents that benefit from the abstraction of water in comparison to those that do not benefit from abstracting water (*competitive rebalancing*);
- The government of Baden-Württemberg invests substantially in maintaining and cleaning water bodies – costs which shall be internalised by the users (*cost recovery*).

As such, the policy objectives represent a mix between the incentive and financing function of the abstraction charge. Following the transposition of the WFD into German federal law, the water abstraction charge can be further seen as the implementation of Article 9 of the WFD. As with the SchALVO, no goals for reaching any of the specific targets of the abstraction charge listed above were quantified (Bergmann and Werry 1989: 7).

5.1.2 *Introducing the Policy Mix*

The SchALVO, which was introduced in 1988 and amended in 2001, curtails standard agricultural practices (*ogL*) in water protection areas. Water protection areas are divided into three zones in which the constraints on agricultural practices differ,

[2]See the legal text introducing the water abstraction charges (Landtag von Baden-Württemberg 1987) as well as in its amendment (Landtag von Baden-Württemberg 2010).

namely, Zones I, II, and III (Mader 2002). To optimise the incentive function and increase the effectiveness of the SchALVO, its amendment further classified these three zones into Low Risk, Problem, and Decontamination areas, depending on their nitrate levels in groundwater and mirroring the objectives of the EC Nitrates Directive (Table 5.2).

As such, the SchALVO now links the immissions and emissions of nitrate. Constraints on standard agricultural practices, as well as compensation payments and control mechanisms, are varying between areas (Table 5.2, LTZ 2010).

Table 5.2 Compensation payments, zone, and area classifications under SchALVO, from 2001

Zone/area	Low risk area	Problem area	Decontamination area
		>35 mg N/l	>50 mg N/l
	<25 mg N/l	OR >25 mg N/l if over the past 5 years nitrate concentrations increased by >0.5 mg N/l	OR >40 mg N/l if over the past 5 years nitrate concentrations increased by >0.5 mg N/l
I (well head): only grasslands or forests are permitted; the application of fertilizers, plant protection products is banned	Compensation payments in zone I only in exceptional circumstances		
II (inner protection zone): in addition to Zone III, Prohibition of the application of manure and sewage sludge; prohibition of animal pens; limited manure spreading and grazing;	Compensation payment for Zone II is only made if the farm holds cattle and can be paid additionally to the compensation payments outlined for Zone II and III		
	Fixed rate (EUR/ha/year) in all areas based on % of agricultural land in Zone II		
	>20 % → EUR 10		
	20–30 % → EUR 40		
	36.50 % → EUR 85		
	<50 % → EUR 160		
II (see above) and III (outer protection zone): Prohibition of tilling of permanent pastures and application of terbuthylazine	No constraints requiring compensation Since 2001 MEKA measures and compensation are allowed	Fixed rate of EUR 165/ha OR Individually set compensation payments based on proof of their economic loss, which range between the fixed rate of EUR 165/ha and the maximum compensation of EUR 200/ha	Fixed rate of EUR 165/ha AND site-specific compensatory payments (EUR 15/ha) OR Individually set compensation payments based on proof of their economic loss, which range between the fixed rate of EUR 165/ha and the maximum compensation of EUR 200/ha

Source: Ministeriums für Umwelt und Verkehr (2001) Verordnung des Ministeriums für Umwelt und Verkehr über Schutzbestimmungen und die Gewährung von Ausgleichsleistungen in Wasser- und Quellenschutzgebieten (Schutzgebiets- und Ausgleichs-Verordnung – SchALVO). Schutzgebiets- und Ausgleichsverordnung für Wasserschutzgebiete (SchALVO)(2001). Stuttgart.

Compensation payments are limited to problem and decontamination areas in Zones II and III. If cattle are held, further compensation may be granted for Zone II. Furthermore, site-specific compensatory payments are only made in decontamination zones. The classifications of these areas are evaluated on an annual basis and are re-categorized if the nitrate levels in the groundwater suggest this is necessary (LTZ 2010).

Compensation payments are conditional upon adhering to the constraints set out in the regulation. A breach of adhering to these constraints is deemed as an administrative offence, while the exceedance of nitrate values in soil is not (Müller 1988). This is, no fines are imposed for surpassing nitrate level thresholds. Rejection of compensation payments does not free the farmer from compliance with constraints (LTZ 2008). Further regulatory instruments, such as the Fertilizers and Plant Protection Act, are underlying the restrictions imposed by the SchALVO. However, unresolved legal concepts of the Fertilizer Ordinance impede its potential impact (Kiefer 2005).

MEKA is a voluntary program for farmers outside of water protection areas in which they would receive compensation for implementing measures that improve environmental services. Farmers can freely choose measures that they deem most appropriate for their operation and location (modular system). In MEKA III, 17 of the 27 measures (63 %) were associated with water quality improvements (IFLS 2010). Each measure is allocated a point score per hectare. The compensation payment is then calculated by multiplying the total points by EUR 10. The measures need to be undertaken for a minimum of 5 years for farmers to be entitled for compensation and the maximum compensation payment is capped at EUR 40,000 per company with the exception of cooperatives (Ministerium für Ernährung und Ländlichen Raum 2008).

The water abstraction charge was first introduced in 1988 by amending Baden-Württemberg's Water Act (*Wassergesetz*) and fundamentally revised in its amendment in 2010 (enforcement in 2011). The amendment aimed to optimise the incentives for conservation and protection of water resources and to incentivize investments by water-intensive industries by introducing offsetting options, simplifying the tariff structure, and offering legal certainty (Landtag von Baden-Württemberg 2010a: 1).

In 1988, the size of the water abstraction charge was based on the origin of the water (surface or groundwater), the amount of water abstracted, and its proposed use (Landkreis Karlsruhe 2010). From 2011 onwards, there were only three cost categories, i.e., surface water, groundwater, and water used by public water supply, and this has facilitated administrative procedures (Table 5.3).

Before the amendment in 2010, exemptions included abstractions below 2,000 m^3/year, abstractors that were exempt from requiring water abstraction permits according to the Federal Water Act or the Water Act of Baden-Württemberg (Kraemer and Jäger 1997: 65), and abstractions below the minimum threshold of EUR 100. Charges for abstractions between 2,000 and 3,000 m^3/year were reduced by 50 %. Water-intensive industries could apply for reductions of a maximum of 90 % if they could prove that the abstraction charge impinged on their competitive position, i.e., profits before taxes were reduced by 5 % due to the water abstraction charge (Bundesverfassungsgericht 2007). Reductions of the charge were made

Table 5.3 Water abstraction charges, 1988, 1998, and 2011 (EUR/m³)

	Cost categories	Original charges (1988, EUR/m³)	Revised charges (1998, EUR/m³) (1)	Revised charges (2011, EUR/m³)
Surface water	Public water supply	0.0256	0.0511	0.051
	Cooling	0.0051	0.0102	0.010
	Irrigation	0.0026	0.0051	/
	Other (incl. production, fisheries)	0.0103	0.0205	0.010
Ground water	Public water supply	0.0256	0.0511	0.051
	Heat production	0.0026	0.0051	0.051
	Other (incl. cooling, irrigation, production, fisheries)	0.0256	0.0511	0.051

Sources: Rott and Meyer 1998; Haug 2007; Landtag von Baden-Württemberg, 2010a
Euro conversion rates from 1998 were applied (EUR1 = 1.95583 DM); (1) the original charges are derived by halving the revised charges, based on the statement by Haug (2007: 45) that charges had doubled in 1998

conditional on water-saving efforts and on substitution of groundwater with surface water where possible.

The amendment of 2010 (*Entgelt für Wasserentnahmen 2010*) led to further exemptions, namely, water for cooling of buildings or irrigation purposes, water used for damage aversion or soil, and groundwater remediation, as well as any water abstractions below 4,000 m³/year. To increase investment incentives, a maximum of 75 % of abstraction charges for surface water could be offset by investment costs for measures which reduce heat pollution, improve the ecology of water bodies, or enable the substitution of groundwater with surface water (§17f). Groundwater charges can be reduced by at most 25 % in specific industries if environmental management systems (EMAS or ISO 14001) are used (§17 g). Further reductions are only possible in the case of particular and atypical burdens (§17h) – these do not include competitive disadvantages caused merely by the abstraction charge (MU 2011).

The *Land* Baden-Württemberg as well as the water suppliers (*Grundwasserdatenbank-Wasserversorgung*) closely monitor the water quality in Baden-Württemberg and use this data to control and assess the measures taken to improve groundwater quality (i.e., SchALVO and MEKA). Alternatively, compliance with the constraints from the SchALVO is monitored on the ground by Rural District Offices who measure nitrate levels (Nmin) from soil samples in autumn. In 2004, soil samples were taken from 40 % of the decontamination areas, 25 % of the problematic areas and 3 % of the low risk areas (Finck and Übelhör 2010). In addition, 5 % of the farms and 20 % of the problematic and decontamination areas are controlled for compliance with restrictions on standard agricultural practices (Fink and Übelhör 2010). Compliance with MEKA measures and eligibility for compensation are monitored by the competent licensing office through site visits.

For the tasks relevant to the water abstraction charge, i.e., the approval process for water abstraction and the official monitoring, the water authorities are responsible. In Baden-Württemberg there are three levels of water authorities: the Ministry of Environment (Supreme Water Authority), Regional Councils (Higher Water Authorities),[3] and the lower administrative authorities, such as the city and county (Lower Water Authorities).[4] Water abstractors are required to hand in their declaration of water abstracted on an annual basis. If this is not done, the charge will be based on estimates from the water authorities (§17b, WEEG 1987).

5.2 Setting the Scene: Challenges and Opportunities

With a GDP per capita of EUR 33,655 in 2008 (StaLaBW 2011), Baden-Württemberg is one of the wealthiest *Länder* in Germany. Its 10,749,000 inhabitants also make Baden-Württemberg one of the more populous *Länder* (StaLaBW 2011). The population density amounts to 301 inhabitants/km^2 (SÄBL 2011). Agriculture was the main land user in Baden-Württemberg in 1988 (49.1 %) and 2010 (45.7 %), experiencing only a 7 % decrease over 22 years. Water protection areas increased significantly over time. In 1985, around 379,000 ha (10 % of the total area) were designated for water protection, while in 2010 they increased to around One million hectares (25 % of total area). Around 360,000 ha within the present water protection zones are dedicated to agricultural practices (Finck and Übelhör 2010).

The main pressures on groundwater arise from diffuse pollution (i.e., nitrate). These can be found in regions dominated by agriculture and are often associated with intensive farming practices. Especially the arable loess soils in the plains of the upper Rhine valley and the Kraichgau are affected (see Map 5.1). Furthermore, groundwater bodies located in the moraine areas of Upper Swabia are also at risk. At the same time, the groundwater reservoirs of the Black Forest and the Swabian Alb show only little contamination (RBMPs). As such, a total of 28 groundwater bodies which make up 19 % of Baden-Württemberg's area are categorized as "under risk" because they show concentrations above 50 mg N/l (see Map 5.1).

According to the River Basin Management Plans (RBMPs) of basins within Baden-Württemberg, the main pressures on surface water include flow regulation and morphological changes, such as a lack of consistent flow, changes in structure of water bodies, backwater in rivers, and water diversions for hydropower and industrial processes. Furthermore, in 50 % of the river basins (Alpenrhein, Oberrhein, and Donau) water abstractions lead to *local* groundwater level reductions (Umweltministerium Baden-Württemberg 2009).

In relation with water use, overall water abstraction increased significantly between 1975 and 1987 by 79 % (LUBW 2010). Afterwards, abstraction levels decreased by 34 % between 1987 and 2007. It is apparent that the energy sector is

---

[3] Regierungspräsidien

[4] Untere Verwaltungsbehörden (Stadt- und Landkreise)

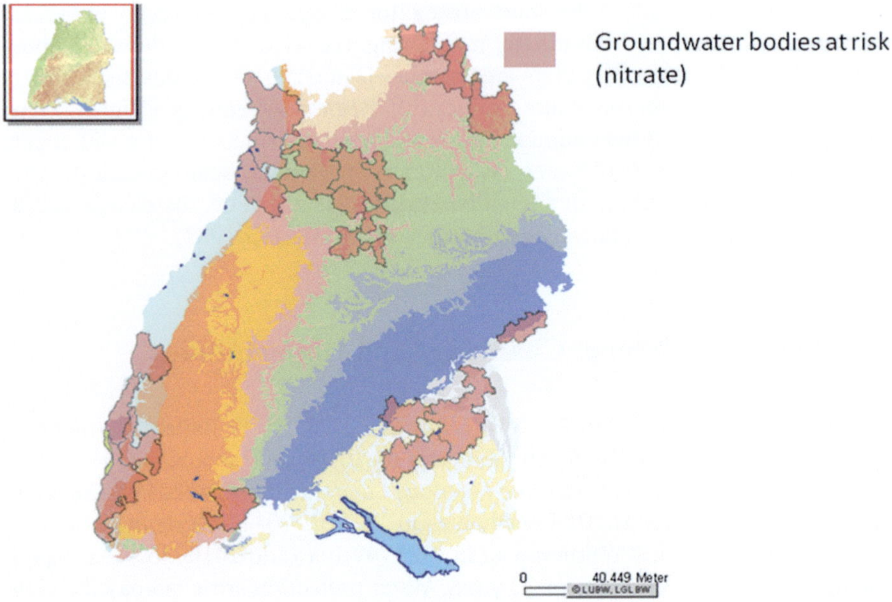

Map 5.1 Groundwater bodies in Baden-Württemberg at risk (>50 mg N/l) (Source: LUBW (2010)

far and away the largest water abstractor in Baden-Württemberg (64 % in 1975, 81 % in 1987, and 77.7 % in 2007) and drove these significant fluctuations in water abstraction. The share of surface water abstracted by the energy sector is constantly 99 % (StaLaBW 2010). With the exception of evaporative and distribution losses, 97 % of the abstracted surface water is returned after its use, mostly to surface water bodies. Aquatic ecosystems are harmed as a result of the higher temperatures of the returned water (thermal pollution) and as a result of residues from coolants (e.g., glycol) (Haug 2007). Water abstraction from agriculture (3.6 mil m^3 in 2007) and services (25.3 mil m^3 in 2007) are comparatively minor.

5.3 The Policy Mix in Action

The introduction of SchALVO in 1988 made compliance with restrictions to the standard agricultural practices, and thus a change in behaviour, compulsory. As nitrate measurements from compliance monitoring of the soil between 1990 and 2008 demonstrate, farmers changed practices in water protected areas, particularly in the early 1990s (Finck and Übelhöhr 2010). Following the amendment, measurements were focused on decontamination and problem areas, and thus are only comparable to a limited extent. Despite the compulsory nature of the SchALVO, 26 % of samples in problem areas (2,678 sites) and 23 % of samples in decontamination areas (952 sites) exceeded the nitrate threshold value in 2010, indicating that not all farmers altered their behaviour. The focus on problem and decontamination areas led to only 38 % of the water protection area being covered by stricter SchALVO restrictions and monitoring. With only 3 % of the low risk area being monitored for compliance with the general restrictions to standard agricultural practices valid in water protection areas (Finck and Übelhör 2010), it was feared that farmers would return to their prior, unrestricted farming practices which do not protect groundwater resources (Kiefer 2005).

However, as the extremely arid year 2003 illustrates, changes in farmer behaviour and weather-related changes in nitrate levels in soils are difficult to distinguish; thus, the impact of the SchALVO cannot be determined with certainty. Contrary to the SchALVO, the MEKA program is voluntary. Changes in behaviour by farmers can be approximated by the take-up of the program measures. The total area in which MEKA measures were introduced grew from MEKA I (815,000 ha, 50 % of agricultural area) to MEKA II and III (900,000 ha, 55 % of agricultural area). For MEKA III 96 % of the targeted area has been achieved between 2007 and 2009. This illustrates a wide acceptance, as MEKA III only ends in 2013. This trend indicates increasing acceptance and willingness to alter farming practices. The main areas in which MEKA measures are being implemented coincide with areas of high nitrate concentrations in groundwater.

In relation with the impact of the abstraction charge, production processes in the energy sector have changed over time, reducing the amount of water required to produce 1 kilowatt-hour (kWh) of energy by 39 %. Water used in the energy sector has fallen as average from 96.7 l/kWh in 1991 to 59.3 l/ kWh in 2007 (StaLaBW 2010). In addition, water productivity (i.e. the value added per m^3 of water used), has increased by 61.3 % in Baden-Württemberg between 1991 and 2007 (StaLaBW 2010).

However, opinions diverge regarding whether these changes in behaviour were caused exclusively by the abstraction charge. For example, a recent study by Fälsch (2011) showed that there has been a substitution effect from industrial self-providers in reaction to the water abstraction charge. The government of Baden-Württemberg also states that the abstraction charge had a clear impact by changing the incentive functions of economic agents (Landtag von Baden-Württemberg 2010b: 6,888). However, other factors, such as higher water and wastewater prices, technological innovation, and the introduction of the fish habitat regulation (VwV-FischgewässerVO 2001), which sets thresholds to the temperature of returned water in Baden-Württemberg, may also have influenced behaviour (Gawel et al. 2011).

5.3.1 Environmental Outcomes

Between 1994 and 2010, there was an overall decrease of 19.5 % (−5.7 mg/l) in nitrate concentrations in groundwater outside of water protected areas, compared to an overall decrease of 15.9 % (−4.3 mg/l) in water protected areas (Fig. 5.1).

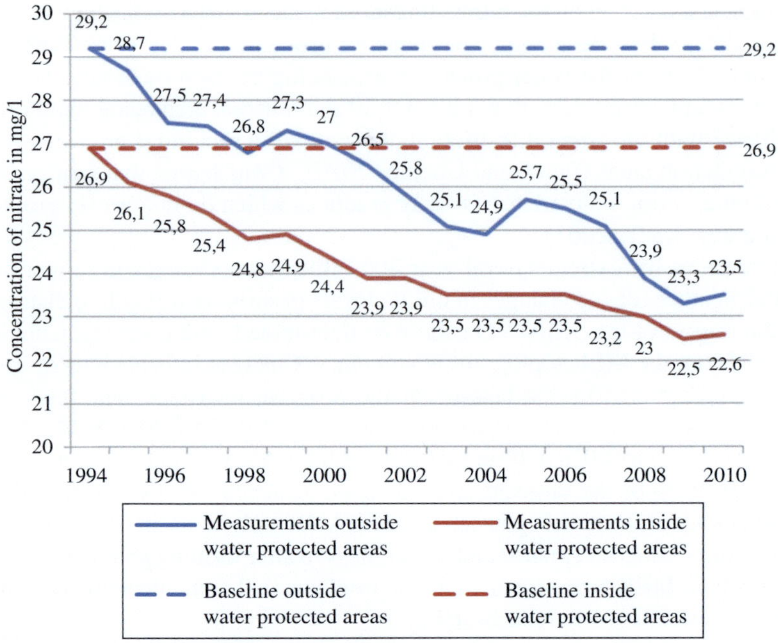

Fig. 5.1 Development of nitrate levels between 1994 and 2010 in and outside of water protected areas and baselines (Source: LUBW 2010: 42); Authors' estimate

When contrasting the change in nitrate concentrations to the baselines of each area, the voluntary MEKA program led to an additional 1.4 mg/l decrease of nitrate (LUBW 2010).

Since the amendment of SchALVO in 2001, decontamination areas have experienced the greatest reduction in nitrate concentrations in groundwater. Concentrations have decreased from 52.1 to 46.5 mg N/l (−10.7 %). Nitrate concentrations in problem areas decreased from 34 to 31.8 mg N/l (–6.5 %). In low risk zones, the levels remained constant at 14.5 mg N/l (LUBW 2010). However, the overall reduction of nitrate concentrations in water protected areas only decreased by 1.3 mg N/l between 2001 and 2010, while it had decreased by 3 mg N/l before the amendment between 1994 and 2001. Thus, while the focus on areas with high nitrate concentrations led to a reduction of concentrations below the thresholds (50 mg N/l), overall the reduction of nitrate concentrations in water protected areas slowed down. This could be explained by the fact that only 38 % of the water protection area was targeted after the amendment and by the low levels of monitoring in low risk areas.

These differing outcomes illustrate that the differentiated restrictions in each area did have an impact on nitrate concentrations. Comparing the reduced pressure from the SchALVO areas with that of the MEKA areas, however, shows that only an additional 13.6 % of reduced nitrate concentrations can be attributed to SchALVO restrictions. It should be noted that other factors, such as differing hydrogeology and thus differing resident time in soils, were not taken into consideration in this analysis.

As a result of reduced pressure from nitrate from agricultural practices, between 2001 and 2010 the percentage of decontaminated and problem areas decreased by 44.3 % and 13.4 % respectively (LTZ 2010).

Total water abstraction has decreased from 7,619 million m^3 in 1987 to 5,015 million m^3 in 2007 (−34 %). As the energy sector was the main driving force behind the increased water abstraction between 1975 and 2007, the behavioural changes described above led to a 37 % reduction in water abstraction between 1987 and 2007. Decreased water abstraction is likely to have a positive impact on pressures outlined in the RBMPs, namely flow regulation and morphological changes, including water diversions for hydropower and industrial processes.

5.3.2 Economic Assessment

This case study describes a policy mix. To achieve a reduction in nitrate concentrations in groundwater, regulatory (SchALVO restrictions) and economic (compensation payments under SchALVO and MEKA) instruments are combined. No regulatory instrument complements the water abstraction charge to reduce water abstractions. Regarding the SchALVO, the exact impact of the economic instrument cannot be singled out.

With the amendment of the SchALVO in 2001, 50 % of the current compensation payments were eliminated, as only targeted areas (i.e. problem and decontamination areas) received compensation payments, rather than all farmers in water protected areas. These savings of EUR 30 million were used to co-finance the MEKA program. CAP payments (pillar II) from the EU co-financed the MEKA program, doubling the total to EUR 60 million (Mader 2002). Thus, the amendment increased the budget for compensation payments from EUR 60 million to EUR 90 million.

Following legal concerns, the revenues from the water abstraction charge are not earmarked for water protection measures, but flow directly into the federal budget of Baden-Württemberg. However, during the introduction of the water abstraction charge and the SchALVO, it was proposed that the revenue, while not earmarked, would be used to finance the compensation payments (Bergmann and Werry 1989: 2; Müller 1988).

Comparing the revenues from water abstraction charges with the expenditures for the compensation payments between 2002 and 2007, it becomes apparent that, although abstraction charges are not legally earmarked to compensation payments, there is a degree of cost coverage. Further, the amendment of the SchALVO took place in a time when the water abstraction charge revenue did not suffice to cover the compensation payments, as in 2000. This may suggest that these cash flows are linked "informally" despite their legal disconnection (Table 5.4). The amendment of the water abstraction charge is estimated to have led to a reduction of revenues from

Table 5.4 SchALVO and MEKA expenses and water abstraction charge revenues, 2000–2007

	Compensation payments					Revenue
Mio EUR	SchALVO (3)	MEKA (total) (4)	MEKA (water protection) (4)	MEKA (water protection) paid by BW	Total compensation payments paid by BW (5)	Water abstraction charge revenue
2000	60 (1)	107.6	84.7	42.35	102.35	93
2001	n/a	128.1	103.1	51.55	n/a	79
2002	22	147.2	117.1	58.55	80.55	98
2003	21.3	147.8	118.7	59.35	80.65	88
2004	21.7	146.7	117.9	58.95	80.65	88
2005	18.7	136	104.5	52.25	70.95	81.1
2006	18.3	112.2	95.8	47.9	66.2	86.5 (6)
2007	18.6	95.2	83.2	41.6	60.2	82

Sources: (1) Müller (1988); (2) Mader (2002); (3) Landtag BW (2008); (4) Personal correspondence with MLR.; (6) Fälsch (2011)

Note: (5) EU payments contribute around 50 % of the MEKA payments; the exact payment for each year should be seen as an estimate. MEKA payments, as part of CAP payments are planned over fixed periods of time (e.g. MEKA II over 1999–2007) so that the height of compensation payments are fixed to a predetermined maximum over this time

water-intensive industries, such as the energy sector, of around EUR 10–11 million (Landtag von Baden-Württemberg 2010a: 3).

A study by IFLS (2010) found that without the agro-environmental MEKA program, farmers would have intensified agricultural production in many instances and, due to economic incentives, would have only adhered to the minimum regulations regarding environmental protection. Compensation payments under MEKA are generally considered to partially and in some cases sufficiently compensate for additional burdens and reduced harvests. However, certain practices, such as the production of biomass and afforestation, are more lucrative to farmers than the agro-environmental compensation schemes. For the compensation schemes to provide a real alternative to these potentially environmentally harmful measures, they need to be expanded and adapted.

Water suppliers, such as the Landeswasserversorgung, feared that the amendment of the SchALVO would reverse incentives for farmers in low-risk and problem areas and lead to increased nitrate pollution in order to receive (higher) compensation payments (Haakh 2001). However, the Nature Protection Association (NABU) rejects this fear, as farmers can barely cover the additional costs and administrative burdens caused by the strict constraints in problem and decontamination areas (Nabu 2011b). The decrease in problem and decontamination areas supports this argument. Further, Haakh (2001) stresses that farmers outside of the problem and decontamination areas only need to follow the general restrictions for water protected areas – restrictions he fears are neither well defined, nor well monitored for compliance. With only 3 % aerial coverage of monitoring (Fink and Übelhör 2010), this may indeed set the wrong incentives. NABU praises the incentives provided by the agro-environmental programs, but criticises the low compensation payments, which in the future are expected to be reduced further due to budgetary constraints (NABU 2011a).

The amendment of the water abstraction charge introduced the option to offset investments which improve water ecology, thus extending the incentive function to ecological measures, rather than to just water savings. The increase in investments related to water protection before the introduction of the water abstraction charge in 1988 and before the enforcement of its amendment by the energy sector (StaLaBW 2011), suggests a correlation and shows an announcement effect, as occurred with the introduction of the effluent tax in Germany in 1976 (Barde and Smith 1997). By analysing the level of the water abstraction charges between 1988 and 2010 for water suppliers, Gawel et al. (2011) found that while the nominal rate remained constant, the real rate decreased by around 35 %. The charge has not been adjusted to inflation – thus the incentive effect is reduced.

Since the amendment, charges for the abstraction of groundwater can be reduced (§17g) by implementing environmental management systems (EMAS or ISO 14001). This also might have a positive effect on risk reduction in the future. Whether a shift from external control to internal environmental management systems empirically increases the awareness of the water abstractors or not remains to be seen.

The split of water abstraction charges paid by industrial sector is mostly shared between the energy sector (40.2 % of total charges paid in 2007) and the public water supply (31.1 %; Landtag von Baden-Württemberg 2010a).

While the public water supply could arguably benefit from decreased nitrate levels in untreated water, as treatment costs would be reduced, clear cost savings have not materialized yet due to the limited change in nitrate concentrations. For the Landeswasserversorgung (LW), one of Baden-Württemberg's main water suppliers, the water abstraction charge comprises 8 % of its operating costs. As tariffs are set to recover all financial costs, the expense is taken on by consumers, with water costs increasing by 8 %.

The regional association for industries in Baden-Württemberg (LVI) states that the water abstraction charges lead to a disproportionate competitive disadvantage, particularly for water-intensive industries, as the surrounding *Länder* do not have this type of charge or, as in the case of Hesse, ceased charging it (LVI 2005). As a result, no new water-intensive industrial plants have been constructed in Baden-Württemberg for a long time – a water-intensive industrial corrugated paper plant, with an investment volume of EUR 500 million, was constructed on the other side of the Rhine in the Rhineland-Palatinate *Land,* which does not charge the abstraction charge (LVI 2005).

The nuclear power plant in Philipsburg (part of EnBW Kraftwerke AG) stated that the liberalisation of the energy market in 1998 increased the competitive disadvantage caused by the water abstraction charge, as costs could no longer be transferred to consumers. Following a law suit demonstrating that the water abstraction charged reduced its profits by more than 5 %, Baden-Württemberg refunded part of the past payments. However, EnBW, which is located in Baden-Württemberg and

Germany's third largest energy supply company, states that the average water abstraction charge still contributes to around 1–2 % of operating expenditures. The amendment of the water abstraction charge was believed to reduce this competitive disadvantage, through the option to offset investment costs. Contrary to LVI's opinion that the water abstraction charge could impede new water-intensive investments, EnBW recently constructed a coal-fired power plant (RDK 8) in Baden-Württemberg (EnBW 2011).

The Ministry of Environment, Climate, and Energy (MECE) in Baden-Württemberg agrees that the "energy location" offers more benefits – such as a central location in the heart of Europe and a high concentration of firms and accredited universities both demanding and supplying services – than the water abstraction charge could outweigh (MU 2011). In addition, sourcing outside of Baden-Württemberg is discouraged by lengthy and extensive administrative procedures necessary to abstract and transport water from neighbouring *Länder* which have not introduced abstraction charges (LW 2011).

The amendment of the water abstraction charge reduces the impact on water-intensive industries while increasing their investment incentives. The public water sector is not expected to be affected, although there may be marginal reductions in charges due to a rounding down of the tariff rate and reduction of the minimum claims limit. At the same time, this amendment will not impact residents directly or indirectly. It is expected that, if the discount options are fully realized, the public budget will decrease by an estimated EUR 10–11 million.

While the agricultural sector only paid a marginal amount of the revenue from the water abstraction charge and was exempted in the amendment, it does benefit from the compensation payments for improved agricultural practices (SchALVO and MEKA). This is perceived, particularly by the water supply industry, as the reversal of the "polluter pays" principle (Müller 1988). While legally the revenues from the water abstraction charge are not earmarked for compensatory payments in agriculture, this perception still remains among other stakeholders.

The compensation payments to farmers, however, are at times perceived to not cover the additional costs (administrative, operational and capital costs) which arise due to production constraints. Further, the annual re-assessment of problem and decontamination areas within the SchALVO, reduce planning security for the farmers and may lead to financial disadvantages (Nabu 2011a).

5.4 The Setting-Up of the Instruments and Consideration of Alternatives

Two legislative changes initiated public discussions on SchALVO and the water abstraction charge. For one, the thresholds of acceptable nitrate concentrations, as stated in the Drinking Water Regulation, were tightened from 90 to 50 mg N/l in 1986. In addition, compensation payments to farmers which were restricted in their agricultural practices by constraints, for example in water protected areas, were made compulsory with the amendment of the Federal Water Law in 1986 (§19(4)).

The *Länder* could decide whether they wanted to implement §19(4) via a centralized model, i.e. the *Land* is responsible for compensation payments to farmers, or via a decentralized model, i.e. the compensation has to occur between the water suppliers and the farmers (Müller 1988).

Given that around 1,000 water companies in Baden-Württemberg were responsible for water supplies and that agricultural activities took place in the around 2,400 water protected areas, the decentralised model did not seem like a viable option. In addition, Baden-Württemberg's history and geography led to very small average farm sizes (in 1987 13.1 ha), which would have increased transaction costs for negotiating compensation (StaLaBw 2008). As strict, area-wide constraints would have been difficult (or impossible) to achieve with the decentralized model, it was decided to introduce the SchALVO in 1988 (Müller 1988).

An array of options was considered to finance the compensation schemes. Following an expert testimony on legal eligibility ("Salzwedel Gutachten"), water abstraction charges crystallized as most promising. This fell in line with the concerns raised in the late 1970s and early 1980s that the current water protection legislation and the *Länder* administrations as a whole were ineffective and not able to fulfil their functions. The choice for water abstraction charges as an economic instrument was in line with the "general movement towards economic and away from regulatory instruments in environmental policy in that time" (Kraemer et al. 1998: 6–7).

The introduction of the water abstraction charge in 1988 was very controversial (Anon 2002). It followed at the *Länder* level after earlier discussions at the federal level in the 1950s and 1960s had failed to impose a federal charge. However, as the Federal Water Act did not provide for abstraction charges, the *Länder* were neither obligated to introduce these charges, nor were they limited in their design if they decided to introduce these (Ginzky et al. 2005).

Initially, the government of Baden-Württemberg intended to earmark the revenues of the water abstraction charges for the compensation payments – the Salzwedel testimony, however, raised serious legal concerns to the legitimacy of this earmarking. Following this, the government of Baden-Württemberg reconsidered the focus of the policy objective of this EPI and diminished its importance as a financing tool for compensation payments (Bergmann and Werry 1989: 2–4). Nevertheless, Müller (1988) states that it is unlikely that Baden-Württemberg would have committed to centralized compensation payments if it had not had the revenues from the water abstraction charge to pay for them.

Baden-Württemberg, in cooperation with relevant water stakeholders, initiated a program to monitor groundwater quality in 1984. Water supply companies supported this undertaking from the beginning by introducing and operating data collection stations and delivering the data to the database for free. In 1992, the water supply companies developed their own groundwater quality database (GWD-WV) in order to increase transparency on water quality levels and monitor and assess the impact of the measures taken to improve groundwater quality (i.e., SchALVO and MEKA) (GWD-WV 2009). These developments facilitated the enforcement of the agro-environmental programs.

The amendment of the EU Nitrates Directive in 1996 tightened the requirements for the "standard agricultural practice" and thus paved the way for the SchALVO amendment in 2001. As the restrictions for farmers were tightened, the focus of measures could be directed to vulnerable zones, without, at least in theory, the deterioration of non-vulnerable zones.

5.4.1 Issues of Implementability

The public was involved in the legislative process of both the introduction of the water abstraction charge in 1987, and its amendment in 2010.

Before the introduction of the SchALVO, water supply companies, such as the Landeswasserversorgung (LW), warned the government about the seriousness of the nitrate problem (LW 2011). However, the entire water supply industry was strictly against the introduction of water abstraction charges to pay for compensation payments for farmers – these were seen as new subsidies for agriculture and a reversal of the polluter pays principle. They suggested strengthening legislation regulating polluters and enforcing it more vehemently (LW 1986). The agricultural sector, on the other hand, supported the idea of compensation payments, as they felt crushed by regulations and restrictions in water protection zones and suffered economic losses as compensation payments did not occur regularly (LW 1986).

Once the water abstraction charge was in force, industries filed constitutional complaints against the lawfulness of water abstraction charges in 1995 (Rott and Meyer 1998). The legislative competence of the *Länder* to introduce water abstraction charges was substantiated by a decision of the Federal Constitutional Court (2 BvR 413/88 and 1300/93). Following this decision, the acceptance of water abstraction charges gradually improved (MU 2011). Nevertheless, several law suits were filed based on differing reductions to the water abstraction charge. As administrations were free to grant reductions up to 90 %, a great heterogeneity in practices developed, which caused discontent throughout the industry.

Several stakeholder groups, among which were the energy industry, manufacturing industry, agriculture, water supply sector, and environmental and user associations, seized the opportunity of public hearings to get involved in the legal process accompanying the amendment to the water abstraction charge in 2010. While the stakeholders belonging to the industry proposed the cancelation of the water abstraction charges, or at least a drastic reduction in the tariffs, the environmental groups lobbied for a drastic increase. Representatives from agriculture approved of the amendment as irrigation practices were made exempt in the amendment due to the small amount of water used. While the majority of the comments by the industry were denied entry into the legal text, the paper, textile, chemical, and energy industries lobbied for and were granted changes regarding the option to offset the water abstraction charge with investments (Landtag von Baden-Württemberg 2010a). In addition, the fee structure and the basis for reductions were changed to establish legal certainty, which had been lacking in the previous version. Both amendments are expected to result in discounts to the industry of around 10–11 Mio EUR annually (of a total revenue of ~80 Mio EUR annually) (LVI 2010). The water supply sector, however, continues to disapprove of the water abstraction charge, on the grounds that water prices reflecting financial full cost recovery suffice as incentives for water users to efficiently use the resource (BDEW 2011). Water companies, however, which abstract most of their water from water bodies which are not endangered by diffuse pollution from agriculture such as the Bodensee water supply company, continue to oppose to the water abstraction charges (BWV 2011).

5.5 Conclusions

The presented policy mix can be seen as a rather flexible tool which is capable of adapting to ex-ante and ex-post situations especially related with the overall performance of the combined instruments to achieve identified goals. The SchALVO was amended in 2001 as a reaction to limited success in reducing nitrate concentrations through voluntary action. The MEKA measures were adapted over time to match the compensation with the burden or losses the measures implied. Furthermore, the (modular) design of the MEKA measures maximizes the flexibility for farmers. Likewise, the water abstraction charge was amended in 2010 to increase the incentives for innovation and sustainable practices and increase legal certainty in administrative procedures.

Fundamentally, and due to the fact that the instruments are interlinked as part of a whole policy mix, it has been a challenge to disaggregate the effects and impacts of the different policy instruments in isolation. Overall, it can be concluded that the MEKA and SchALVO measures have been considerably successful in reducing groundwater nitrate concentrations in Baden-Württemberg. However, it can be assumed that the success would have been higher if monitoring activities had been expanded and enforcement measures, such as fines for non-compliance, had been imposed. On the other hand, strict enforcement is difficult when monitoring the impact of agricultural practices is done by measuring the nitrate levels in soil, since concentrations are aggravated by the impact of climatic conditions.

While the water abstraction charge internalises the *environmental and resource costs*, the compensation payments for farmers arguably contradict the *polluter pays principle*, both concepts which are set out in Article 9 of the WFD. Legal certainty and clarity regarding reduction schedules for the water abstraction charge appeared to be crucial for increasing acceptability among industries (e.g. energy, chemical and paper) and decreasing transaction costs, particularly legal costs, for all stakeholders. Furthermore, the option to offset investment costs for ecologically-friendly measures against the abstraction charge further increased acceptance among the industry and was perceived as compensation for any competitive disadvantage the charge might have caused. The perception that revenues are being used to finance measures which improve water quality (i.e. MEKA and SchALVO) increased the acceptability of water supply companies which depend on water sources endangered by agriculture. Finally, experience with these measures in Baden-Württemberg has shown that transaction costs can be reduced by introducing joint applications for compensatory measures (e.g., for MEKA and SchALVO) and by harmonizing administrative procedures to already existing economic or regulatory instruments (e.g., the water abstraction charge was linked to existing procedures of the effluent tax).

References

Anon. (2002). *Der Wasserpfennig*. Stuttgarter Zeitung online. http://content.stuttgarter-zeitung.de/stz/page/285089_0_4612_-stichwort-der-wasserpfennig.html. Accessed 27 June 2011.

Barde, J. P., & Smith, S. (1997). Do economic instruments help the environment?. *The OECD observer,* No. 204. Paris: OECD.

BDEW. (2011). *Verordnung über die Erhebung eines Entgelts für die Entnahme von Wasser aus Gewässern des Landes Sachsen-Anhalt*. Wasserentnahmeentgeltverordnung des Lan-des Sachsen-Anhalt – WasEE-VO. Entwurf. Berlin, 01 Nov 2011.

Bergmann, E., & Werry, S. (1989). *Der Wasserpfennig: Konstruktion und Auswirkungen einer Wasserentnahmeabgabe*. Forschungsbericht 10201116, Umweltforschungsplan des Bundesministeriums für Umwelt, Naturschutz und Reaktorsicherheit, Wasserwirtschaft, UBA-FB 88-035. Berlin: Umweltbundesamt.

Bundesverfassungsgericht. (2007). Beschluss BVerwG 7 B 53.06 – VGH 8 S 314/03. Bundesverfassungsgericht. http://www.bverwg.de/enid/0,3a97fa655f76696577092d0964657461696c093a09636f6e5f6964092d0938343138093a095f7472636964092d093133333232/Entscheidungen/Entscheidung_8n.html. Accessed 30 Sept 2007.

BWV. (2011). *Wasserentnahmeentgelt*. Bodensee-Wasserversorgung. http://www.zvbwv.de/de/wasserentnahmeentgeld.html. Accessed 9 Nov 2011.

EnBW. (2011). Personal communication with Dr. Bernhard Schneider. November 11, 2011

Fälsch, M. (2011). *Lenkungserfolg und Weiterentwicklung bestehender Wassernutzungsabgaben – Wasserentnahmeentgelte*. Germany: Universität Leipzig.

Finck, M., & Übelhör, W. (2010). *Ergebnisse aus 20 Jahren Schutzgebiets- und Ausgleichsverordnung (SchALVO)*. Landinfo 5/2010.

Gawel, E., Köck, W., Kern, K., Möckel, S., Fälsch, M., Völkner, M., & Holländer, R. (2011). *Weiterentwicklung von Abwasserabgabe und Wasserentnahmeentgelten zu einer umfassenden Wassernutzungsabgabe*. Umweltbundesamt, (UFOPLAN) 370926201.

Ginzky, H., Bothe, R., & Richter, S. (2005). Wasserentnahmeentgelte, Ökonomische und verfassungs- und europarechtliche Aspekte. *GWF Wasser/Abwasser, 146*(2), 114–121.

GWD-WV (2009). Ziele, Grundwasserdatenbank – Wasserversorgung. grundwasserdatenbank.de. Accessed 28 Oct 2011.

Haakh, F. (2001). *Die Novellierung der SchALVO*. LW Schriftenreihe 2001 – Beitrag 4.

Haug, S. (2007). Wasserbedarf in Baden-Württembergbei 5,3 Milliarden Kubikmeter pro Jahr. *Statistisches Monatsheft Baden-Württemberg, 5*, 44–48.

IFLS. (2010). *Halbzeitbewertung – Maßnahmen und Entwicklungsplan Ländlicher Raum Baden-Württemberg 2007–2013 (MEPLII)*. Stuttgart: Ministerium für Ernährung und Ländlichen Raum Baden Württemberg.

Kiefer, J. (2005). *Die SchALVO – eine zentrale Regelung als Alternative*? In Landwirtschaft und Grundwasser, Dokumentation der 18. Wissenschaftliche Fachtagung, Universität Bonn, 9. Juni 2005.

Kraemer, R. A., & Jäger, F. (1997). Institutionen der Wasserwirtschaft in Europa: Länderberichte. In F. N. Correira & R. A. Kraemer (Eds.), *Eurowater* (pp. 13–188). Berlin: Springer Verlag.

Kraemer, R. A., Strübin, M., & Hansen, W. (1998). *Money flows: economics of water supply and sewerage in Germany*.

Landkreis Karlsruhe. (2010). *Informationen zum Wasserentnahmeentgelt*. Amt für Umwelt und Arbeitsschutz.

Ministerium für Ernährung und Ländlichen Raum (2008). *Agrarumweltprogramm des Landes Baden-Württemberg – MEKA III*.

Landtag von Baden-Württemberg. (1987). Gesetzenwurf der Landesregierung: Gesetz zur Änderung des Wassergesetzes für Baden-Württemberg (Entgelt für Wasserentnahmen), Drucksache 9/4237 eingegangen am 18.03.1987. Stuttgart: Staatsministerium Baden-Württemberg.

Landtag von Baden-Württemberg. (2008). Antrag der Abg. Bernd Murschel u.a. GRÜNE und Stellungnahme des Umweltministeriums, Nitratbelastung im Grundwasser, Drucksache 14/3325.

Landtag von Baden-Württemberg. (2010a). *Gesetzentwurf der Landesregierung: Gesetz zur Änderung der Vorschriften über das Wasserentnahmeentgelt*.

Landtag von Baden-Württemberg. (2010b). Plenarprotokoll 14/97 vom 31.07.2010. Stuttgart: Landtag von Baden-Württemberg.

LTZ. (2008). *Die Schutzgebiets- und Ausgleichs-Verordnung Praktische Umsetzung im Ackerbau und auf Grünland,* Landwirtschaftliches Technologiezentrum Augustenberg, Merkblätter für die umweltgerechte Landbewirtschaftung Nr 20 (4. Auflage). Karlsruhe.

LTZ. (2010). *SchALVO Nitratbericht – Ergebnisse der Beprobung 2010,* Landwirtschaftliches Technologiezentrum Augustenberg, Ministerium für Ländlichen Raum, Ernährung und Verbraucherschutz, Baden-Württemberg. Stuttgart.

LUBW. (2010). *Grundwasserüberwachungsprogramm – Ergebnisse der Beprobung 2010*. LUBW Landesanstalt für Umwelt, Messungen und Naturschutz Baden-Württemberg Referat 42 – Grundwasser.

LVI. (2005). *Belastungen der Industrie durch Wasserpfennig nicht mehr hinnehmbar*. Standpunkte 02/2005, Landesverband der Baden-Württembergischen Industrie.

LVI. (2010). *Geschäftsbericht 2010, O*stfildern. Landesverband der Baden-Württembergischen Industrie

LW. (1986). *Umstrittener Wasserpfennig – Auflage oder Entschädigung?* LW Schriftum Nr 165.

LW (Landeswasserversorgung). (2011). Personal communication. November 14, 2011

Mader, T. (2002). *Ziele und Inhalte der SchALVO – der Schutzgebiets- und Ausgleichsverordnung von Baden-Württemberg*. Workshop – Kooperation zwischen Landwirtschaft und Wasserversorgung in der EU, 10–12 Apr 2002, Universität Dortmund.

Müller, E. (1988). Der Wasserpfennig in Baden-Württemberg aus der Sicht der Wasserwerke. *GWF Wasser Abwasser, 129*, 11.

Nabu. (2011a). *Baumann: Fördersätze sind viel zu gering*. Naturschutzbund Deutschland. http://baden-wuerttemberg.nabu.de/themen/landwirtschaft/aktuelles/14154.html. Accessed 14 Nov 2011.

Nabu. (2011b). Personal communication with Matthias Strobl. November 2011.

Rott, U., & Meyer, C. (1998). Aufkommen und Verwendung von Wasserentnahmeentgelten in der Bundesrepublik Deutschland. *Wasser und Abfall, 139*(12), 772–782.

SÄBL. (2011). *Gebiet und Bevölkerung-Statistische Ämter des Bundes und der Länder*. http://www.statistik-portal.de/Statistik-Portal/de_jb01_jahrtab1.asp. Accessed 27 June 2011.

StaLaBW. (2008). *Durchschnittliche Betriebsgröße der landwirtschaftlichen Betriebe in Baden-Württemberg seit 1979*. http://www.statistik.baden-wuerttemberg.de/Landwirtschaft/Indikatoren/AS_betriebsgroesse.asp. Accessed 14 Nov 2011.

StaLaBW. (2010). Wasserwirtschaft in Baden-Württemberg, CD aus der Reihe Statistische Daten 03/2010. Statistisches Landesamt Baden-Württemberg.

StaLaBW. (2011b). Bruttoinlandsprodukt und Bruttowertschöpfung in Baden-Württemberg, Statistisches Landesamt Baden-Württemberg. http://www.statistik.baden-wuerttemberg.de/VolkswPreise/Landesdaten/LRtBWSjewPreise.asp. Accessed 4 July 2011.

MU (Ministerium für Umwelt, Klima und Energiewirtschaft Baden-Württemberg). (2011). Personal communication with Dr. Axel Hollenbach. November 2011

Umweltministerium Baden-Württemberg. (2009). *Bewirtschaftungspläne inkl. Maßnahmenprogramm für das BG Alpenrhein/Bodensee*. Hochrhein, Oberrhein, Neckar, Main, Donau. http://www.um.baden-wuerttemberg.de/servlet/is/63582/. Accessed 29 Aug 2011.

VwV – Fischgewässer VO. (2001). *Fischgewässerverordnung*, VwV Verwaltungsvorschrift des Ministeriums für Umwelt und Verkehr zum Vollzug der Verordnung des Ministeriums für Umwelt und Verkehr über die Qualität von Fischgewässern – VwV-Fischgewässerverordnung.

Wasserentnahmeentgeltgesetz (WEEG). (1987). Gesetz zu Änderung des Wassergesetzes für Baden-Württemberg (Entgelt für Wasserentnahmen) vom 27. Juli 1987, GBl.1987 Nr. 10, pp 224–228.

Chapter 6
The Danish Pesticide Tax

Anders Branth Pedersen, Helle Ørsted Nielsen, and Mikael Skou Andersen

Abstract This chapter analyses the Danish pesticide tax (1996–2013) on agriculture which was introduced as an ad valorem tax in 1996, doubled in 1998, and redesigned in 2013 as a tax based on the toxicity of the pesticides. The Danish pesticide taxes probably represent the world's highest pesticide taxes on agriculture, which makes it interesting to analyse how effective they have been. The analysis demonstrates the challenges of choosing an optimal tax design in a complex political setting where, additionally, individuals in the target group have different rationales when making decisions on pesticide use. It also demonstrates that a small first, green tax step over time might develop into a better tax design.

Keywords Pesticide tax • Price elasticities • Behavioural responses • Effectiveness • Reimbursement

6.1 Introduction

Denmark's landscape is dominated by agriculture. In 1995, the year before the pesticide tax was first introduced, 66 % of the land use was agriculture and in 2014 it remains so (Statistics Denmark 2011, 2014). In 1999, OECD (1999: 3) concluded that there was a concern for nutrient and pesticide discharges from agriculture in Denmark. Meanwhile, Denmark was and is one of very few countries where the population has the privilege of consuming largely untreated tap water due to high water quality, making treatment unnecessary. In contrast to most other countries, the Danish water supply for drinking water purposes is sourced entirely from groundwater (GEUS 2010; Aarhus University 2011). This fact has contributed to the

A.B. Pedersen (✉) • M.S. Andersen
Department of Environmental Science, Aarhus University,
Frederiksborgvej 399, 4000 Roskilde, Denmark
e-mail: abp@envs.au.dk; msa@dmu.dk

H.Ø. Nielsen
Department of Environmental Science, Aarhus University,
Grenåvej 12, 8410 Rønde, Denmark
e-mail: hon@envs.au.dk

M. Lago et al. (eds.), *Use of Economic Instruments in Water Policy*,
Global Issues in Water Policy 14, DOI 10.1007/978-3-319-18287-2_6

development of a strong norm among Danes for having untreated tap water. According to an expert involved in the 1995 political processes regarding introduction of a pesticide tax this norm was shared by the politicians involved; pesticide pollution of drinking water was considered unacceptable, while there was less focus on the negative effects of pesticides on biodiversity (Interview, Ministry of Taxation 2011).

Prior to the 1996 pesticide tax, a general pesticide fee (3 % of the wholesale price of pesticides) had been in force, but the purpose of this tax was only to recover the administrative costs associated with the approval of pesticides, and it had no effect on pesticide use, nor was it expected to (Ministry of Taxation 2004; Andersen et al. 2001). Furthermore, some information and command-and-control policy instruments were in force prior to 1996 (Pedersen et al. 2011), but these didn't deliver the expected reduction in pesticide use.

The new tax was levied on sales and aimed to reduce use of approved pesticides to contribute to achievement of one of the objectives of the government's 1986 Pesticide Action Plan – a 50 % reduction of pesticide use (Pedersen et al. 2011). The tax revenue was fully reimbursed to the agricultural sector (ibid). An ex-ante impact assessment showed that the tax would reduce the use of pesticides by 8 %, assuming a price elasticity of demand of –0.5 and a price increase of 15 %. If the tax were to lead to development of more alternative (mechanical) pest protection methods, a total of 10 % reduction could be expected. If a more conservative price elasticity was used, a 5 % reduction could be expected, according to assessment, but it was underlined that uncertainties were high (Minister of Taxation 1995; L 44 1997/1998).

It soon became clear that the policy instruments included in the 1986 Pesticide Action Plan would not achieve the objective of a 50 % reduction in pesticide use, although the Ministry of Taxation assessed that the pesticide tax 'probably' had an – unspecified – effect on pesticide use. Consequently, the Danish Parliament decided to double pesticide taxes as an average across types as of November 1998; tax rates on fungicides, herbicides and growth regulators were more than doubled while the increase in tax rates on insecticides was lower, (see Table 6.1) (L 44 1997/1998; Ministry of Taxation 2004).

Ex-ante modelling predicted that the new tax rates would reduce pesticide use by 8–10 % from 1998 to 1999 (assuming a price elasticity of –0.75), compared to a situation without tax increases. The Ministry of Taxation estimated the elasticity of

Table 6.1 Danish pesticide tax 1996–2013 (% of retail price, exclusive VAT and other taxes)

	Period	
Pesticide type	1996–1998	1998–2013
Insecticides	37	54
Fungicides	15	33
Herbicides	15	33
Growth regulators	15	33

Source: Minister of Taxation (1998)

further tax changes to be within the range of −0.5 to −1.0, and that a projected 35 % decrease in the price of grain would reduce pesticide use by another 10 % (L 44 1997/1998). In total, a reduction of 18–20 % was expected from 1998 to 1999, which would result in pesticide use corresponding to a Treatment Frequency Index (TFI) just below 2.0 (L 44 1997/1998). The TFI represents the average number of pesticide applications on cultivated areas per calendar year in conventional farming (based on total cultivated area and total pesticide sales in Denmark), assuming use of a fixed standard dose, and is used as a standard measure of total pesticide use. A 1999 expert committee further assessed that the economically rational level of pesticide use for farmers overall, after the tax increase, would amount to a TFI of 1.7. In accordance with this, the government raised its level of ambition in the succeeding 2004–2009 Pesticide Action Plan, expecting the 1998 pesticide tax, in combination with some voluntary policy instruments, to reduce pesticide use to a TFI of 1.7 (Pedersen et al. 2011, 2012a). The reduced use of pesticides was expected, 'in the short or the long term', to reduce pesticide residues in crops, water courses, lakes, ground water, soil and rainwater and thereby to lower the risk of environmental damage and negative health effects (L 44 1997/1998). The tax rates of 1998 were in force until 2013, when the tax was redesigned as a tax based on the toxicity of the pesticide instead of the price of the pesticide (see below).

One of the arguments for differentiating the 1996/1998 tax among types of pesticides (see Table 6.1) was that the costs per treatment vary quite a lot for different types of pesticides. A differentiation of the tax would therefore approximate a tax-per-treatment principle. The tax was charged to manufacturers and importers who then incorporated it into the product price. All manufacturers/importers were obliged to register with the tax authorities. Taxed products had to be marked with a special label designed by the authorities. This special label indicated the tax category and the maximum price of the product, the argument being that this system precluded the possibility of registering the product at a low price (and a low tax) before selling it at a higher price without a higher tax. Customs and taxation authorities were obliged to control manufacturers and importers (Ministry of Taxation 1998). The tax also applied to other pesticide users such as private home owners and horticulturists (in the analysis below, the focus is on agriculture). The tax revenue – also the part of the revenue collected from pesticide use among private home owners – was fully reimbursed to the agricultural sector primarily through a lowering of the land tax and through different types of support (e.g. subsidies for organic agriculture and protection of the water environment) (Ministry of Taxation 2004; Interview Ministry of Taxation 2011).

6.2 Setting the Scene: Challenges, Opportunities and EPIs

The introduction of the 1996 pesticide tax took place against a background of failure to reach the aims of the Danish pesticide policy with the previous (regulatory and informational) policy measures and a general Danish move towards a green tax reform, shifting the tax burden from income taxes to environmental taxes (Ministry of Taxation 2001). Thus, an expert committee had paved the way for the tax with a 1992 report proposing a reform that would include, among others, more environmental taxes on water, energy and transportation in order to encourage work and discourage consumption (Ministry of Taxation 2001: 47).

As mentioned above, expectations were that the tax could reduce pesticide residues in crops, water courses, lakes, ground water, soil and rainwater and thereby lower the risk of environmental damage and negative health effects. However, the tax design was not optimal from an environmental viewpoint, as it was not based on the toxicity of the pesticides (OECD 1999: 3) (see discussion of this below).

All Nordic countries (Denmark, Finland, Iceland, Norway and Sweden) have introduced pesticide levies on agriculture (Danish Competition Authority 2006: 253). Furthermore, a few other OECD countries, e.g. Italy, France and some North American states (e.g. British Columbia and Washington) have introduced pesticide levies on agriculture (OECD and EEA 2014). However, the average Danish tax level seems to have been substantially higher than tax rates in other countries (OECD and EEA 2014).

In connection with the Danish implementation of the EU Water Framework Directive (EC/60/2000) the pesticide tax was totally redesigned in 2013. The EU Water Framework Directive (WFD) prescribes a 'good chemical status' in surface waters and, in principle, a no-pollution-at-all standard for groundwater, although in practice the principle is defined as minimum anthropogenic impact in both surface waters and groundwater (European Commission 2011). In the Danish river basin management plans – produced to comply with the WFD – pollution from pesticides is listed as a source of pressure on groundwater and drinking water (Danish Nature Agency 2011). In order to achieve the objectives of the WFD, then, an effective tax design is imperative. The redesigned tax now reflects the environmental harm of the chemical compounds (measured by their environmental behaviour and their negative effects on human health and the environment (Danish Parliament 2012)) rather than the sales price of the product. Furthermore, average tax levels have been raised. The aim of the tax redesign was to increase farmers' economic incentive for using pesticides with low risk for human health and the environment. The effects of the reformed tax could not yet be assessed by the end of 2014, partly because statistics for 2013 were not yet available, partly because the tax was introduced in July of 2013 and therefore did not directly affect pesticide use for the 2013 season. Moreover, farmers appear to have hoarded chemicals in 2012, the year prior to the introduction of the tax, see Fig. 6.1 below. In fact, in 2012 pesticide purchases were significantly higher than pesticide use, a statistic which is also being collected as of 2012 (Danish Environmental Protection Agency 2013b). This implies that the effect of the tax may not be accurately assessed for the first couple of years following implementation.

6.3 The Pesticide Tax in Action

The introduction of the relatively high Danish pesticide tax in 1996 reflects in part a growing focus during the late 1980s and early 1990s on reducing pollution from agriculture, coupled with a strong norm related to untreated drinking water and a general move to replace high income taxes with green taxes. At the same time agricultural organizations were as per tradition invited to participate in negotiations about the design of the tax, and the choice of an ad valorem tax with reimbursement to the agricultural sector was in line with agricultural interests given that they were under pressure to accept a tax of some form. Even so economic models predicted that the tax would achieve the necessary reduction in pesticide use. However, farmers did not respond to the price signal to the degree expected.

6.3.1 The EPI Contribution

6.3.1.1 Environmental Outcomes

This section (and Sect. 6.3.2) focuses primarily on the response of the economic agents, i.e. farmers' use of pesticides – partly because a behavioural response, or lack thereof, by definition translates into changes, or lack thereof, in pressures and impacts on the water-related ecosystem and partly because studies on the environmental effects of the pesticide tax are lacking.

Measuring the exact effect of the pesticide tax on pesticide use is complicated by the fact that the Danish pesticide policy employs a mix of policy instruments – a common challenge for EPI's assessed in this book. The first Danish Pesticide Action Plan (1986) relied mainly on regulatory and information measures, but these were later supplemented with economic instruments such as the pesticide tax and voluntary agri-environmental schemes (Pedersen et al. 2011). As mentioned above, it was expected that the new tax rates in combination with a projected decrease in the price of grain would reduce pesticide use to a level of a TFI just below 2.0 in 1999 (see above). The development of the Danish TFI is illustrated in Fig. 6.1.

The figure for 1985 is an average of the years 1981–1985. For the years 1997–2012 the figures are a product of Danish EPA's so-called 'new method' for calculating TFI. The switch of calculation methods in the late 1990s meant that the TFI figure calculated was a bit higher (in the interval 0.07–0.27 for the years 1997–2012) compared to when the old method was used.

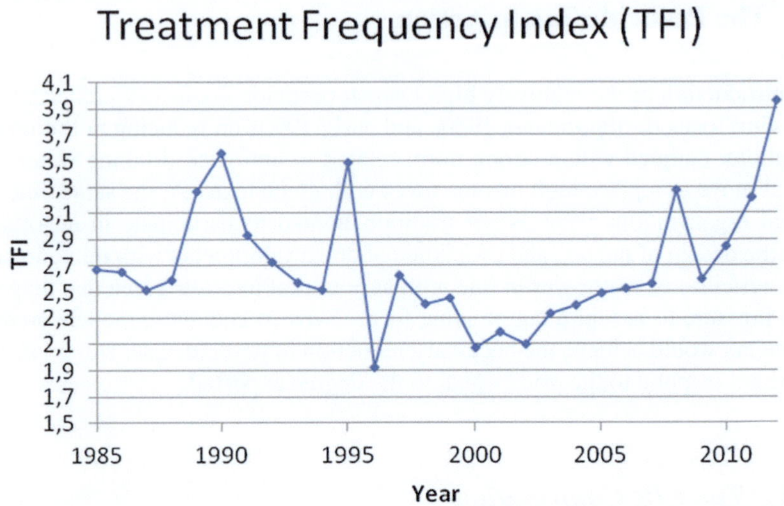

Fig. 6.1 Danish treatment frequency index (1985–2012) (Sources: Index made by Christina Bøje (Danish EPA) based on yearly EPA reports. The years 2007–2012 are corrected with the newest figures from Danish Environmental Protection Agency 2013; the 1981–1985 average is from Danish EPA (1998))

In the period before the introduction of the tax (1981–1995) the TFI hovered at around 2.5 (except for 1989, 1990 and 1995). In 1996, when the pesticide tax was first implemented, the TFI dropped to the lowest level (1.9) for the entire period 1981–2012. Much of the explanation for this decrease appears to be that farmers had hoarded pesticides in 1995 (TFI 3.5) in anticipation of the tax (Statistics Denmark 1997). In 1997–1999, pesticide use was back at a level around a TFI of 2.5 despite the doubling of tax rates in 1998. Consequently, the expectation of a TFI just below 2.0 was not met in 1999, despite the twin incentives of decreasing grain prices and increasing pesticide prices that year. By 2000 pesticide use did drop to a TFI level of 2.0, but since then the TFI gradually rebounded to a level around 2.5. In four of the last 5 years for which statistics are available (2008–2012) measured TFI has been well above 2.5. In 2012, a new 'record' was reached with a 3.96 TFI, possibly, again, due to a hoarding effect in anticipation of the redesigned pesticide tax to be implemented in 2013.

The assessment of the pesticide tax must also take into account changes in the external context that may have counteracted the pesticide tax. While the price on pesticides for most years has remained at the 1996 level, it did decrease during some years, e.g. 2005–2008. When the price decreases, so does the nominal value of the tax. The grain price has been fluctuating considerably (e.g. it was very high in 2007, but lower every year between 1997 and 2006 compared to 1995–1996) (Ørum et al. 2008:103; Pedersen et al. 2012a). Higher grain prices may have stimulated preventive spraying in some crops some years. The composition of crops also affects pesticide use and therefore the TFI – different crops need different treatment. However, the development in the composition of crops on Danish farms in the years 1996–2001 led to a *decrease* in the actual need for pesticides estimated to be 0.08 in the

TFI (Ørum 2003). For the period 2003–2007, the development in the composition of crops has not substantially changed the need for pesticides (Ørum et al. 2008: 105). The occurrence of new pests in Denmark, in particular more insects, stimulated by unusually mild Danish winters in some years might have influenced the use of pesticides, although while a popular argument among farmers this has not yet been systematically documented. Finally, an increase in the amount of winter crops combined with a poor crop rotation at approximately 50 % of the farms with winter crops has increased the need for herbicides (Ørum et al. 2008). Such changes would alter the economically optimal level of the TFI from the original estimate of 1.7 (Ørum et al. 2008), although the impacts, as outlined, exert pressure on the TFI in either directions, increasing or decreasing the TFI in any given year. Thus for 2007, Ørum et al. (2008) calculated the economically optimal TFI level to be 2.08 – and this figure may be too low, as the estimate was calculated before the exceptionally high price level for grain that year were known.

With the pesticide use currently well above 3 (the 3 year-average for 2010–2012 was 3.34, according to Danish Environmental Protection Agency 2013a), clearly the Danish mix of policy instruments has failed to deliver on the objective of reducing pesticide use to a level of 1.7 TFI. In a 2010 assessment, the Danish Economic Councils (2010: 158f) concluded that the 1998 tax has failed to give the farmers incentives to reach the 1.7 target – this despite the fact that Danish pesticide tax levels are the highest in the world according to the Danish Competition Authority (2006: 253). The explanation for the poor effect of the tax, according to the Danish Economic Councils, is an inelastic demand for pesticides – apparently, the expectations of the Ministry of Taxation regarding the elasticity (see above) were too optimistic. This conclusion is further supported by a study of pesticide decisions among Danish farmers, showing that for about half of the farmers price incentives were not a dominant factor in decisions on pesticide (Pedersen et al. 2011). The implication is that tax levels must be quite high for the tax to have the desired effect for a significant share of farmers.

No ex-post evaluations have assessed specifically whether the pesticide tax has delivered the expected reductions in the use of pesticides, namely a 5–10 % reduction (for the 1996 tax) and an additional 8–10 % reduction by 1999, following the rate increases in the 1998 tax (see above). The trajectory of the TFI alone indicates that the tax has only a small effect on the use of pesticides (this lack of effect will be discussed further in Sect. 6.3.2). Consequently, the environmental effects will likely be quite small, too. It is conceivable that the developments in grain prices (increases some years) as well as pesticide prices (decreases some years) have counteracted the taxes, obscuring an actual tax effect. But while this conclusion might hold for 2007 and 2008, which saw abnormal price developments, the pattern for the first half of the decade does not appear to support such a conclusion (Pedersen et al. 2012a: 10). Moreover, sharp ups and downs in grain prices in the last half of that decade do not match the continuous upward trajectory of pesticide use.

6.3.1.2 Economic Outcomes

A government analysis of pesticide policy instruments concluded that, *in general*, ad valorem taxes are cost effective policy instruments for reduction of pesticide use – although, this statement was not based on an empirical assessment of the cost effectiveness of the pesticide tax (Ministry of Environment et al. 2007: 17).

Needless to say, farmers being the target of the tax are therefore to some extent burdened by the tax. However, the revenue is fully reimbursed to the sector. Until 2003, the revenue was reimbursed minus the revenue from the old wholesale tax (see Sect. 6.2.2) primarily through a lowering of the land tax by 0.43 %. The remaining part of the revenue was channelled into the yearly Finance Act, where the Ministry of Food, after negotiations with the agricultural organisations, reimbursed the revenue to purposes within the agricultural sector. In 2003, the reimbursement system was changed, and it was decided to reimburse a fixed percentage (83 %) of the revenue to a lowering of the land tax. The remaining 17 % are distributed to different activities in the agricultural sector through the Ministry of Food and the Ministry of Environment. Between 2001 and 2008, total revenue has varied between DKK 359 and 423 mill (Dansk Landbrug 2007). While the sector as a whole is reimbursed, each individual farmer is still faced with an incentive to reduce his use of pesticides in order to reduce marginal costs, assuming he applies optimising principles to pesticide decisions.

6.3.1.3 Distributional Effects and Social Equity

The agricultural sector is the main sector affected by the Danish pesticide tax. Farmers who have reduced their use of pesticides due to the tax might hypothetically have experienced positive health effects. Use of pesticides in Denmark was assessed by a 1998 committee not to constitute a large threat to farmer health, and epidemiological analyses have detected no long-term health effects among farmers from occupational exposure to pesticide levels resembling current Danish use of pesticides (Bichel Committee 1998). However, 25 % of the Danish farmers hold the perception that their health risk of spraying pesticides is large or very large (Pedersen et al. 2011).

The pesticide tax has had some *distributional effects* within the agricultural sector. These effects were analysed before the implementation of the pesticide tax in 1996. Given market characteristics, pesticide prices are decided based on the product's use value for the farmers. While a pesticide tax does not increase the use value of the pesticide for the farmer, producers and suppliers will probably have to carry part of the tax burden (Minister of Taxation 1995).

In a 2006 analysis, the pesticide tax was deemed among the ten most costly regulations within the jurisdiction of the Ministry of Taxation, measured upon the burden induced on the businesses. This was due to a complex administrative system The average burden of this system is estimated to be DKK 21,000 per year per manufacturer/producer. The system is criticized for being too costly and inflexible.

Furthermore, it reduces competition, because the maximum price of the product has to appear on the label (Danish Competition Authority 2006: 254). When the tax was redesigned in 2013, the labelling system was no longer necessary and therefore cancelled.

Furthermore, ex-ante analyses showed geographic disparities in the tax due to the tax level, the reimbursement system as well as differences in crops. E.g. land prices differ in different regions of Denmark. Consequently, farmers living in areas with high land prices would get a higher amount of money through the reimbursement scheme than farmers living in areas with relatively low land prices.

The new 2013 pesticide taxes will affect different types of farmers differently, as the farmers use pesticides with different risk profiles. E.g. strawberry producers might experience decreasing pesticide prices, while potato producers might experience increasing prices (Danish Environmental Protection Agency, undated). In the mid-2000s an average farm of about 165 ha spent DKK 100,000–150,000 per year on pesticides (Danish Competition Authority 2006).

6.3.2 The EPI Setting Up

6.3.2.1 Institutional Set-up

The introduction of the pesticide tax in 1996 took place against a general move towards a green tax reform (Ministry of Taxation 2001). Even so, the introduction of the pesticide tax met with opposition. While the Social Democrat-led government proposed the tax with reference to the polluter pays principle (Ritzaus Bureau 30.11.1995), agriculture argued that it would weaken the competitive position of Danish agriculture, while the right-wing opposition parties argued that they were against allowing polluters to pay for their actions rather than to ban dangerous pesticides (Ritzaus Bureau 1.12.1994). In the end, the government also leaned on the EU which strongly espoused the polluter pays principle (Ritzaus Bureau 30.11.95).

An important aspect of the institutional setting is a strong network involving farmers organizations and the Ministry of Agriculture (Daugbjerg and Pedersen 2004), which affected the design of the pesticide tax both in 1995 and 1998. The government established a commission of high-level civil servants to produce a proposal for a pesticide tax, but with the mandate that the tax had to be put together so as not to diminish the international competitiveness of agriculture and so that revenues were reimbursed to agriculture (ibid: 234).

The pesticide tax did not change existing institutions directly related to pesticide policy, but it did change the land taxes as these were lowered in order to allow for a pesticide tax. Moreover, the pesticide tax led to the establishment of a new institution, a fund to administer the earmarked tax revenues, led by a board in which agricultural interests have the majority, while consumer and labour interest organizations are also included (Promilleafgiftsfonden 2011).

6.3.2.2 Transaction Costs and Design

When the tax was originally conceived in the 1990's, a tax based on toxicity was discussed in the government, particularly among the Ministry of Taxation, the Ministry of Environment and the Ministry of Agriculture (Interview, Ministry of Taxation 2011). The Ministry of Taxation preferred a tax based on the toxicity of pesticides, but according to the Environmental Protection Agency (EPA) it was impossible to establish such a tax because it was impossible to rank the different types of negative effects of pesticides (on groundwater, fish in watercourses, biodiversity in windbreaks etc. etc.) (Interview, Ministry of Taxation 2011). The Ministry of Agriculture preferred an ad-valorem-tax to a per-unit-tax because such a tax would confer a smaller share of the tax burden on farmers and a larger share on producers/importers, while the full revenue was reimbursed to the agricultural sector – thereby ensuring a net benefit for the sector. Furthermore, agriculture would also get a reimbursement of the tax revenue paid by private home owners (Interview, Ministry of Taxation 2011). This model was finally chosen. The tax design was not optimal from an environmental viewpoint. On the other hand, the average tax level has, to our knowledge (see also Danish Competition Authority 2006), for many years constituted the world's highest pesticide tax, representing a most likely case for a behavioural effect. Moreover, the formulation of the tax may serve to illustrate a rather classic path from economic text book into the real world of interests and politics as well as practical constraints on how to measure toxicity.

When the tax was introduced some transaction costs were assessed. Using sales as the tax base was expected to minimize inspection costs and administrative costs, due to the relatively few import and production companies compared with the number of retailers (Minister of Taxation 1995). It was estimated that non-recurrent expenses to the labelling system, information and computers would be DKK 2.1 mill. (1995). Monitoring costs were unknown. Operational costs were estimated at DKK 1 mill. for pressing and sending out of the price labels, but could be underestimated – in 2006, one of the two largest chemical companies estimated their labelling costs to be between DKK 1.5 and 2.0 mill. per year (Landbrugsavisen 2006).

This system was considered one of the ten most burdensome regulations for the companies within the jurisdiction of the Ministry of Taxation (see Sect. 6.3.3). The labelling system also imposed inflexibility on prices as labels were printed months in advance of sales. One company informed that it had to put labels on 300,000 products every season (Danish Competition Authority 2006). For instance, when world market prices decreased, the companies had to put new labels on the products (Interview, chemicals and feed company, August 2011).

Additionally, there were operational costs for the fund administering the earmarked funds.

6.3.2.3 Implementability

The Danish pesticide tax was a national tax and therefore not a flexible instrument in the sense that the tax could be adapted to local particularities. However, the tax was flexible in the sense that farmers could determine whether to pay the tax or to reduce their pesticide use. As for the policy process agricultural interests enjoyed a privileged position in the policy community while environmental and other groups at the time worked more at the periphery of the policy areas, when the tax was introduced (Daugbjerg and Pedersen 2004; Interview, Ministry of Taxation 2011; Interview, Danish Water and Wastewater Association 2011). Needless to say, agricultural organisations and farmers were against the introduction of the tax and were fighting it in the media, as well as other arenas. However, the policy design, particularly the reimbursement of the tax revenue through land taxes and the establishment of a new institution administering the revenue, reflected the wishes of agriculture and eased the implementation (Interview, Ministry of Taxation 2011).

An important barrier for the implementability of the pesticide tax seems to be that contrary to what is normally assumed in economic modelling not all farmers are profit maximizers. A 2011 Danish study based on a survey with 1.164 farmer respondents systematically analysed the most important economic and non-economic barriers in the decision patterns of Danish farmers regarding plant protection (Pedersen et al. 2011, 2012b; Christensen et al. 2011). One of the main findings of the study, which applied cluster analysis, was that approximately one third of the Danish farmers attach greater weight to obtaining physical yield than to prices on pesticides and crops, when they make decisions. These farmers primarily optimise physical yield (crops). On the other hand, around half of the farmers focus more on prices. They optimise economic yield. In other words, only about half of the farmers respond to price incentives in the manner assumed in ex-ante analyses of pesticide taxes. The diminished focus on prices is motivated by the professional satisfaction gained from producing the highest yield possible, while for farmers who are neither profit nor crop optimizers the explanation may be that relatively small price changes may not command adequate attention in a complex decision situation (Nielsen 2009). The analysis indicates that farmers who are more focused on optimising physical yield (and less on prices) are less responsive to increases in pesticide taxes and other types of economic instruments than the farmers in the price-oriented cluster. These differences do not appear to reflect underlying structural characteristics, as the farmers in the two groupings are alike with regard to structural variables such as farm size and distribution across plant, cattle and pig production (Pedersen et al. 2011, 2012; Christensen et al. 2011; Nielsen 2009).

Additionally, Ørum (2003) and Ørum et al. (2008) demonstrate that while a TFI of 1.7 is economically optimal for farmers, according to calculations, within a TFI interval between 1.7 and 2.0, farmers' economic outcome would not vary much. The implication – emphasised by the authors – is that behavioural changes would not happen automatically, but requires 'strong(er) incentives', for instance through a pesticide quota system or higher pesticide taxes (ibid). Furthermore, structural developments in Danish agriculture exhibit consistently increasing farm size. The share of farms larger than 75 ha increased from 8 % in 1989 to 25 % in 2009 (Statistics Denmark 2011: 243). A 2003 estimation indicated that larger farms (150–200 ha) tend to use 15 % more pesticides than smaller farms (50–80 ha) corrected for crop composition and location (Ørum 2003).

Current levels of illegal imports are impossible to estimate but every now and then illegal pesticide transports are uncovered by the authorities (Ministry of Environment 2011a). In December 2011, the Danish Ministry of Environment revealed the most severe example of illegal import of pesticides to date. An importer of pesticides was reported to the police for illegal import and resale of 45 tonnes of pesticides from Germany in the period 2006–2009. A second company and 44 farmers and horticulturists were reported to the police in the same case (Ministry of Environment 2011b).

All sector policies affecting the prices of crops and pesticides can reinforce/reduce the expected effects of the pesticide tax. A prime example is the EU Common Agricultural Policy (CAP), which previously revolved around product support rather than producer support, providing incentives for larger production and potentially reducing the effect of the pesticide tax. An example of the CAP affecting pesticide use is the dramatic decrease in fallow fields in recent years following the European Union 2008 abolishment of the requirement for arable farmers to leave 10 % of their land fallow to allow the farmers to maximise their production potential (European Commission, undated). Another example is the trend towards moving of measures from the CAP's single payment scheme to the rural development scheme.

6.4 Conclusion

The Danish pesticide tax was implemented in 1996 and the tax rate doubled in 1998. No ex-post evaluations have assessed specifically whether the 1996 pesticide tax has delivered the predicted 5–10 % reduction in pesticide use or whether the doubling of the tax rate in 1998 has delivered an additional 8–10 % reduction, as also predicted. The trajectory of the treatment frequency index (TFI) alone indicates that the tax has only a very small effect, at best. It is conceivable that the developments in grain prices (increases some years) as well as pesticide prices (decreases) have counteracted the taxes, obscuring an actual effect of the taxes. But while this may hold for 2007 and 2008 with abnormal price developments, the pattern for the first half of the decade does not appear to support such a conclusion. Nor has the development in the composition of crops substantially changed the need for pesticides. However, poor crop rotation at some farms and the appearance of new pests have increased the use of pesticides some (Ørum et al. 2008).

One reason for the small effects might be that about one third of Danish farmers can be considered to be less responsive to economic policy instruments than the main share of farmers, as the former focus more on optimizing yield than on prices on pesticides and crops (see Pedersen et al. 2011, 2012b). Professional pride in producing a large crop appears to drive the behaviour of these farmers rather than tweaking their profits. Therefore, a pesticide tax does not give these farmers as strong an incentive to change behaviour as it does the farmers who are more focused on optimizing economic yield. This is not to say that the crop yield optimizers would not respond to a stronger economic incentive, they are also businessmen, but it does corroborate and explain the rather low price elasticity on pesticide taxes and suggests that for these farmers taxes would have to be increased to well above economic optimization levels to have a significant impact on behaviour.

Overall, the Danish pesticide policy instrument mix can be considered a failure, as the policy mix has fallen considerably short of delivering on the policy objective of a TFI of 1.7, which was predicted based on ex-ante modelling. In fact, pesticide use has risen considerably over the years.

As for cost effectiveness of the pesticide tax no precise assessment has been undertaken. However, a government analysis of policy instruments to fulfil the aims of the Danish pesticide policy concludes that, in general, ad valorem taxes (such as the Danish pesticide tax) are cost-effective policy instruments for reduction of the use of pesticides (Ministry of Environment et al. 2007: 17). However, this rests on an assumption that the taxes are effective, which has not been demonstrated. Transaction costs of the pesticide tax were assessed ex ante to be quite small.

The tax has led to some distributional effects within the sector. For instance, farmers who grow crops with a higher pesticide need and farmers living in regions with lower land values will, on average, experience a poorer net result than other farmers.

Many farmers hold the opinion that the pesticide tax is unfair and represents just another burden reducing their income. Furthermore, importers and producers of pesticides found the price label system connected to the tax to be costly, a perception which was supported by a 2006 analysis concluding that the price label system was among the ten most costly regulations within the jurisdiction of the Ministry of Taxation. When the tax was redesigned in 2013, the labelling system was cancelled as the tax was no longer an ad valorem tax.

The agricultural sector is the main sector affected by the pesticide tax. However, the full revenue is reimbursed to the sector – primarily through lower land taxes – what eases the economic burden. This reimbursement model was the result of intense exchange/negotiations between agricultural organisations and three ministries, when the tax was designed.

The design may not have been optimal when the tax was designed in the 1990s given that the tax rate was based on price instead of on toxicity (OECD 1999). However, it's introduction in 1996 represents an important first step, and the design was improved in 1998, when the tax rates were doubled. Furthermore, the ad valorem tax (1996–2013) might have made it politically feasible to implement a redesigned pesticide tax in 2013 based on the toxicity of the pesticides (and with quite high tax rates from a comparative perspective). The new tax will most likely have an effect on pesticide use, but it remains a challenge that some Danish farmers do not react to price incentives in to the degree or in that manner economic modelling predicts.

References

Aarhus University. (2011). *Grundvandets kvalitet.* Available from http://naturogmiljoe.dmu.dk/vand/33/. Accessed 12 Sept 2011.

Andersen, M. S., Dengsøe, N., & Pedersen, A. B. (2001). *An evaluation of the impact of green taxes in the Nordic countries*. Copenhagen: Nordic Council of Ministers.

Bichel Committee, (1998). *Rapport fra hovedudvalget*. Copenhagen: Ministry of Environment.

Christensen, T., Pedersen, A. B., Nielsen, H., Mørkbak, M. R., Hasler, B., & Denver, S. (2011). Determinants of farmers' willingness to participate in subsidy schemes for pesticide free buffer zones – A choice experiment study. *Ecological Economics, 70*, 1558–1564.

Danish Competition Authority. (2006). *Konkurrenceredegørelse 2006*. Copenhagen.

Danish Economic Councils. (2010). *Økonomi og Miljø 2010*. Copenhagen.

Danish Environmental Protection Agency. (1998). *Bekæmpelsesmiddelstatistik 1997*. Copenhagen.

Danish Environmental Protection Agency. (2013a). *Bekæmpelsesmiddelstatistik 2012*. Copenhagen.

Danish Environmental Protection Agency. (2013b). *Bekæmpelsesmiddelstatistik 2012: Forbrug lavere end indkøb*. Available at http://mst.dk/service/nyheder/nyhedsarkiv/2013/nov/bekaempelsesmiddelstatistik-2012-forbrug-lavere-end-indkoeb/. Accessed 5 Nov 2013.

Danish Environmental Protection Agency. (undated). *Hvad betyder pesticidafgiften for landmandens omkostninger?* Copenhagen.

Danish Nature Agency. (2011). *River basin management plans*. Available at http://www.naturstyrelsen.dk/Vandet/Vandplaner/Se_vandplanerne. Accessed 15 Aug 2011.

Danish Parliament. (2012). *Forslag til Lov om ændring af lov om afgift af bekæmpelsesmidler*. Approved 12 Jun 2012.

Dansk Landbrug. (2007). *Dansk landbrug i tal 2007*. Copenhagen.

Daugbjerg, C., & Pedersen, A. B. (2004). New policy ideas and old policy networks: Implementing green taxes in Scandinavia. *Journal of Public Policy, 24*(2), 219–249.

European Commission. (2011). *Introduction to the new EU Water Framework Directive*. Available at http://ec.europa.eu/environment/water/water-framework/info/intro_en.htm. Accessed 12 Sept 2011.

European Commission. (undated). *Health check of the Common Agricultural Policy*. Available at http://ec.europa.eu/agriculture/healthcheck/index_en.htm

GEUS. (2010). *Grundvand – Status og udvikling 1989–2009*. Copenhagen: GEUS.

Interview Chemicals and Feed Company. (2011). Employee (anonymous), interviewed August 2011 (unpublished).

Interview Danish Water and Wastewater Association. (2011). Employee (anonymous), interviewed November 2011 (unpublished).

Interview Ministry of Taxation. (2011). Interview with an expert in the 1995–1996 negotiations (anonymous), interviewed November 2011 (unpublished).

L 44 1997/98. L 44 (som fremsat): Forslag til lov om ændring af lov om afgift af bekæmpelsesmidler og lov om ændring af forskellige punktafgiftslove (Afgiftsforhøjelse m.v.).

Landbrugsavisen. (2006). *Klart ja til at undvære prismærker på pesticider*. 24 Mar 2006.

Minister of Taxation. (1995). *Forslag til lov om bekæmpelsesmidler.* Lovforslag nr. L 186. Fremsat den 7. marts 1995 af skatteministeren (Carsten Koch).

Minister of Taxation. (1998). *Forslag til Lov om ændring af lov om afgift af bekæmpelsesmidler og lov om ændring af forskellige punktafgiftslove (Afgiftsforhøjelse m.v.)* (LSF 44, approved 23 Jun 1998).

Ministry of Environment. (2011a). *Landmand politianmeldt for import af ulovlige sprøjtemidler*. Available at www.mst.dk. Accessed September 2011.

Ministry of Environment. (2011b). *Danmarkshistoriens største svindelsag med sprøjtemidler afdækket*. Available at www.mst.dk. Accessed 12 Dec 2011.

Ministry of Environment, Ministry of Finance and Ministry of Food. (2007). *Rapport fra et tværministerielt udvalg. Analyse af virkemidler til opfyldelse af Pesticidplan 2004–2009 mål om en behandlingshyppighed på 1.7*. Copenhagen.

Ministry of Taxation. (1998). *Bekendtgørelse af lov om afgift af bekæmpelsesmidler (LBK nr. 798 af 09.11.1998)*. Copenhagen: Ministry of Taxation.

Ministry of Taxation. (2001). *Indkomstopgørelser i ikke-erhvervsmæssig virksomhed – Hvorfor skatteministeret afviste udlevering af visse dokumenter i 10-mandsprojektsagerne – Omlægning af personskatterne 1987–2002*. Copenhagen: Ministry of Taxation.

Ministry of Taxation. (2004). *Rapport om Muligheden for at omlægge pesticidafgiften til en afgift på behandlingshyppighed*. Copenhagen: Ministry of Taxation.

Nielsen, H. (2009). *Bounded rationality in decision-making: How cognitive shortcuts and professional values may interfere with market-based regulation*. Manchester: Manchester University Press.

OECD. (1999). *Denmark – Conclusions and recommendations (OECD environmental performance reviews)*. Paris: OECD.

OECD & EEA. (2014). *Database on instruments used for environmental policy*. Available at http://www2.oecd.org/ecoinst/queries/. Accessed 3 Sept 2014.

Ørum, J. E. (2003). Opdatering af Bicheludvalgets driftsøkonomiske analyser (Working report no. 20). Copenhagen: Environmental Protection Agency.

Ørum, J. E., Boesen, M. V., Jørgensen, L. N., & Kudsk, P. (2008). *Opdateret analyse af de driftsøkonomiske muligheder for en reduceret pesticidanvendelse i dansk landbrug – en beskrivelse af udviklingen fra 2003–2008*. Copenhagen: University of Copenhagen.

Pedersen, A. B., Christensen, T., Nielsen, H., & Hasler, B. (2011). *Barrierer i landmændenes beslutningsmønstre vedrørende ændret pesticidanvendelse (Bekæmpelsesmiddelforskning fra Miljøstyrelsen nr. 134 2011)*. Copenhagen: Danish Environmental Protection Agency.

Pedersen, A. B., Nielsen, H. Ø., Andersen, M. S. (2012a). WP3 EX-POST case studies: The Danish Pesticide Tax (Deliverable D3.1: EPI-WATER – Evaluating Economic Policy Instruments for Sustainable Water Management in Europe, EUFP7-project).

Pedersen, A. B., Nielsen, H., Christensen, T., & Hasler, B. (2012b). Optimising the effect of policy instruments: A study of farmers' decision rationales and how they match the incentives in Danish pesticide policy. *Journal of Environmental Planning and Management, 55*(8), 1094–1110.

Promilleafgiftsfonden. (2011). Vedtægter for promilleafgiftsfonden. http://www.promilleafgiftsfonden.dk/~/media/promilleafgiftsfonden/Vedtaegter.ashx. Accessed 14 Oct 2011.

Ritzaus Bureau. 30.11.95. *EU siger ja til dansk pesticidafgift*.

Ritzaus Bureau. 1.12.94. *Farlige pesticider skal forbydes, ikke beskattes*.

Statistics Denmark. (1997). *Miljø*. Copenhagen: Statistics Denmark.

Statistics Denmark. (2011). *Statistical yearbook 2011*. Copenhagen: Statistics Denmark.

Statistics Denmark. (2014). *Statistical yearbook 2014*. Copenhagen: Statistics Denmark.

Chapter 7
Subsidies for Drinking Water Conservation in Cyprus

Maggie Kossida, Anastasia Tekidou, and Maria A. Mimikou

Abstract This study investigates four subsidies for drinking water conservation initiated in Cyprus in 1997, namely: the construction of domestic boreholes for garden irrigation, the connection of a borehole to toilet cisterns for flushing, the installation of domestic grey-water recycling systems, and hot water recirculators. The policy objective on launching these incentives, presented here as an Economic Policy Instrument (EPI), was to reduce drinking water demand in households, partly supplied by desalination, especially during drought periods. Thus, the focus of reducing drinking water consumption was not directly linked to an overall reduction of the domestic water consumption. From 1997 to 2010 a total of 13,172 subsidies have been granted, amounting to EUR 5.5 million, resulting in a cumulative saving of 12.42 mio m^3 of water. The overall performance of this EPI is subject to uncertainty, while its overall usefulness as an EPI is questionable due to externalities, mainly related with its impact on the overall domestic water consumption and the exploitation of regional groundwater resources.

Keywords Cyprus • Drinking water conservation • Subsidies • Drinking water demand • Boreholes • Water recycling

7.1 Introduction

The EPI investigated in this study (subsidies for drinking water conservation in Cyprus) was initiated in 1997 by the Water Development Department (WDD), focused in the beginning on subsidies to construct domestic boreholes for garden irrigation and connecting a borehole to toilet cisterns for flushing. These were

M. Kossida (✉) • A. Tekidou • M.A. Mimikou
Laboratory of Hydrology & Water Resources Management,
Department of Water Resources and Environmental Engineering,
School of Civil Engineering, National Technical University of Athens,
5, Iroon Polytechniou str, 157 80 Athens, Greece
e-mail: mkossida@chi.civil.ntua.gr; atekidou@chi.civil.ntua.gr;
mimikou@chi.civil.ntua.gr

M. Lago et al. (eds.), *Use of Economic Instruments in Water Policy*,
Global Issues in Water Policy 14, DOI 10.1007/978-3-319-18287-2_7

followed in 1999 by additional subsidies to install domestic grey-water[1] recycling systems, and hot water recirculators[2] later on. During the same period (1997) public water supply of desalinated water had been introduced as a source for domestic water, with the purpose to reduce the deficit resulting from the growing demand. The rationale of the WDD on launching this EPI was to save valuable drinking water from the distribution network in households; Part of this water was now coming from desalination and is thus too costly (in terms of production and supply) to be used for gardens and toilet flashing, especially during drought periods. From 1997 to 2010 a total of 13,172 subsidies have been granted (of which 59 % for new boreholes, 34 % for connection of boreholes to toilets, 6 % for recirculators and 1 % for grey-water recycling systems installation). The total calculated amount of euros paid for those subsidies is about EUR 5.5 million. The vast majority (61 %) of the subsidies were given in households of the Nicosia water district, 13 % in Lemessos, 10 % in Ammohostos, 9 % in Larnaka and 9 % in Pafos water districts.

Prior to 1997 the water policy was much focused on increasing water supply and exploiting every drop of water ("not a drop to be lost in the sea"), thus lot was invested in dam infrastructure and increasing their capacity (i.e. the average 1980s storage capacity has doubled in the 1990s) (Kotsila 2010). At the same time though, precipitation trends have been decreasing, thus the water policy in the early 2000 has been shifted towards alternative water supplies, efficient water use and conservation; sustainability has not though been paid much attention yet. The current EPI was run in parallel with a bundle of additional measures that included reduction of leakage through restoration of the networks, progressive block tariffs, meter installation, water saving campaigns etc., in an attempt of the WDD to tackle the increasing per capita consumption and water scarcity problems. Thus, the business as usual baseline has been going through a major transformation (Charalambous et al. 2011; I.A.CO 2011).

7.2 Setting the Scene: Challenges, Opportunities and EPIs

Cyprus has a typical Mediterranean climate with mild winters, long, hot and dry summers, and short autumn and spring seasons. The average annual rainfall is about 500 mm, with a high spatiotemporal variability (ranging from 300 up to 1,100 mm), while 2–3-year drought events are often observed (Kossida et al. 2012). Evapotranspiration is high and corresponds to 80 % of the rainfall. Cyprus has been identified as one River Basin District for the purpose of the Water Framework

[1] Grey-water is defined here as domestic wastewater from laundry, dishwashing and showers.

[2] Hot water recirculators pull hot water from the water heater while they send back (at the same time) cooled-off water creating a closed loop. These systems conserve water (no wasting of water while waiting for the hot water to arrive to the tap) and use little energy.

Directive, and is subdivided into 9 hydrological regions made up of 70 watersheds (MANRE 2005). The area under government control contains 47 watersheds. In terms of land use, arable land and permanent crops are dominant (48 %), followed by forests (44 %). Wetlands and water bodies account for 0.4 % only (MANRE 2010). The most important economic sector is the tertiary, both in terms of economic output (81 % of the GDP) and employment (72 %), showing upward trends. The agricultural sector (primary), on the contrary, has experienced downward trends.

Cyprus has experienced many drought episodes varying from below normal precipitation (81–90 % normal) to severe drought (≤70 % normal) (WDD 2009). The long term annual average (LTAA) precipitation from 1901 to 1970 was 541 mm, while the LTAA from 1971 to 2009 has fallen to 463 mm (EEA 2011). The volume of water falling over the total surface area of the free part of Cyprus (5,800 km^2) is estimated at 2,750 mio m^3, but only 10 % (275 mio m^3) is available for exploitation, since the remaining 90 % returns to the atmosphere as direct evapotranspiration. The net rainfall is distributed between surface and groundwater storage with a ratio 1:3 respectively. From the groundwater storage approximately one-third flows out into the sea.

Cyprus water abstraction (205 mio m^3/year on average since 1998) comes from groundwater (75 %) and surface water (25 %), while additional water is supplied by desalination (24 mio m^3/year on average since 1998), water reuse and emergency water transfers (e.g. in 2008 from Greece). About 52 % of this abstracted water is provided to the users by the Public Water Supply System (PWSS) while the remaining 48 % through self-supply (agriculture is the dominant user of self-supplied water). The 2008 annual water use per capita was 276 m^3 (or 755 l/cap/day). The main water user is agriculture (59 %), followed by domestic (30 %), tourism (5 %), livestock (3 %), and industrial (3 %) (MANRE 2010). Cyprus has experienced many drought episodes and water scarcity situations, with its groundwater resources being over-exploited and its water stress conditions reaching critical levels. Based on calculations of the Water Exploitation Index (WEI), which is here defined as the percentage of total annual abstraction of the 30 years-LTAA availability of water resources, Cyprus has been extremely water stressed since 1998 (WEI >40 %) with its groundwater resources being most stressed. Comparing the surface and groundwater exploitation indices separately we observe that the groundwater is much overexploited (95–127 %), while surface water exploitation is below 40 % (10–34 % demonstrating an overall increasing trend), and thus leveraging the WEI to unsustainable conditions (Kossida 2010).

Under this context, the specific policy objective of the EPI was drinking water conservation, especially since desalinated water was a major part of the domestic supply: substituting valuable drinking water from the distribution network in households that is too costly to be used for gardens and toilet flashing, especially during drought periods. Secondary objectives related to water security, especially in periods of drought, and overall water saving.

7.3 The "Subsidies for Drinking Water Conservation" in Action

The WDD subsidies target new installations at household level, which are located within the boundaries of any water district and connected to the Municipal and Communal PWSS. Four subsidies for domestic water saving have been launched gradually from 1997 to 2010:

1. Construction of borehole for the irrigation of household gardens (EUR 700) in 1997
2. Connection of the borehole with the toilet cisterns (EUR 700) (applicable also for schools, office premises, shops, institutions etc.) in 1997
3. Installation of a grey-water recycling system (EUR 3,000) (applicable also for schools, military camps, public buildings, gyms, hotels etc.) in 1999
4. Installation of a hot water recirculator (EUR 220) in 2007

The above rates are applicable from 2009 onwards; lower rates were initially set and gradually increased. The rationale behind the EPI was based on the fact that water used for flashing and garden irrigation constitutes a major micro-component of the domestic water use with a significant share in the consumption, and the same applies to laundry, dishwashing and shower water that can be recycled. Nevertheless, no detailed study prior to the launch of the subsidies has been identified that assessed their impact and effectiveness or identified a logical basis on how the subsidy amount has been set. The only prior application was a pilot study on grey-water recycling in seven establishments in Nicosia that was run for 1.5 years prior to the subsidy as experimental work (Kambanellas 2007).

All subsidies were granted by the WDD following an application submission by the beneficiary and two site inspections. Regarding enforcement, although a cap of 250 m^3 groundwater abstraction per year was imposed to the new boreholes, the water meters were not monitored by the WDD for compliance. Additionally, neither inspection of the installations after start-up or other safeguarding mechanisms, nor any follow-up survey to assess the EPI's effectiveness were implemented. Only one follow-up study has been identified to assess the actual performance on boreholes for garden irrigation. In 2007–2008 extreme drought influenced the beneficiaries into heavily applying for the subsidies (increase of 170 % of the number of subsidies awarded) probably driven from their will to secure water.

7.3.1 The EPI Contribution

7.3.1.1 Environmental Outcomes

Among the four subsidy categories, constructing boreholes for garden irrigation received high response (59 %), while 34 % where given for connecting a borehole to toilet cisterns, 6 % for installing hot water recirculators, and only 1 % for installing grey-water recycling systems. By looking at the temporal evolution of the

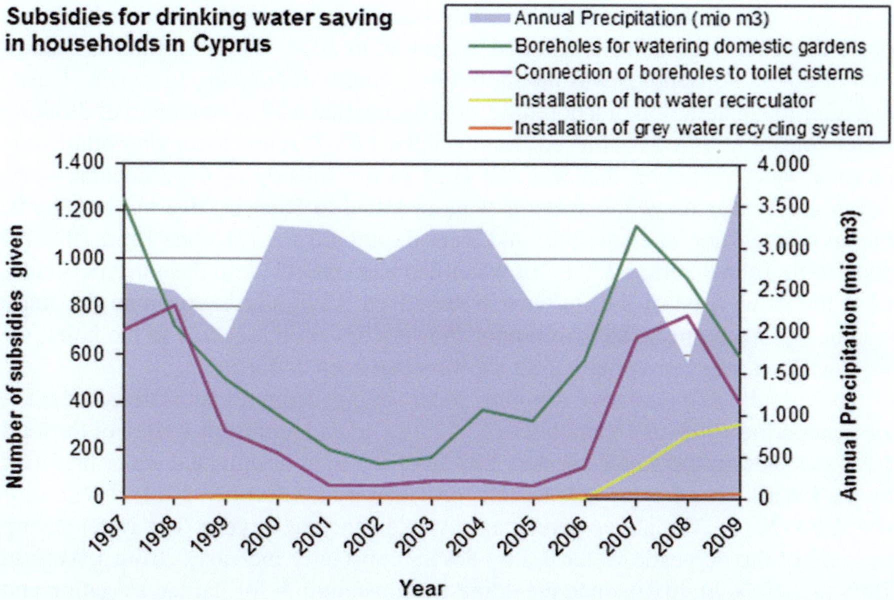

Fig. 7.1 Number of subsidies given per category as compared to annual precipitation (mio m^3) for the period 1997–2009 (Source: Compiled by the authors. Data provided by the WDD in I.A.CO Ltd (2011) and EEA (2011))

number of subsidies as compared with the respective precipitation (Fig. 7.1), we can observe the following pattern: the number of subsidies paid increased in periods of low precipitation/drought events (e.g. 2007–2008), while it declined during periods of relatively high precipitation (e.g. 2001–2004).

In order to assess the effectiveness of the EPI to reduce the pressure on domestic drinking water supply (policy objective), calculations of the total volume of water saved have been made based on the number of subsidies granted and assumptions on the potential savings induced by each subsidy category as listed below (I.A.CO Ltd 2011; Kambanellas 2007):

- On an average four-person family consumption of 600 l/day, a share of 30 % is used for outdoor purposes. Thus, using groundwater from boreholes for irrigation can cover this demand.
- On an average four-person family consumption of 600 l/day, a share of 27 % is used for flashing. Thus, supplying of borehole groundwater to toilet cisterns can cover this micro-component of use.
- Hot water recirculators can save up to 60 m^3/year of water.
- Laundry, dishwashing and shower effluents account for up to 50 % of the household water use. The operation of a grey-water recycling system can divert these volumes of water for outdoor use or for flashing (average saving 240 m^3/year).

Post-evaluation data that would allow the direct estimation of the water savings are not available, and thus the proxy calculations cannot be properly assessed for their accuracy. Only one follow-up study has been identified to assess the actual performance on boreholes for garden irrigation: in 2004, drinking water consumption of 20–30 households was monitored in a suburb of Nicosia, 12 months before and after the installation of a borehole, concluding that a 27 % reduction of drinking water consumption was achieved. Kambanellas (2007) refers to another pilot study on grey-water recycling that was run prior to the subsidy as experimental work. Seven grey-water recycling systems were installed in Nicosia (five in households, one in a hotel, one in a stadium) and were monitored for 1.5 years (mid-1997 till end-1998). In that period 220 m^3 of water had been recycled. In the current calculations the value used of 240 m^3/year water saved is slightly higher than the study results, yet since only water from pool showers has been recycled in the hotel, we would expect a higher volume if all showers had been connected.

The calculated cumulative drinking water savings from all subsidies during the 14-year period 1998–2010 amount to 12.42 mio m^3 and represent 1.50 % of the total 1998–2010 domestic water use and 3.37 % of the total desalinated water provided by the PWSS (data for the calculations provided by I.A.CO Ltd 2011; WDD website; EEA 2011). The above percentages vary from year to year: The water saving as share of the domestic water use by PWSS constantly increases (from 1.04 % in 1998 to 2.10 % in 2010) since the domestic consumption for garden irrigation and toilet flashing (the two dominant subsidies) is now substituted by self-supplied groundwater (boreholes). The water saving as share of to the desalinated water provided by the PWSS is variable, with the maximum being observed in 2007 (4.69 %) and the minimum in 2002 (2.34 %). As desalinated production significantly grows after 2007 this share is further decreasing (Fig. 7.2). It has to be emphasized that the

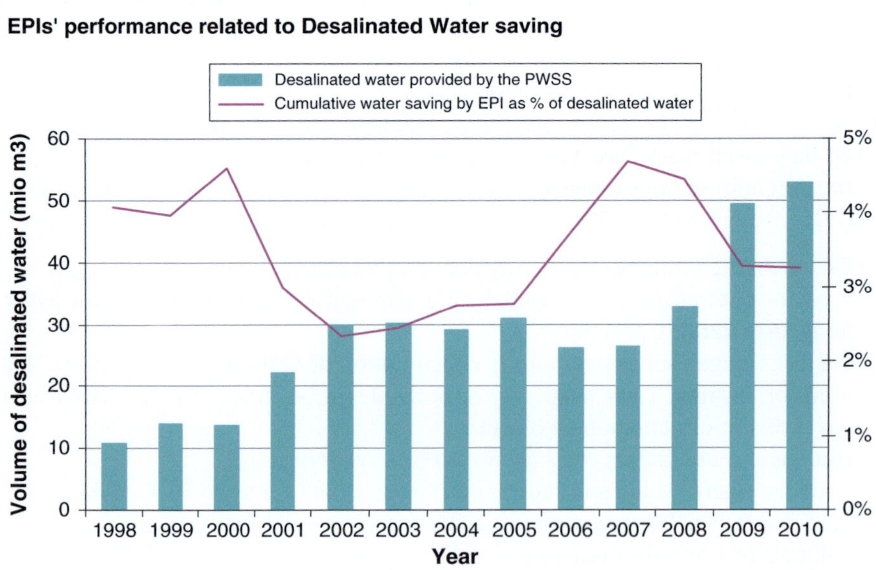

Fig. 7.2 EPI's performance related to desalinated water supply (Source: Compiled by the authors. Data provided by the Water Development Department (WDD) in I.A.CO Ltd 2011, and the WDD website)

calculation of cumulative water savings was performed by adding to a current year the savings that would also occur from all the subsidies of the previous years. This assumes that the past installations (i.e. boreholes, recirculators, grey-water recycling systems), as result of previous years' subsidies, are operational and fully functional every year, and maintained properly so that they can render the predicted estimated savings (e.g. pumps in old boreholes are maintained, old recirculators are working etc.).

Although the EPI introduced savings in the drinking water supplied by the PWSS, its impact on the total domestic water use cannot be comprehensively assessed. Assuming that the recirculators and grey-water recycling systems have resulted in overall saving of domestic water consumption, the same cannot be concluded for the boreholes' subsidy since the availability of free groundwater (no pricing) may have led the beneficiaries to over-pump and irrationally use excess water. The rational or irrational use of the boreholes (no monitoring and enforcement was implemented) relates to the individuals' behaviour (education, awareness, incentives, water saving culture). Furthermore, the borehole abstractions may have put additional pressure on the groundwater resources. WDD stated that groundwater levels and geology were considered in the evaluation of the applications, and that the aquifers where subsidies were approved are marginal and of poor quality and thus practically not exploitable for may uses. Nevertheless, a comprehensive study on the cumulative effect of the boreholes (given especially the fact that many illegal wells do exist on the island) in the different districts should probably have been undertaken prior to the launch of such measures in order to assess its environmental sustainability. Currently, no such assessment can be concluded, except that, on the positive side, this subsidy has in some way allowed the government to have an idea of the number of domestic boreholes as it acts as an incentive for people to follow the procedure of applying and registering their borehole (as opposed to drilling it illegally).

It is reasonable to assume that the induced water savings would be substituting part of the desalinated water supply. Thus, they can also be translated to equivalent energy savings (due to the decrease in desalination production needs) and corresponding CO_2 emissions reduction. Desalination at the current water production (47.7 mio m^3/year) implies a total electricity consumption of 217 GWh/year (Manoli 2010). Based on the Cyprus Energy Efficiency Report 2001, 762 gCO_2 emissions are generated per KWh produced. Thus, the total CO_2 emissions generated from the desalination plants energy consumption account for 165,199 tones CO_2/year. Each m^3 of water produced by desalination requires on average 4.5 KW (Manoli 2010), thus 3.43 $KgCO_2$ are generated per m^3 of water produced. The subsidies granted saved in total 12.42 mio m^3 of water, and assuming this volume would have come from desalination they resulted in a total 55,891,080 KWh of energy saving and 42,601 tons of CO_2 emissions saved for the entire period, or 3,277 tones/year on average. Acknowledging that pumping from the garden boreholes and the operation of recirculators consume energy as well, the net savings are in fact somehow lower.

7.3.1.2 Economic Outcomes

The payments provided for each subsidy were not kept constant throughout the implementation period; the subsidies paid varied among and within the intervention category, resulting thus in different costs for the WDD every year. It is not evident that the updates of the subsidies were based on specific studies or monitoring of the effectiveness of the EPI, but rather on ad-hoc or spontaneous reaction of the WDD. Similarly a cost-benefit analysis previous to the launch of the measure or an ex-ante comparison with alternative measures has not been performed (at least to the best knowledge of the authors). The total calculated amounts of euros paid in subsidies from 1997 to 2010 is about EUR 5.5 million (of which 59 % for new boreholes, 24 % for connection with toilets, 3 % for recirculators, 4 % for recycling) (Kossida et al. 2013). These payments do not represent the total cost of the EPI since transactions costs (e.g. costs derived by the field inspections) are not included.

To assess the cost-effectiveness of the EPI, the unit cost of each m^3 of drinking water saved has been calculated for each subsidy type and year, and has additionally been compared with the selling prices of water from desalination plans (as formulated in 2009). To obtain this ratio (balanced cost), the total cost of the subsidy each year has been divided with the cumulative water saved from the subsidies granted during the current plus all previous years, based on the assumption that the past interventions continue to be exploited by the beneficiaries (Fig. 7.3). To further assess the net amount of euros paid each year for additional new savings, the total cost of subsidies of each year has been divided with the additional savings generated explicitly that year. This was done in order to get a better insight on cost recovery per subsidy type and time period (Fig. 7.3). The overall average cost per m^3 saved from all the subsidies during the whole 1997–2010 period is EUR 0.43 (Kossida et al. 2011). At the beginning of the implementation, the EPI comes at a high cost, e.g. subsidies provided for connection to toilet cisterns in 1997 and 1998 result in EUR 2.83 and EUR 1.52 paid per m^3 water saved respectively (note that the investment cost in these calculations is considered as a cost only in the year when the investment was made). As the EPI implementation progresses and water saving is accumulating over the years (benefit of previous investments), the unit cost is decreased to as low as EUR 0.10/m^3 (years 2001–2005). A time frame of about 3 years was thus required for the EPI to become cost-effective as compared to the selling prices of the Desalination Plants and water tariffs. It has to be noted that during that period the amount paid per subsidy was kept at low levels (EUR 170 for the boreholes, EUR 340 for the grey-water recycling). From 2006 onwards the unit cost has highly and abruptly increased, reaching values higher than the desalinated water selling prices. The maximum is observed in 2007, where unit costs are in the range of EUR 2.5/m^3 and continue to be high and above desalinated water selling prices for the following years. This change is probably due to the fact that the payments were significantly increased (EUR 700 for the boreholes, EUR 1,700 followed by EUR 3,000 for the grey-water recycling systems), as well as the number of subsidies given (dramatic increase of 100–400 % in some categories). Apparently, as Cyprus was facing severe drought conditions during that period, the applications submitted

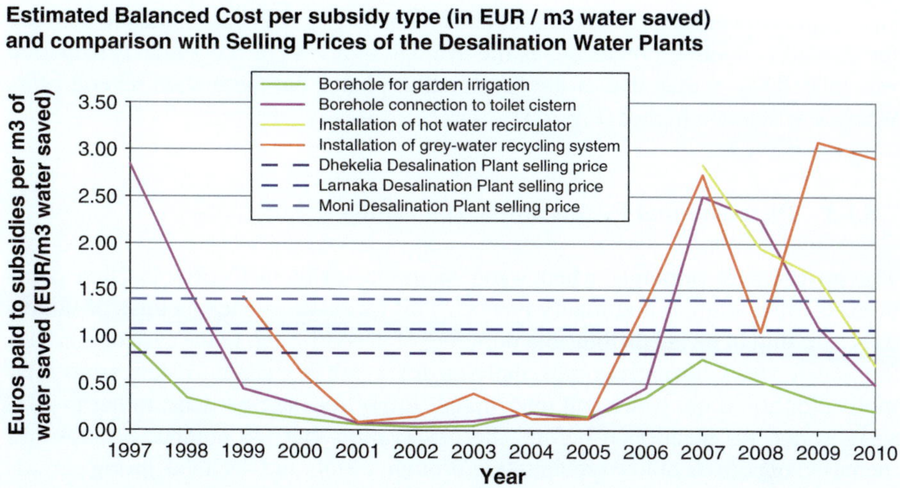

Fig. 7.3 Cost-effectiveness of EPI (*top*: Balanced cost per subsidy type (in EUR/m^3 of drinking water saved; *bottom*: Net ratio of EUR paid in subsidies every year per new added/m^3 of drinking water saved) (Compiled by the authors with data provided by the WDD in I.A.CO Ltd 2011)

were probably much more than in the previous years, leading us to conclude that the EPIs probably did not induced a change of behavior towards water conservation, but rather acted as a mean to individuals to secure domestic water using alternative free resources (they did thus decreased water supply risk), and people might have after all implement these measures even if the subsidies were not available. Looking further at the net cost of additional new savings generated every year, we can observe that after 2004 this becomes disproportionally high, implying that the increases in the amounts paid were probably too high (subsidies should probably have kept at lower rates). Thus, it is not clear whether the EPI contributed to increase the overall economic efficiency, as the average unit cost of the 1997–2010 period was indeed lower than that of the desalination plants, but there were several years where it was much higher (Fig. 7.3).

7.3.1.3 Distributional Effects and Social Equity

The government principle when water shortages arise in Cyprus is "first come humans, then animals and finally plants". This rationale creates feelings of unfairness resulting in illegal drilling and pumping of groundwater. In the case of this EPI, the beneficiaries' incentives into applying for a subsidy seem to stem from their motivation to secure water and interrupted supply for their gardens, rather to conserve water. As mentioned before, the observation that the number of subsidies increased in periods of low precipitation/drought events and declined during periods of relatively high precipitation (Fig. 7.1) possibly conveys a message on the individuals' responsive behaviour (rather than proactive).

Social inequalities can arise from the subsidies for boreholes: during dry periods when water supply is cut regularly and while some people are suffering from water shortages, others may water their gardens, causing aggravation. Furthermore, it brings up questions on environmental cost recovery and whether money should be granted to people as they are already benefiting from acquiring an additional "free" water supply. On the other hand, in an interview (Cyprus Mail 2008) WDD senior staff defended that licenses to drill boreholes are given every year and a large number of new boreholes were dug in 2008 (year of acute water crisis) causing hardship and inconvenience for those who could not afford their own borehole, thus the subsidy may have created opportunities for these people.

Additional conflicts may rise by the farmer's community. Although stated by the WDD that boreholes were approved on the basis that they were exploiting marginal aquifers of urban centers and of poor quality unsuitable for other users, public proof of evidence was lacking and thus farmers could assume that the drawdown may affect nearby irrigated agriculture and their wells' capacity. Finally, given the process of the borehole subsidy, conflicts may arise between the WDD (executive level) and the Local District Offices (end-users level).

7.3.2 The EPI Setting-Up

The institutional set-up in Cyprus is built in three levels (Aeoliki Ltd 2009): a policy level (cooperation among four Ministries, namely the Ministry of Agriculture, Natural Resources and Environment (MANR&E), the Ministry of the Interior, the Ministry of Finance and the Ministry of Commerce and Industry); an executive level (with responsible actors being the WDD of the MANR&E for planning, designing, constructing, operating and maintaining water works, and the District

Administration (DA) of the Ministry of Interior implementing/enforcing water laws) (Government of Cyprus 2010); an end-user level (local organisations like the Municipal Water Boards, the Village Water Commissions, the Irrigation Divisions and Associations, the Sewerage Boards).

The design, implementation and enforcement of the EPI were carried out solely by the WDD. This entails evaluation of applications, inspections prior and after the installation. The work load required substantial time, man-power and money, and probably created the incapability to monitor (e.g. the borehole meters) and follow up on the effectiveness of the measure. If responsibilities had been better shared among the executive and end-user levels (e.g. monitoring carried by the Local District Office), the implementation might have been more successful: better selection of the beneficiaries based on specific additional local criteria (i.e. loose conditions when it comes to the selection of beneficiaries for borehole drilling are reported by some water officers, Charalambous et al. 2011), stronger enforcement of the EPI's constraints (i.e. respect of the groundwater abstraction cap), monitoring and assessment of its impacts and benefits that would allow update and re-design of the EPI. Regarding the construction of boreholes, the Local District Office was involved in granting a drilling permission, but not in the actual evaluation process; it was acting rather as an additional intermediate agent who was gathering paperwork to forward it to the WDD, burdening thus in a sense the process.

Transaction costs have been identified in relation to the design, implementation and monitoring and enforcement. With regard to the design of the EPI, no engineering or economic assessment studies have been identified prior to its implementation, with the exception of subsidies for the installation of grey-water recycling systems (Kambanellas 2007). Five years of research (1985–1991) and 2 years of experimental work (1997–1998) on a pilot scale led to launching this subsidy in 1999. Thus, design costs related to costs paid to researchers for designing the pilot study, the purchase and installation of seven systems in Nicosia, lab costs, and field trips expenses (assuming the labour cost of the involved WDD officers was included in their salary). Based on the Citizen's Charter Report (WDD 2005), a series of actions had to be undertaken from the time of application until the subsidy is paid to the beneficiary (submission of application, preliminary inspection, approval, installation, final inspection, grant). Implementation costs are thus generated by the need for field inspection (two to three times is total) and the interaction between the WDD and the DO in the case of boreholes. These extra labour costs generated for the technicians and the officers can be covered by their salary, yet transaction costs are still evident and associated with opportunity costs in this case. The instrument had provisioned the installation and monitoring of water meters in the boreholes. Nevertheless, monitoring and control activities have not been identified. Control of the borehole meters would imply field trips (and thus associated expenses), and monitor of the house meters to assess water savings would imply interaction with the DO, thus labour costs if additional personnel is required to run the assessment.

The implemented EPI was aligned with the prevailing laws and policy setting, while no barriers linked to other policies could impede its implementation. In terms of flexibility, subsidies themselves are flexible and can be adjusted to local conditions;

adequate planning is though required in the designing phase, as well as a follow-up on their effectiveness that can allow re-design and post-implementation adaptation when conditions change. Nevertheless, this has not happened in this case as a uniform approach has been applied, regardless the local particularities. Although the amount paid for subsidies have been updated from 1997 to 2010 these adjustments have not been based on a post-implementation review.

Regarding the selection of beneficiaries for borehole drilling, loose conditions were reported by some water officers (Charalambous et al. 2011). During the extreme drought of 2007–2008, the number of subsidies paid drastically increased (amounting to an investment cost of about EUR 2.5 million for the 2 years), demonstrating the fact that external factors probably led to spontaneous and poorly thought reaction in terms of economic efficiency (both due to the increased numbers of subsidies awarded, as well as to the increased grant per subsidy paid). A total of 3,504 subsidies were given for construction of new boreholes and connection to toilet cisterns, and 419 for installation of recirculator and grey-water recycling systems. The resulting unit cost for every m^3 of drinking water saved with these investment costs of the years 2007–2008 reached EUR 2.5 in some cases (e.g. for the subsidies regarding the connection of cisterns to boreholes and the installation of recirculators). The EPI had not provisioned for measures to monitor the achievement of policy objectives and to avoid negative effects.

For financial matters the WDD has to consult with the MANR&E, the Planning Bureau for the authorization of funds and expenditure, the Ministry of Finance and the Accountant General for finance and tenders and the Loan Commissioners for loans for subsidized projects. It is also monitored from the Audits Office and has to justify any change from the original contracts for water development works. This process of obtaining the release of the funds can be tedious, requiring much time and effort. The WDD is bounded on the government procedures for all its actions. That could also be problematic and most importantly time consuming for the procedures, and might have been the root of poor planning of the EPI in terms of grants awarded per subsidy type and their respective updates.

Finally, regarding the EPI and sectoral policies, no specific barriers linked to other policies that posed problems to the successful implementation of the EPI have been identified. On the other hand, the EPI, and specifically the subsidies for boreholes may have put additional pressure on the groundwater resources with negative effects on the environment. Although it was stated by the WDD that the aquifers where subsidies were approved are marginal and of poor quality and thus practically not exploitable for may uses, no pressure and impact analysis. This goes against environmental policies, in this specific case the Water Framework Directive. WFD is intended both to safeguard drinking water supplies and to prevent ecological damage. Similarly, among the goals of the WFD and Groundwater Directive is the good chemical status of the groundwater, and thus with the borehole subsidies the WDD could further deteriorate the groundwater bodies (since less quantity could results in less dilution), when in fact they should try to improve it.

7.4 Conclusions

From 1997 to 2010 a total of 13,172 subsidies have been granted. By looking at their temporal evolution in comparison with the respective precipitation, it is observed that subsidies pick-up in periods of low precipitation (drought events), conveying a message that the motivation of the beneficiaries was securing uninterrupted water supply for their gardens, rather than conservation, and their behaviour was reactive rather than proactive.

The fact that enforcement by the WDD was non-existent, and thus no regular monitoring of the boreholes' meters has been implemented, weakened the EPI's performance and its overall benefits. On the positive side, since the water saved from the subsidies would have originated from desalination, equivalent energy savings and corresponding CO_2 emissions reduction have been induced, estimated to a total of approximately 56 million KWh of energy saving and 3,277 tons of CO_2 emissions/year on average.

The overall average cost per m^3 of drinking water saved from all the subsidies during the whole 1997–2010 period is EUR 0.43 (based on the assumptions and necessary proxies made in this study). Additional transaction costs have not thought been assessed. At the beginning of the implementation, the EPI comes at a high cost, (e.g. EUR 1.52–2.83/m^3 in 1997–1998) since the investment cost is considered as a cost only in the year when the investment was made and water savings have not yet accumulated. As the EPI implementation progresses and water saving is accumulating over the years, the unit cost is decreased as low as EUR 0.10/m^3 (years 2001–2005). A time frame of about 3 years was thus required for the EPI to become cost-effective as compared to the selling prices of the Desalination Plants and water tariffs. From 2006 onwards the unit cost has abruptly increased, reaching values higher than the desalinated water selling prices. This change is due to the fact that the payments were significantly increased, as well as the number of subsidies awarded, supporting evidence that its cost-benefit clearly relates to the design parameters.

The overall performance of the EPI is subject to uncertainty. While drinking water conservation has likely been achieved, all results are based on proxy calculations, (due to lack of proper monitoring), and thus subject to bias. At the same time, there is no clear evidence that an overall reduction of the domestic water consumption has been achieved. The selection of boreholes as a subsidy creates ambiguity, regarding the adverse impacts on groundwater and the irrational use of a free water supply (thus resulting in an overall increase if domestic water use). Weaknesses in the design (no impact assessment prior to implementation, no research behind the selection of the amounts paid, etc.) and enforcement of the EPI (no monitoring and follow-up) cause reservations regarding its effectiveness. There is no evidence that the implementation of the EPI would have been enacted even if the negative net benefit was recognised, yet the subsidies that related with the boreholes (two out of the four subsidy categories) could have been redrawn due to strong arguments by environmentalists (since these were the ones who also received strong criticism after implementation).

In parallel to the subsidies, the WDD had launched a bundle of demand reduction measures: awareness campaigns, water reuse, water pricing, water metering installation, leakage reduction. Thus, it is difficult to decouple the actual effect of the investigated EPI and the savings that are explicitly attributed to the subsidies. While the EPI was aligned with the prevailing laws and policy setting, and it has a flexibility potential to be adjusted to local conditions, public participation, inclusion of stakeholders and collective design were not pursued. If incorporated, these could have brought up issues of social equity, possible unsustainability of the measure as such, and useful suggestions for re-design and enhancement. Additionally, the whole process was much centralised, whereas if a rational partition of responsibilities had been foreseen (i.e. carrying of the inspection by the Local District Office) the burden would have been shared and thus enforcement and follow-up might have been possible allowing in turn real ground evaluation of the EPIs effectiveness.

For this EPI to be successful some key enabling factors and preconditions need to apply. Adequate design, prior to the implementation of measures, based on field research, survey, impact assessment and pilot applications, is essential. This design process needs to be collective, seeking public participation and involvement of the stakeholders in order to allow for the identification of issues of social equity and unsustainability (e.g. in relation to the amounts granted, the expected response, etc.). Enforcement and monitoring, that will allow the timely collection and analysis of data to assess the performance and re-evaluate the original design are further needed. A share of responsibilities among the competent authorities is essential during the implementation phase. Involving regional authorities that could (a) convey local knowledge on the specific prevailing conditions, and (b) perform the inspections, can allow the proper adaptation of the subsidies, while reducing the burden and cost from the central agent. Awareness rising and targeted education of the beneficiaries must have a central role. It is critical that they, as end-users, understand that their main incentive should be water conservation as opposed to saving money from their water bill or securing uninterrupted watering of their gardens, avoiding thus irrational use.

References

Aeoliki Ltd. (2009). *Institutional framework and decision making practices for water management in Cyprus*. INECO project report. Accessed 15 Sept 2014.

Charalambous, K., Bruggeman, A., & Lange, M. A. (2011). *Policies for improving water security, the case of Cyprus*. Project Report, WP4 Deliverable, CLICO FP7 project, March 2011.

Cyprus Energy efficiency report, January 2011. Accessed 15 September 2014.

Cyprus Mail. (2008). *Water authority defends new borehole subsidies*. http://www.cyprusedirectory.com/cyprusguide/cyprus.aspx?ID=8990. Accessed 15 Sept 2014.

European Environment Agency (EEA). (2011). *WISE-SoE reporting on water quantity, waterbase – water quantity*. Accessed 15 Sept 2014.

Government of Cyprus. (2010). *Unified water management law (in Greek)*. Official Government Gazette, Nicosia, Cyprus. Accessed 15 Sept 2014.

I.A.CO Ltd. (2011). *Report on the assessment of water demand management measures (in Greek)*. MANRE, February 2011.

Kambanellas, C. A. (2007). *Recycling of grey water in Cyprus*. Sustainable Water Management 1-2007, Water Development Department, Nicosia.

Kossida, M. (2010). Towards a Water Scarcity & Drought Indicator System (WSDiS). WS&D Expert Network. Helsinki, 30 Sept 2010.

Kossida, M., Tekidou, A., & Mimikou, M. (2011). *Subsidies for drinking water conservation in Cyprus*. Deliverable D3.1 – Review reports, WP3 EX-POST Case studies, EPI-Water project.

Kossida, M., Kakava, A., Tekidou, A., Iglesias A., & Mimikou, M. (2012). *Vulnerability to water scarcity and drought in Europe*. Thematic assessment for EEA 2012 Report. ETC/ICM technical report 2012/3. ISBN 978-80-85087-13-0.

Kossida, M., Tekidou, A., & Mimikou, M. (2013). *Assessing the role of subsidies as an economic policy instrument for water conservation: An ex-post evaluation in Cyprus*. In IWA 3rd conference on water economics, statistics, and finance, 24–26 Apr 2013, Marbella.

Kotsila, P. (2010). *The socio-environmental history of water development and management in the Republic of Cyprus*. MS thesis, Universitat Autònoma de Barcelona. Accessed 15 Sept 2014.

Manoli, A. (2010). *Desalination in Cyprus*. Presentation to the Spanish Cypriot Partnering Event, Nicosia.

Republic of Cyprus, Ministry of Agriculture Natural Resources and Environment (MANRE). (2005). *Water framework directive (2000/60/EC)*. EU Summary Report Article 5 & 6. Accessed 15 Sept 2014.

Republic of Cyprus, Ministry of Agriculture, Natural Resources and Environment (MANRE). (2010). *Water framework directive – reporting sheets on economics*. Nicosia, February 2010. Accessed 15 Sept 2014.

Republic of Cyprus, Ministry of Agriculture, Natural Resources and Environment (MANRE), Water Development Department (WDD) Website. http://www.cyprus.gov.cy/moa/wdd/Wdd.nsf/index_en/index_en?OpenDocument. Accessed 15 Sept 2014.

Water Development Department (WDD). (2005). Citizen's Charter', Press and Information Office, Republic of Cyprus 359/2005 – 2.000, ISBN 963-38-347-5. Accessed 15 Sept 2014.

Water Development Department (WDD). (2009). *Special report 2.1, economic analysis of water use, calculation of total cost of water services, identification of current levels of cost recovery) (in Greek)*. Accessed 15 Sept 2014.

Chapter 8
Residential Water Pricing in Italy

Jaroslav Mysiak, Fabio Farinosi, Lorenzo Carrera, Francesca Testella, Margaretha Breil, and Antonio Massaruto

Abstract This chapter analyses the residential water pricing system in Italy and reviews the empirical outcomes of water tariffs in the Po-River Basin District (P-RBD), and especially in the Emilia Romagna administrative region (RER). The tariff system is imbedded in a composite regulatory framework governing the water supply and sanitation (WSS) services that was instituted in the 1990s. The scope of the review embraces both the outcomes of the WSS reform and the accomplishments of the per-capita and social water tariff variant introduced in RER, along with the service performance criteria meant to encourage better service provision and conservation of water resources. Starting from 2011 the regulation of the water tariffs has been progressively reorganized. As the reorganisation is not yet fully realised, and our analysis concentrates on the ex-post review and assessment, we concentrate on the water tariff system in place until 2012.

Keywords Residential water tariffs • Emilia Romagna • Social tariff • Water pricing reform • Abrogative referendum

---

J. Mysiak (✉) • F. Farinosi • L. Carrera • M. Breil
Fondazione Eni Enrico Mattei (FEEM) and Euro-Mediterranean Centre on Climate Change (CMCC), Isola di San Giorgio Maggiore, 30124 Venice, Italy
e-mail: jaroslav.mysiak@feem.it; fabio.farinosi@feem.it; lorenzo.carrera@feem.it; margaretha.breil@feem.it

F. Testella
Department of Political Sciences, Communication and International Relations, University of Macerata, Piazza Strambi n.1, Macerata, Italy
e-mail: francesca.testella@unimc.it

A. Massaruto
Dipartimento di Sscienze Economiche e Statistiche, Università di Udine, Via Tomadini 30a, 33100 Udine, Italy
e-mail: antonio.massarutto@uniud.it

M. Lago et al. (eds.), *Use of Economic Instruments in Water Policy*,
Global Issues in Water Policy 14, DOI 10.1007/978-3-319-18287-2_8

8.1 Introduction

A residential water tariff is a price that domestic users pay for water supply and sanitation (WSS) services; that is abstraction, storage, potabilisation, conveyance, wastewater collection and treatment. Water tariffs may be designed and structured so as to encourage water conservation and greater water use efficiency; with tangible environmental benefits. In doing so, water pricing may pursue multiple policy goals, seemingly at odds but reconcilable in principle: *water use efficiency*, that is avoiding wasteful use of water; *allocation efficiency*, thus maximising overall society's benefits from water uses; *financial viability*, meaning ability to compensate capital, skills and technology needed to ensure water services and sanitation; and *social equity*, usually referring to the affordability of the water service as a public interest good.

The EU Water Framework Directive (2000/60/EC), the flagship of Community water-related policies, compels an *adequate* contribution of the different water uses, including the households, to the *recovery of the costs* of water services. What is an *adequate* level of cost recovery is left to the discretion of the EU Member States (MS), based on the '*social, environmental and economic effects of the cost recovery as well as the geographic and climatic conditions of the region or regions affected*' (Directive 2000/60/EC). This requirement has not been fully translated into Italian WSS regulation. As a results, the water tariff system plays a limited role.

In this chapter we analyse water tariff system in Italy and the tariff variant introduced in Emilia Romagna administrative region (RER). The tariff system is a part of a comprehensive legislative and regulatory framework that determines the organisational and management structure of the service provision, and the competences and jurisdiction of the respective authorities. The framework had been laid down in the law 36/1994 (so-called Galli law), later incorporated into the law 152/2006 (so-called Environmental code). According to this system, the central government exercised authority over the conceptual design of water tariff system, whereas the power of articulating the water tariff structure and levels was delegated to lower authorities. The water services are organised within water supply and sanitation (WSS) districts (the so-called optimal territorial areas, ATOs). According to the Environmental code, the water tariffs were designed as a price-cap system in relation to the quality of service, amortisation of physical capital, costs of maintenance, and return to capital investments. The price-cap refers to the difference between real and reference operational costs which cannot exceed 30 %. The remuneration of invested capital, set to 7 % of the envisaged investment capital of the water utility, has been at the centerstage of the public abrogative referendum (June 2011). The referendum responded to a 2009 law requesting that water services are either commissioned to entirely private or public-private companies. In the latter case the private constituent should account of at least 40 % of company's capital. The referendum succeeded both to block what has been labelled as 'privatisation' of WSS, and to abolish the return to capital investments as a part of the WSS tariff method. Starting from 2011, the authority over water tariffs design has been partially transferred to the Authority for energy, gas, and water services who initiated, as a transitory measure, a new tariff method. The changes of tariff systems after 2011 are not subject of our analysis, both because the new system is not yet finalised and reorganisation is not yet fully realised, and our review concentrates on the ex-post assessment of empirical evidence.

8.2 Setting the Scene: Challenges, Opportunities and EPIs

Italy is characterised by abundant but unequally distributed renewable water resources. Besides, the relative high climate variability is likely to be further reinforced as a result to medium- to long-term effects of human-induced climate change. The Po-River Basin District (P-RBD) is one of the eight river basin districts (RBDs) established under the EU Water Framework Directive (2000/60/EC) and the legislative decree 152/06 which transposes the WFD into national legislation (the so-called Environmental Code). It is the largest single river basin (RB) in Italy, and an engine of economic growth. The per-capita gross domestic product in the 26 provinces comprised by P-RBD ranges between 21,000 and 38,000 PPS (purchasing power standards) and is above the EU average for all but a few provinces. The administrative Region of Emilia Romagna (RER) situated in the North-East of Italy and partially included in the Po-River Basin District (P-RBD). Emilia Romagna extends over 22,445 km^2 and is home to 4,432,500. The Region includes nine districts (Provinces), nine WSS districts (ATOs) and intersects seven primary water basins among which the most important is the Po-River Basin.

Annual average precipitation within the P-RBD is nearly 1,200 mm, or around 78 billion m^3. Civil water use accounts for around 12 % of the water withdrawals in the river district. The main source of water withdrawal are aquifers in the upstream part of the district whereas several provinces in the downstream part withdraw water from the surface sources and the Po river itself. The city of Ferrara, situated close to the river outlet, is supplied by 72 % from the Po river (ATO Ferrara 2006). The long term average discharge of the river at Pontelagoscuro is 1,540 m^3/s whereas the water abstraction for public water supply varies between 0.9 and 1.2 m^3/s. In summer 2007, river discharge at Pontelagoscuro was as low as 168 m^3/s, barely above the minimum environmental flow of 150 m^3/s, which exemplifies the vulnerability of the WSS provision.

The population in P-RBD amount to 17 residents (+6 % compared to 2001) mostly concentrated in small towns below 25,000 residents. Within the river district, the cities with above 100,000 resident are 11, with total population amounting to 3,400,400 inhabitants (or 20 % of the whole P-RBD population). According to the demographic projections, the population is expected to increase by 7–26 % by 2050. The average domestic water consumption in the main towns is highly heterogeneous, ranging from 240 l/day/inhabitant (l/day/pc) in Lodi to 132 l/day/pc in Reggio nell'Emilia (average 197 l/day/pc). The lowest consumption is typical for the Emilia Romagna region (RER) situated in the downstream part of the basin. The registered water losses are 21.6 % on average across the major town in the P-RBD, and ranging between 34.5 % (Torino) and 7.25 (Aosta).

The average tariff per capita are highest in the RER, whereas the citizens of other major regions comprised in the P-RBD (Piedmont, Valle d'Aosta, Lombardy) pay relatively less. The 2011 data (Federconsumatori 2011) shows that Reggio nell'Emilia is the town with highest average water prices (EUR 2.24[1]/m^3), while Milano's residents pay the lowest tariffs (EUR 0.67/m^3).

The RER government modified the method to determine the water tariffs by the regional decree 49/2006. The method introduced performance factor (PCn) that allows to 'penalise' water utilities not encouraging enough the final consumers to conserve water, while rewarding those who manage to do so. The regional decree 49/2006 introduced the obligation to connect the water tariff to the number of household members. The ATO Bologna fulfilled the obligation by implementing the so-called 'per-capita' tariffs (PCT). The PCT was experimentally introduced in five municipalities in 2008 and fully applied starting from 2009. The tariff is applied only to domestic water uses and includes a fix and a variable component, both dependent on the number of household members.

The domestic water supply is priced with fixed and volumetric components, the latter based on *increasing block tariffs* (IBTs). The tariff is set to recover financial costs of the service to some extent, that is investment costs, operational and management costs, and administrative and support costs. The environmental and resources costs are not included, contrary to what is required by the Water Framework Directive (WFD). RER deploys 'social tariff', subsidised by other user groups, in response to the affordability of household water services. The water tariff is connected to the quality of the service provided, assessed using a set of environmental and service performance indicators.

8.3 The Water Tariffs System in Action

8.3.1 The EPI Contribution

8.3.1.1 Environmental Outcomes

According to the latest available data,[2] the total water withdrawals[3] in RER declined by 1.6 % between 2005 and 2008. With exception of Modena, the withdrawal declined in all ATOs situated in the Emilia part of the region, and increased in the Romagna part, likely as a result of seasonal water demand of attractive touristic

[1] These tariffs are calculated based on reference consumption of 200 m^3/year by a family with two children. Hence the tariffs contain a higher share of the more expensive volume-price block.

[2] The data from the latest water census (published in 2014 and referring to water consumption in 2012) is not yet available in a disaggregated form (per WSS basin and major cities). The data used in our analysis refer to the water censuses in 2008 and before.

[3] This data refers to water withdrawn by water utilities serving specific WSS district (ATOs) and is not necessary indicator of water consumption, as significant volumes of water are transferred between WSS districts.

attractive place along the North Adriatic Sea. The observable changes range between –23 % in Piacenza to +15 % in Modena. Bologna, Parma and Ferrara ATOs show a reduction of –11, –3, and –1 % respectively. Overall, the water withdrawals for public distribution in RER amount to 121 m^3/year per capita (ISTAT 2009a), which is less than the national average (198 m^3/year per capita). The variation in the withdrawals per capita span from 48 m^3/year (Ravenna ATO) to 184 m^3/year (Forlì-Cesena).

Across the P-RBD, the major cities with highest reduction of water consumption includes Parma (–35 % over the period 2000–2011; from 201 to 137 l/day/pc), whereas only Cremona increased the consumption per capita (+3.5 %, from 203 to 211 l/day/pc). On average the water consumption in the P-RBD amounts to 197 l/day/pc.

Households' per-capita water consumption in the district towns in RER is commonly lower than in other cities within the river basin. The highest per-capita consumption is registered in Piacenza (78 m^3/year/person) and the lowest Forlì-Cesena (51 m^3/year/person). Also with respect to losses in water pipeline system RER performs better than most of the other regions. Compared to national average (32 %) and worst performer (Puglia, 47 %), the RER loss rate (24 %) is lower by one and three quarters respectively. Within RER the losses span between 18 (Forlì-Cesena) and 30 % s (Ferrara) (ISTAT 2009a).

Normally, the quantity of water withdrawn is negligible in the basin's water budget. However, during the recent drought spells in 2003 and 2006–2007, the preventive reduction of the domestic water consumption had sizeable effects (ARPA Regione Emilia-Romagna 2006). In the Romagna part of the region, supplied from the Ridracoli dam, the water shortage reached even more critical levels, triggering the declaration of state-of-the-emergency in May 2007.

The riverine ecosystems along the river network and the delta benefit from the combined effect of reduced water consumption in agriculture, industry and domestic sectors. Po-River Delta is one of the most valuable wetlands in Italy and a biodiversity hotspot – NATURE 2000 site – of European importance. The Delta is undergoing lasting changing under significant anthropogenic pressures, sea level rise and sea water infiltration upstream for a considerable distance from the mouth. Hence, the Po-River Delta is extremely sensitive to reduced river flow (RER 2009).

Decree 152/06 specifies the requirements put on quality and coverage of wastewater treatment, in compliance with the Council Directive 91/271/EEC concerning urban wastewater treatment. In RER, 2,163 wastewater plants served about 6.2 million PE[4] (81.6 % coverage) (ISTAT 2009a). The coverage of domestic users increased from 64.2 % PE in 2005 to 67.3 % (+2.9 %). The number of urban agglomeration below 2,000 PE without a wastewater treating (WTT) system in 2008 was still high (1,609). However, the number of larger settlements (>2,000 PE) not connected to treatment plant is only 21, down from 179 in 2005. According to the State of the Environment in RER, the quality of surface water bodies has not

[4]Person equivalent (PE) is a quantity of biodegradable organic substances from the civil use discharged in 24 h and corresponding to biochemical demand of oxygen equal to 60 g per day.

improved notably between 2004 and 2008. This is because agriculture remains the major source of pollution and reduced point pollution is not easily discernible in the quality of water bodies.

8.3.1.2 Economic Outcomes

Compared to the situation before 1994, the reform of the water services and sanitation had helped to modernise WSS, and reduce fragmentation in both service provision and water tariffs in place. Between 2001 and 2010, the number of water utilities operating in the RER went down from initial 157 to 18. The number of tariff basins – areas applying the same tariff structure and levels, was reduced from 214 to 37 (Table 8.1).

The reform however did not ensure level of investments necessary into extension and modernisation of water infrastructures. In 2007, the average annual per-capita investment in WSS amounted to EUR 37.00 (min-max range EUR 19–117) (CONVIRI 2008). According to the only study available, this is by far too little (Massarutto 2011). Most of the investments are designated for new infrastructure, whereas improvement of the existing infrastructure is dedicated only some 37 %. These shares tend to be opposite among the developed countries with high WSS connectivity (40 % for new infrastructures and to 60 % for maintenance of existing infrastructure) (CONVIRI 2008). According to (CONVIRI 2008), the new investments are financed predominantly from the collected revenues (46 %) and public transfers (21 %). Own capital investments and loans are represented by 11 % and 14 % respectively.

Table 8.1 Evolution of the water services and sanitation sectors in Emilia Romagna region (RER) between 2001 and 2010

			2001		2005		2010*	
ATO		POP 2006	WU	TB	WU	TB	WU	TB
1	Piacenza	278,224	30	47	28	30	2	3
2	Parma	420,077	26	47	26	40	4	7
3	Reggio Emilia	501,364	2	2	2	2	2	2
4	Modena	670,098	32	32	4	5	3	5
5	Bologna	954,682	50	50	4	7	2	8
6	Ferrara	353,303	2	2	2	2	2	2
7	Ravenna	373,449	5	5	1	3	1	5
8	Forlì-Cesena	377,993	8	8	1	3	1	3
9	Rimini	294,074	2	21	1	2	1	2
	Total ER	**4,223,264**	**157**	**214**	**69**	**94**	**18**	**37**

Source: Online sources of the italian Statistical Bureau (www.istat.it), own elaboration
Note: *POP 2006* population living in the different ATOs in 2006, *WU* number of water utilities operating in the RER, *TB* number of tariff basins
*Domestic tariffs only

Table 8.2 Actual and planned investment in ATO Ferrara

	HERA	CADF	TOTAL
Population (2006)			353,304
Aqueduct length (km)	2,420	2,264	4,684
Sewage system length (km)	928	905	1,833
Investments 2005–2007 (EUR)	25,872,000	14,039,041	39,911,041
Investments 2008–2012 (EUR)	53,074,000	20,100,000	73,174,000
Investments 2012–2024 (EUR/year)	10,000,000	4,300,000	1,300,000

Source: ATO 6 Ferrara (2007)

Table 8.3 Actual and planned investment in ATO Bologna

	HERA
Population (2008)	960,343
Aqueduct length	8,801 km
Sewage system length	3,504 km
Investments 2004–2006	EUR 82,000,000
Investments 2007–2009	EUR 108,000,000
Investments beyond 2010	EUR 194,720,565

Source: ATO Catchment Area Plan

In each ATO, water supply and sanitation services are commissioned to one or more water utility for the period up to 30 years. ATO Bologna commissioned the service until 2021 to HERA Group S.p.A; ATO Ferrara commissioned the service until 2024 to HERA Group S.p.A. and CADF S.p.A.; and ATO Parma commissioned the service to IREN S.p.A., Montagna 2000 S.p.A., Salso Servizi S.p.A. and Emilia Ambiente S.p.A (RER 2010). The two largest water service providers in RER (Hera and Iren) are multi-utility corporations with large turnover. Business diversification influence positively company's ability to access credits. The Tables 8.2 and 8.3 show the planned investments in the ATO Ferrara and ATO Bologna.

Over the period 1999–2008 Aosta and Sondrio registered the highest reduction of water losses (−72.3 e −62.9 % respectively) while Cuneo and Asti registered a substantial increase of losses (+184 e +102 % respectively).

The RER included an economic incentive for water utilities to reduce water losses and improve the quality of the services; the co-called performance factor (PCn). The PCn is determined by two sets of indicators with respect to quality of the service (e.g. unplanned service disruption, customer satisfaction, call centre service), and environmental performance (e.g., water losses and per-capita water consumption) (RER 2006).

The current tariff systems in Italy led to a great differences in water prices across the ATOs (Federconsumatori 2011). Calculated for a representative level of households' water consumption (200 m^3/year), the water bills across districts' capital range from around EUR 0.58/m^3 (Milan) and EUR 2.39/m^3 (Florence)

Table 8.4 Share of cost components in the water price

	Bologna (EUR/m³)	Ferrara (EUR/m³)	Parma (EUR/m³)
Operating costs	0.019	0.025	0.049
Maintenance	0.042	0.043	0.062
Compensation for the invested capital	0.059	0.050	0.093
Investments in water treatment structures	0.119	0.119	0.205
Investments in water losses reduction	0.091	0.089	0.148

Source: RER (2005)

(Federconsumatori 2011). In 2010, average price of water in the tree district town analysed in this study was well above the national average: Bologna EUR 1.51/m^3; Parma EUR 1.91/m^3; and Ferrara EUR 2.03/m^3 (Federconsumatori 2011). In principle, water bills in Italy are lower than in most other European countries. These differences lie in the incomplete amortisation of water pipeline systems initially build using public money.

In 2005, the Water Conservation Plan estimated the incidence of some of the costs into the total amount of the tariff for each ATO in the Emilia Romagna region (Table 8.4).

8.3.1.3 Distributional Effects and Social Equity

The price of WSS increased substantially since the introduction of the Galli law. Yet compared to other European countries, Italy is still among the countries spending a relatively small proportion of household incomes on water service. However, the number of families which spend more than 3 % of their income for water is on the rise (AUTORIDSRU 2011).

Between 2001 and 2010, the average prices paid by households for water services rose by 66.7 % in Italy and by 68 % in the RER (Table 8.5). In some districts the price increase toped 200 %. To compare, from 2001 to 2007 the net household incomes increased only by 17 % in Italy and 14 % in RER (ISTAT 2009b).

There have been some attempts to define the highest socially acceptable share (SAS) of cost of water service in terms of household incomes, originating from studies on impacts of privatization of water services in 1980s and early 1990s in UK and Wales. Fitch and Price (2002) for example set the SES to 3 %, drawing on the measure of fuel poverty (>10 % of household income). The average cost of water service in Italy does not yet reach a level of concern, but raising poverty and related problems of access to services are being raised.

Poverty indicators show that on average 15.2 % of households in Italy and 9.5 % of households in the Region Emilia Romagna are considered poor according to the EUROSTAT indicator of deprivation. The number of households facing difficulties in paying bills for services (including water and heating), 10.6 % in the national average and 4.6 % in the Emilia Romagna Region, is especially high among single

Table 8.5 Average water charges (Euro per typical annual consumption of 160 m^3) in the Region Emilia Romagna (RER) in 2001 and 2010. National average for 2001 based on an annual consumption of 150 m^3 (AUTORIDSRU 2011)

	2001	2010	Difference
	EUR	*EUR*	%
Italy	**135**	**225**	
Piacenza	67	205	205.97
Parma	**135**	**274**	**102.96**
Reggio Emilia	160	295	84.38
Modena	113	205	81.42
Bologna	**152**	**189**	**24.34**
Ferrara	**186**	**284**	**52.69**
Ravenna	173	267	54.34
Forli-Cesena	196	270	37.76
Rimini	155	239	54.19
Minimum value RER	67	189	182.09
Maximum value RER	196	295	50.51
Medium value RER	**149**	**250**	67.79

parent households and elderly people. In these statistics, water consumption is not considered as a separate indicator. In 2009, 10.6 % of Italian households and 4.6 % of those in the Region of Emilia Romagna were facing problems in providing for adequate heating of their dwellings (AUTORIDSRU 2011). The same report estimates that in 2009, water bills amounted to 0.5 %, for waste collection to 0.6 % and heating to 3 % in terms of household incomes.

The resolution for the regional government n. 560/2008 adopted guideline for the application of social tariff as a way of protecting low-income households. The subsidised water tariffs are offered to all households below a certain threshold, determined with an indicator of wealth ISEE (*indicator of comparable economic conditions*, ISEE[5]). For the territory of the whole region, there is a single threshold that specifies the economically and socially most marginalised and vulnerable households. A second threshold is variable and is determined by each AATO. It specifies households exposed to less extreme economic and social hardship. The social tariff is financed through the application of higher water tariffs (up to 1 %) applied to wealthier consumers. Facing the second highest water tariff in RER, the ATO Ferrara was the first one to apply the social tariff (resolution n. 5 of 17 December 2007). In 2008, the water tariffs were increased 0.5 % and the proceeds collected were designated to co-finance the water consumption by disfavoured clients, elderly citizens and physically impaired persons. ATO Parma adopted the social tariff in 2009 (resolution n. 15 of 22/12/2009) (Tables 8.6 and 8.7).

The collected funds for social tariffs amounted in 2009 to EUR 59,075 in Bologna, EUR 193,088 in Ferrara and, in 2010, ca. EUR 300,000 in Parma.

[5] In Italian, *Indicatore Situazione Economica Equivalente.*

Table 8.6 Example of social tariffs in the selected ATO

ATO	Most marginalised groups (ISEE) (EUR)	Less marginalised groups (ISEE) (EUR)	Price increase for other users (%)
Ferrara	<2,500	2,500–5,000	0.5
Parma	2,500–5,000	2,500–5,000	

Table 8.7 Number of households- beneficiaries of social tariffs in 2009

	Bologna		Ferrara		Parma[a]	
	No of households	% of all households	No of households	% of all households	No of households	% of all households
First income band	643	0.2	555	0.3	2,400	1.2
Second income band	2,150	0.5	1,593	1	7,100	3.6
Total	2,793	0.7	2,148	1.3	9,500	4.8

[a]Values for Parma refer to 2010, the first year of the tariff in this area (AUTORIDSRU 2011)

The quality of the water supply and sanitation services is regularly evaluated in terms of customer satisfaction. Generally, the communication of water authority yields medium level of satisfaction, whereas price level receives lowest scores. Some areas within RER display a higher degree of dissatisfaction (AUTORIDSRU 2011). Half of the consumers does not drink tapped water or only or rare occasion, complaining "bad taste" (AUTORIDSRU 2011).

8.3.2 *The EPI Setting Up*

8.3.2.1 Institutions

Water and sanitation (WSS) service in Italy are regulated by the law 152/2006. The service is organised within the WSS districts (so-called *optimal territorial areas* or ATOs) that in RER coincide with the boundaries of lower administrative districts (provinces). Until recently, each ATO was governed by an autonomous regulatory authority (*ATO Authority*, AATO). In 2010, these authorities were dismantled and their competences transferred to regional administrations. Each ATO is managed according to a plan (the so-called *optimal territorial area plan*, hereafter PA) that specifies priorities and future investments within the WSS basin, and specifies the water tariffs.

Article 154 of the Environmental Code (law 152/2006) equals water tariffs to compensation for water services and connects them to quality of water and water services, amortisation of physical capital, costs of maintenance and return to capital investments. Until 2011, the water tariff system was based on the so-called *'nor-*

malised method' (NM) introduced in 1996. Using the NM, the AATO determined the reference tariff within their jurisdiction. This in turn are translated into actual tariffs by taking into account organizational model of the management, water quantity and quality, the level of quality of water service, financial plan, and actual costs of the management. Typically, water tariffs for residential water use employ three blocks: the first is subsidised, second is regular and third penalises excessive water use. The tariff contains a fixed and a variable component of water supply, purification fee and sewage fee.

The Region Emilia Romagna (RER) transposed the law 36/94 by the regional law (RL) n. 25 of 6 September 1999.[6] In order to incentive water conservation, while respecting social equity aspects, the tariff blocks could be varied according to territorial criteria, users' type and volume of consumption.

The RL of 14/04/2004 n. 7 modified the RL 25/99 in a way that was at odds with the provisions of the law 36/94: it assigned the regional government the task of defining the water tariffs, while taking into account the recommendations of an expert commission established for this purpose, and the results of consultations involving syndicates, and key economic and social players. Among others, the tariff had to include incentives to use natural resources efficiently. Subsequently, the resolution n. 5749 of 16 April 2004 established an expert commission whose task was it to revise NM and make recommendation with respect to the reference tariff. In 2006, the regional government's presidential decree (DPRG) n. 49 of 13 march 2006 (modified successively by the DPRG n. 274 of 13/12/2007) adopted a tariff method for the integrated water service. The innovation of tariff system introduced in RER include among other the promotion of high quality service and water conservation through the water tariffs, higher flexibility with respect to the price-cap, and the option to disentangle the water supply and waste water discharge tariffs, more adequate remuneration of the invested capital.

The article 2 of the RL 10/2008 instead assigns the task of specifying the reference tariff to the regional government who is also asked to develop an economic and financial plan of integrated water service. The Constitutional Court, with the sentence 29/2010, ruled unconstitutional the two articles mentioned above. The Constitutional Court argued that the protection of the environment and the guarantee of market competition are of exclusively competence of Central State. The Court affirmed that the aims of water tariff discipline are to protect the environment and to apply a uniform tariff system in all the country without any difference among the various Regions. The regional government argued that the RL 10/2008 acted in order to prevent the specification of water tariffs in a fragmented way, individually for different ATO. With a circular PG2010.0103608 of 13/04/2010 the Directorate General for Environment of the RER confirmed the validity of the tariff method introduced by the RL 49/2006 (along with subsequent modifications).

The Water Conservation Plan of RER foresees water tariffs that incentive water conservation. The DPRG 49/2006 introduced the obligation that within 5 years, or

[6] Later modified by RL n. 27 of 21/10/2001, n. 1 of 28/01/2003, n. 7 of 14/04/2004 and n. 10 of 30/06/2008.

at the time of the first revision after 1/12/2007, the tariffs have to consider the number of household components (art. 10, comma 5). The ATO Bologna introduced another change, the so-called 'per-capita' tariffs (PCT), experimentally in five municipalities in 2008 and since 2009 in the whole territory of competence. The tariff is applied only to domestic water uses and includes a fix and a variable component, both dependent on the number of household members. The tariff is organised in five blocks, the first two of which are subsidised, the third is standard one, and the last two are penalising the excessive water use. The pro-capita tariffs are specified in five blocs: the first two subsidised, and the last two penalising the high water use.

8.3.2.2 Transaction Costs and Design

At the state level, the costs of regulation of water tariffs include the operative costs of the overseeing agency. The agency is set to collect the data about tariffs applied by ATOs across the country, verify the compliance with the state regulation, revise regularly the tariff system, and produce annual reports about the state of WSS in Italy. Since 1994, the agency changed twice, incurring further costs due to reorganisation and restructuring. The *Vigilance Committee for Water Resources* (CO. VI.RI.) was initially established in 1994 and abolished in 2009. Its successor, the *National Commission for Water Resources*, was abolished 2011. Later, the advisory and compliance control tasks have been assigned to the *National Agency for Water Resources Vigilance*.

At the level of the WSS districts (ATOs), the transaction or institutional costs are internalised through water tariffs and born by the consumers. These include costs of negotiated agreements among the participating municipalities, and the operational costs of the Authority of ATO. In addition, the costs of regional vigilance committee or tariff commissions such as that established in RER by the resolution n. 5749 of 16 April 2004.

Large proportion of the transaction costs are impaired by the litigation costs. Between 2008 and 2010, the Constitutional Court had intervened several times with respect to the water supply and sanitation service (sentences 335/2008, 246/2009, 307/2009, 29/2010, 142/2010 e 325/2010).

8.3.2.3 Implementation

The governance regime of water supply and sanitation in Italy is based on cooperative arrangements between state and regional governments. The centrally governed water tariff system in place until 2012 was a result of a negotiated agreement, and subject of a periodic review conducted in collaboration with the Ministry of Environment and the Ministry of Finance. The vertical disaggregation of regulatory competences respects the subsidiarity principle and power division between state and regions. At the level of an ATO, the constituting municipalities cooperate for the sake of coordinated and more efficient water service provision.

In RER, the governance regime is a result of a constructive public debate. The regional legislation is a result of an extensive consultation between the regional authority and social stakeholders. In 2004, the social water tariff was negotiated between regional authorities and labour unions (CGIL, CISL and UIL), resulting in production of a guidance document and pro-capita tariff later codified in the regional law.

On 12–13 June a citizen initiated referendum was held in Italy to partially abrogate the law 166/09 (so-called Ronchi law), decree 133/2008 and legislative decree 152/06 (the so-called Environmental Code) referring to the public water supply. Two out of four quest of the referendum address the public water services. The first quest addressed the article 23bis of legislative decree 133/2008 concerning the privatisation of public services with economic relevance, modified by Law 166/2009. Since 1999, public water services were entrusted to public (in-house) or private companies – water utilities. The legislative decree 133/2008 put higher burden on commissioning water supply and sanitation to in-house public water utilities, encouraging greater private sector participation. The law 166/2009 went further and requested that by December 2011 water services are either commissioned to entirely private or public-private companies. In the latter case the private constituent should account of at least 40 % of company's capital. The public water utilities were admitted only in transitional mode or in situations in which the market mechanism is either inefficient or useful.

The second quest sought abrogation of article 154 of legislative decree 152/06, determining the return on invested capital (ROIC) by the normalised method (NM). The ROIC provides incentive to invest into modernisation of water infrastructure, modernising the water services and making them more reliable. The normalised method for tariff determination (NM) set the ROIC to 7 %. Before the referendum, the Constitutional Court backed the ROIC by ruling that public water service was essentially an economic service (judgment n. 325/2010).

The referendum reached quorum and both quests, as well as the additional two not referring to the water services, were approved by the public ballot. The abrogation of article 154 of legislative decree 152/06 concerning ROIC has uncertain legal outcomes. Unaffected by the referendum is also the article 117 of the legislative decree 267/00 requesting an adequate compensation of the invested capital based on prevailing market conditions.

8.4 Conclusions

The WSS reform in the 1990s reorganized the water service and set out for a more efficient and harmonised water service provision. The reform had helped to reduce fragmentation in both service provision and water tariffs in place, as shown by the evidence collected. Although the available data is patchy and rife with uncertainty of many kinds, a decreasing trend can be observed in water abstraction/consumption pro-capita and water pipeline leakage. Similarly, the household access to WSS has steadily improved. RER performs better than the national average in all environmental outcomes, with a high variability across the WSS districts (ATOs). The price of a cubic metre of water and wastewater services, adjusted for inflation, increased significantly over the past years. Compared to other OECD countries, the water price adjusted by purchasing power parities is still low (OECD 2009), the main reason being that the initial capital investments borne by the central state are not amortised in the current tariff systems. On the downside, the tariff system has not guaranteed necessary investments into extension and modernisation of water infrastructures. The planned investments in water infrastructure are by far too low in order to guarantee a sustainable and reliable water services. The failed attempt to reinforce participation of public sector in WSS provision introduced a regulatory uncertainty discouraging from investments. The water utilities will have access to external sources of finance, such as loans, only if a sufficient and reliable stream of revenue is ensured.

Empirical evidence shows that water pricing is a suitable tool for encouraging water conservation and demand management. Water is a social good whose service provision can be governed by economic instruments. The recognition of right to water as a fundamental human right is not at odds with the participation of private sector in the water service provision. The access and affordability of water can be reconciled with water pricing in several ways. In RER, it is managed by social tariffs whose costs are distributed among the wealthier consumers. Alternatively, it could be managed either by income support (connected or not to water consumption), or by facilitated payments. See OECD (2009) for further discussion of both.

The extent of litigation with respect to regulatory authority over water supply and sanitation services underlines the unresolved issue of power sharing between the state and regions. Given the large economic and social disparity across the administrative regions, more flexibility and discretion is warranted at the regional level in order to adapt water pricing schemes to specific environmental and socio-economic conditions. The performance factor introduced in RER is an example of regulatory innovations that are worth to pursue. However, it should be based on a simple set of service quality indicators that can be easily collected and assessed. The water tariffs system in Italy and elsewhere is not shielded from political interference. The current water pricing regulation blurs the distinction between the regulator and regulatee. On the one hand, local governments of municipalities assembled in a single WSS district play a part in water services regulation and tariff specification. On the other hand, it is common that the water utilities to which the WSS is commissioned are controlled by local governments.

References

ARPA Regione Emilia-Romagna. (2006). Monitoraggio e caratteristiche idrologiche della magra estiva del Fiume Po nell'anno 2006.

ATO Ferrara. (2006). Piano di conservazione della risorsa, Agenzia di Ambito per i Servizi Pubblici, Ferrara.

ATO Ferrara. (2007). Piano d'Ambito per il Servizio Idrico Integrato annualità 2008 – 2024. Ambito Territoriale Ottimale n. 6, Ferrara.

AUTORIDSRU. (2011). Relazione annuale sullo stato dei servizi idrici, di gestione dei rifiuti urbani e sull'attività svolta. Autorità regionale per la vigilanza dei servizi idrici e di gestione dei rifiuti urbani, Bologna.

CONVIRI. (2008). Rapporto sullo stato dei servizi idrici. Comitato per la Vigilanza sull'Uso delle Risorse Idriche, Roma.

Federconsumatori. (2011). IX Indagine nazionale a campione sulle tariffe del Servizio Idrico Nazionale. Federazione Nazionale Consumatori e Utenti.

Fitch, M., & Price, H. (2002) Water poverty in England and Wales. Chartered Institute of Environmental Health. http://www.cieh.org/library/Knowledge/Environmental_protection/waterpoverty.pdf. Accessed 15 Sept 2014.

ISTAT. (2009a). *Censimento delle risorse idriche ad uso civile*. Roma: Istituto Nazionale di Statistica.

ISTAT. (2009b). *Il reddito disponibile delle famiglie nelle regioni italiane*. Roma: Istituto Nazionale di Statistica.

Massarutto, A. (2011, unpublished). *La riforma della regolazione del servizio idrico in Italia.*

OECD. (2009). *Managing water for all – An OECD perspective on pricing and financing* (p. 147). Paris: OECD Publishing.

RER. (2005). *Rapporto sulle attività di smaltimento delle acque reflue urbane e dei fanghi*. Bologna: Regione Emilia Romagna.

RER. (2006). *Metodo tariffario per la regolazione e la determinazione della tariffa del servizio idrico integrato in Emilia-Romagna*. D.P.G.R, n. 49/2006. See also Bollettino Ufficiale della Regione Emilia-Romagna 2-1-2008 – parte seconda – N.1.

RER. (2009). *Relazione sullo stato dell'ambiente della Regione Emilia Romagna*, Bologna.

RER. (2010). *Relazione annuale sullo stato dei servizi idrici, di gestione dei rifiuti urbani e sull'attività svolta Anno 2010. Autorità regionale per la vigilanza dei servizi idrici e di gestione dei rifiuti urbani Regione Emilia Romagna*, Bologna.

Chapter 9
Water Tariffs in Agriculture: Emilia Romagna Case Study

Michele Vollaro, Laura Sardonini, Meri Raggi, and Davide Viaggi

Abstract The chapter presents changes in the irrigation tariff system of the irrigation district Tarabina, in the Emilia Romagna Region, Italy. In order to improve the management of the irrigation water resources (distribution of water and related costs), in 2006 the users voluntarily replaced the area-based payment (a financial instrument) with a volumetric tariff (EPI) and introduced a set of formal rules. In the following years, a reduction in water use at district level has been observed. Such an outcome has aroused a particular interest in studying the contribution of the volumetric tariff, intended as an EPI, on the reduction of water use. The capability of such an EPI in reducing the amount of water used in agriculture would strengthen the policy intentions of the EU of implementing measures that induce a more efficient use of water resources. Based on a counterfactual analysis, it has been found that the introduction of the volumetric tariff induced a reduction, on average, of about 50 % of the water used for irrigation along with a reduction of about 70 % of the costs for the non-irrigators. Such findings suggest that EPIs, associated to other instruments, such as site-specific regulations, might improve their effectiveness and pursue multiple policy goals.

Keywords Irrigation water management • Marginal pricing • Volumetric tariff

9.1 Introduction

The chapter reports the water management experiences of an users-based irrigation organization in Emilia Romagna Region and aims at assessing, through a qualitative approach, the relative performances in terms of improvements in water allocation

M. Vollaro (✉) • L. Sardonini • D. Viaggi
Department of Agricultural Sciences, Alma Mater Studiorum University of Bologna, viale Fanin 50, 40127 Bologna, Italy
e-mail: michele.vollaro@unibo.it; laura.sardonini@unibo.it; davide.viaggi@unibo.it

M. Raggi
Department of Statistical Sciences, Alma Mater Studiorum University of Bologna, via delle Belle Arti 41, 40126 Bologna, Italy
e-mail: meri.raggi@unibo.it

M. Lago et al. (eds.), *Use of Economic Instruments in Water Policy*, Global Issues in Water Policy 14, DOI 10.1007/978-3-319-18287-2_9

and relative costs. The case represents an interesting example of improvements in the governance of irrigation water that took place in the irrigation district Tarabina, in which a voluntary change in the tariff system, from a unique area-based payment to a composite tariff accounting for the quantity of water used, set up by the users to resolve distributional issues in the quantity and costs of irrigation water, have implied a remarkable reduction of water use. Although the choice of implementing volumetric tariffs has not essentially been a response to changes in the availability of irrigation water resources, this particular experience *de facto* demonstrates the potential of improvements in water management (water pricing and metering) as effective adaptation strategies aimed at improving the management of water resources by the employment of an Economic-Policy Instrument (EPI).

9.2 Setting the Scene: Challenges, Opportunities and EPIs

Incentive pricing is the instrument envisaged by the Water Framework Directive (WFD) in art. 9 for inducing (i) the full-cost recovery of the water services, including the environmental and resource costs, and (ii) a more efficient use of the water resources, concurring to the environmental objectives, in the context of the application of the Polluter Pays (PPP) and User Pays (UPP) Principles. The adoption of pricing has been highly recommended also by the *Blueprint* (EC 2012), which is an orientation document about water policy at EU level that focuses also on quantitative aspects of water resources.

The case study is located in the South-East of the Emilia Romagna Region and is part of the district managed by the Land Reclamation and Irrigation Board "Romagna Occidentale" (LRIBRO). The focus of the study is the introduction of an incentive pricing instrument (volumetric tariff system) in a sub-area of LRIBRO.

Although the diffusion of pricing mechanisms across the EU is mostly related to environmental and/or quantitative issues, the adoption of a volumetric tariff in the irrigation district Tarabina is the governance response to an intentional correction of the repartition of water costs and allocation among district members. Indeed, many members, especially non-irrigators, considered the area-based tariff as an unfair pricing system, but also many irrigators were not able to stand anymore to repeated increases in the tariff level. The change to a volumetric tariff system represented, therefore, a solution for improving fairness among non-irrigators and an instrument for inducing self-regulation in the use of irrigation water among irrigators.

This specific incentive pricing has been chosen among a set of other possible instruments mainly because the irrigation district is served by a network of pressure pipes, but also for fulfilling the requirement provided by the art. 11 of WFD, which recognizes pricing as a "basic" measure, namely minimum requirements to be complied with. Moreover, the Tarabina Management Committee (TMC), in agreement with the LRIBRO authorities, adopted a set of formal rules in order to provide the best management ground for the implementation of the incentive pricing. This innovative governance and institutional setting is in line with the indications of the WFD, which provides River Basin Authorities (RBAs) with the opportunity of creating *ad hoc* policy mixes by envisaging "supplementary" measures concurring to the environmental objectives of the Directive. However, an aspect to be considered is the specific context in which the EPI has been implemented. Indeed, the district Tarabina is a relatively small area which covers about 700 ha and includes approximately 50 farms (one of them is a cooperative and covers more than an half of the total surface). Moreover, the area hardly suffers from water scarcity because the irrigation water is delivered by the Canale Emiliano Romagnolo[1] (CER) for the means of a long-term contracts of water supply with LRIBRO. Such contextualization has represented a favorable ground for the adoption of an incentive pricing system based on water metering, especially in virtue of the relatively low costs of implementation, both at administrative level and for farm-level adaptation of irrigation facilities.

Despite the specific context considered, incentive pricing instruments are usually adopted for inducing users to profitably self-regulate the consumption of a good (behavioural change/collective action) (Cross 1970) in order to promote the realization of one or more social outcomes (e.g., reduction in pollution, adoption of water saving technologies…) (Rogers et al. 2002; Ward and Pulido-Velazquez 2009), especially improvements in allocation efficiency of available resources. Indeed, incentive pricing instruments have been envisaged by the WFD on the basis of the dynamic relations between quality and quantity of water resources (an increase in quantity induce increase in quality, *ceteris paribus*) for concurring to the objective of improving the environmental status of water bodies. However, the effectiveness of an EPI cannot be evaluated solely on the performance of the pursued outcome, because its implementation might produce second-order effects or affect other factors not properly or directly considered during the design stage, such as, e.g., the ability of the EPI of not debilitating economic development, the effort to avoid unfairness in the distribution of economic and financial burdens among members of the society and of the economic sectors (avoid social conflicts). This is especially true/valid for EPIs, like incentive pricing, which operate through the internalization of water uses' costs. Based on such considerations, this case study proposes to analyse the effects on water use of an incentive pricing instrument that has been designed for correcting the cost distribution of irrigation among users. Such quantitative aspects of the outcomes of the incentive pricing instrument have been assessed by the means of a counterfactual analysis, based on a performance's comparison with respect to the "twin" irrigation district Selice in which the tariff system has remained unchanged.

---

[1] CER is one of the most important water infrastructures in Italy. It delivers water from the Po River to supply agriculture (mainly) and industrial uses in the south eastern areas of the Emilia Romagna Region.

9.3 The Volumetric Tariff in Action

The agricultural area of the district is served by a network of pressured pipe system that was built by the national government in the early 1980s. Such type of infrastructure allowed for an autonomous administration of the district, called "cost center", such to keep the accounting system of the district Tarabina separated from the general administration of the LRIBRO. The definition of Tarabina as a "costs center" involved the introduction of a management committee (TMC) (farmers elect seven members out of a total of nine). Data on land use and the crop mix in the Tarabina area are not available from statistical sources (due to lack of information at the appropriate scale), but qualitative information was made available by the technical staff of LRIBRO, consulted through direct interviews. They stated that the main specialization in the Tarabina area is horticultural crops and that heterogeneous crop mixes are present at the farm level based on combinations of other crops, such as seed for industrial uses, cereals and fruit (peaches, kiwis, apricots, plums). Data on water uses and tariff paid by district members are available at aggregate (district) level up to 2011 and at farm level (for irrigators) since 2006. However, the staff of LRIBRO cannot release such data because of privacy restrictions on the use of such information. Data on M&O costs are available at aggregate level.

At the outset of the irrigation system, a flat-rate (*per ha*) tariff system was adopted (representing a minimum contribution, equal for all members, to the maintenance and operational (M&O) costs of the district). In 2005, the TMC proposed to change the pricing system, supporting those farmers who complained of excessive water tariff increases (from EUR 20/ha in 1983 to EUR 155/ha in 2005 for all farmers, both irrigators and non-irrigators). The solution was identified in shifting towards the adoption of a volumetric tariff, implemented through the installation of water meters, by charging water users according to the actual applied quantity of irrigation water and by the collateral adoption of a formal set of rules, needed for governing the new EPI. The majority of farmers decided to adopt the new volumetric tariff system. Its introduction was first tested in 2005 and definitely adopted in 2006.

The new pricing is called "trinomial", since the tariff is the sum of three components:

- A fixed component (EUR/ha): paid by both irrigators and non-irrigators, representing a payment quota for M&O costs;
- A volumetric component (EUR/m^3): representing the actual water use, quantified by water meters and paid by irrigators only to recover the costs of the resource and its delivery;
- A variable component (EUR/ha) introduced to recover all the remaining costs related to water use (not covered by the previous two quotas); this part is charged in the next business year and includes additional costs such as non-ordinary interventions, unmetered water use and M&O costs, and is paid by irrigators only.

Figure 9.1 represents the rationale of the ex-post analysis, performed in order to clarify which were the preconditions of EPI introduction, the EPI and which are the main effects to be analyzed:

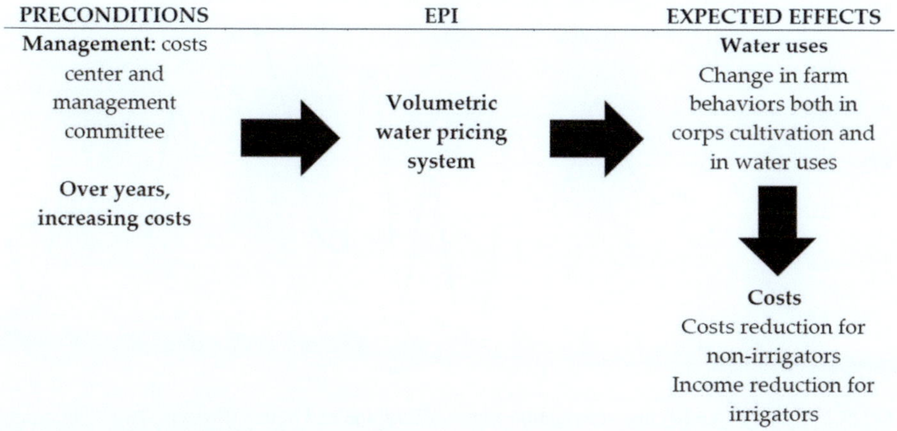

Fig. 9.1 Rationale under the implementation of the volumetric water pricing system in Tarabina (Source: Own elaboration)

9.3.1 The EPI Contribution

The contribution brought about the implementation of the volumetric tariff in Tarabina can be better assessed by implementing a counterfactual analysis based on the performance of the irrigation district Selice, instead of focusing on time-differences within the same district Tarabina. The "cost center" Selice is considered the twin of Tarabina since it is identical as regards the agricultural and infrastructural characteristics. Selice neighbors Tarabina from the South border and its plain agricultural land of about 1,300 ha is shared among 42 farms that receive water from the CER.

Since 1983, the contributive system in Selice is regulated by a monomial areal tariff and the district has its own formal set of rules for the management of irrigation infrastructures and water resources. Given the close vicinity to Tarabina, the weather conditions in Selice can be considered as yielding the same effects on water use borne by Tarabina. Indeed, by exploring the linear trends in water use as shown in Fig. 9.2, Tarabina records a marginal increase close to 24 m^3/ha per year until 2005, thereafter it shows a null tendency, while Selice shows a marginally increase of 21 m^3/ha per year over all the considered period.

9.3.1.1 Environmental Outcomes

In the context of the environmental outcome, the main result of the EPI implementation, judging by the responses of the agents involved, is the reduction of the global amount of water used by farmers in the irrigation district Tarabina. In the period prior to the introduction of the EPI, the distribution of water use was particularly variable, as noted in Fig. 9.3, with a general increasing trend and an average consumption of about 440,000 m^3.

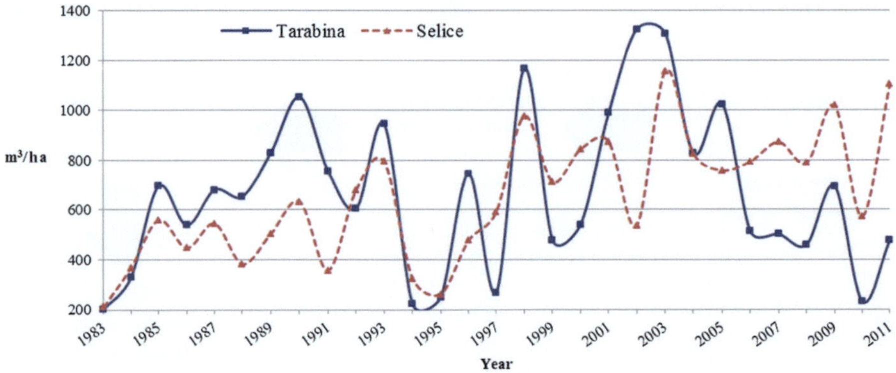

Fig. 9.2 Unitary (per ha) use of irrigation water in Tarabina and Selice (Source: Own elaboration on LRIBRO data)

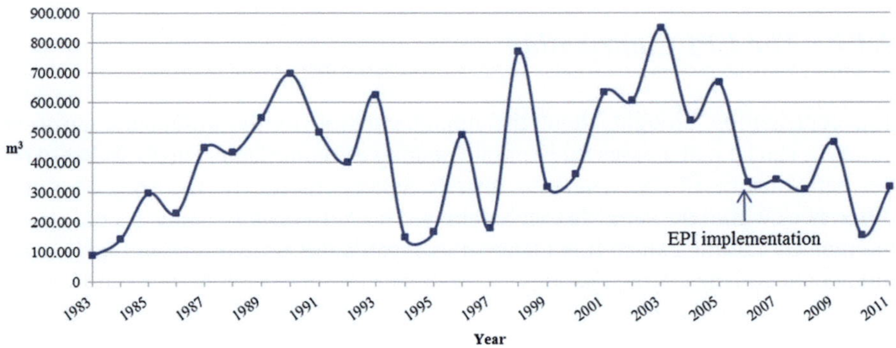

Fig. 9.3 Water use distribution in Tarabina between 1983 and 2011 (Source: Own elaboration on LRIBRO data)

After the introduction of the volumetric tariff in 2006, the distribution seems to follow a more stable trend with an average level of 320,000 m^3, about 30 % lower than the one registered in the previous period. The variability in water use is likely linked both to climatic factors (such as rainfall and temperature) and the water requirements of crops that differ from year to year. However, by looking at the variation in water use in the twin irrigation district Selice, for which the weather effects can be considered the same as in Tarabina, the average consumption of water changed from about 835,000 to 1,100,000 m^3, an increase of about 32 %. In terms of water quality or pollution problems, the EPI implementation did not bring about any change, as the water in this area is good enough for irrigation. Moreover, from a social perspective, the EPI is not likely to have clear effects in terms of environmental pressures, as previous studies show that they are poorly related to changes in water use (at least in the relevant use interval) (Raggi and Viaggi 2009).

Two main consequences, related to changes in individual behavior, have been observed with respect to pressures on water-related ecosystems after the introduction of the EPI. The first is the elimination of "chiari" (lake advocated for hunting activities) and the second concerns energy use. Nevertheless, another aspect regarding possible crop changes should deserve particular attention. However, according to LRIBRO technical staff, crop cultivation did not change on the majority of farms.

With respect to the first aspect, the use of water meters discouraged the non-agricultural uses of water, such as the "chiari" that are ponds used for recreational purposes, in particular hunting activities. Before the EPI implementation, the "chiari" were filled at least two or three times each summer with large quantities of water (up to 200,000 m^3). The new tariff system triggered an incentive to reduce the amount of water used to fill the "chiari", but, however, the abandonment of such behavior brought about, as a direct ecosystem consequences, a significant reduction in the number of birds. Indeed, the other aspect to be considered is that the purpose of "chiari" is not to provide environmental improvements.

The second aspect concerns the variation in energy use, measured as the total costs for energy services. The data available cover the period 1983–2011, but have serious limitations in assessing the effect of the introduction of volumetric pricing in 2006 (the data covers only 5 years of the EPI implementation and data related to 2002 and 2004 are missing). In Fig. 9.4, the trend related to energy costs is shown. Up until 1993, the trend is that of increasing costs; in the subsequent 10 years the behavior is rather variable and after 2006 it shows a stability.

By relating the energy costs to the total amount of irrigation water delivered (energy costs per unit of irrigation water), an increasing trend of EUR 0.0027/m^3 per year is observed, while Eurostat data on energy prices for industrial purposes increased by EUR 0.0017/Kwh per year.[2] An estimate of energy consumption at

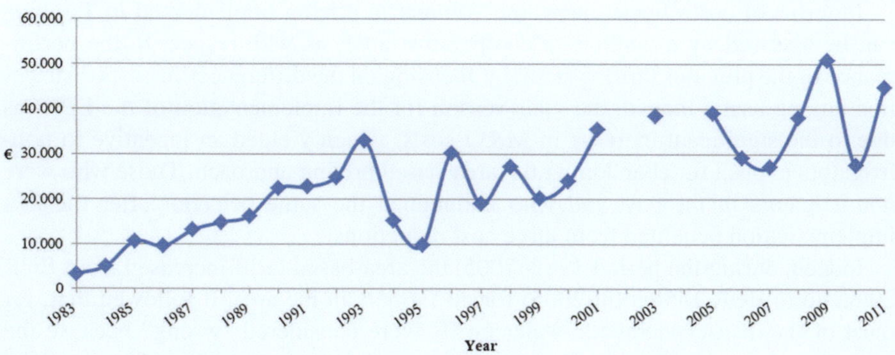

Fig. 9.4 Energy costs in Tarabina between 1983 and 2011 (Source: Own elaboration on LRIBRO data)

[2]The observed period goes from 1991 to 2011, according to the availability of Eurostat data on unitary energy prices (EUR per kilowatt-hour) for industrial purposes in Italy, including levies and taxes.

district level would be necessary to isolate the effects of both water consumption and inflation on the total energy costs. However, given that the computed trends present a difference in the order of millesimals, the yearly fluctuations of energy costs in the Tarabina district could be partially attributed to the variation in yearly water consumption. Therefore, it may be that one of the outcomes of the EPI is a stabilization in energy costs, which implies a benefit in environmental terms due to a more accurate use of energy and, maybe, a reduction in fossil fuel inferred by the increases in unitary energy costs since 2008 (EUR 0.0028/Kwh per year).

9.3.1.2 Economic Outcomes

The EPI in this context was not compared to alternative water pricing systems at the time of its selection, but its implementation was decided upon by implicitly comparing it to the existing area-based pricing system. The choice of the EPI was quite obvious because of the presence of a pressure pipe system, in principle allowing for an easy installation of water meters and related reporting procedures. The shift from an area-based to a volumetric water pricing system was hence identified as the 'best' solution by and for the users in the area. The change of water pricing system was also supported by the good characteristics of the hydraulic system and the small geographic area covered.

Compared to the previous area-based system, the EPI contributed to economic efficiency both in terms of water allocation among farmers and overall water use. The shift in the pricing system resulted in water re-allocation between users in terms of quantity used, in particular providing incentives to use less water for farmers with lower marginal value of water (that would have used more water in an area-based system, in which the marginal cost of water is zero).

In terms of cost-effectiveness, the volumetric pricing implemented in Tarabina can be assessed by a qualitative comparative analysis with respect to the performance of the previous tariff system, by focusing on the differences in costs distribution among users. Indeed, the main reason for the implementation of the EPI was due to the significant increase in M&O costs, which yielded an incentive to non-irrigators to push for abandoning the area-based pricing approach. Those who were non-irrigators in the past and who maintained the same behavior after the EPI implementation benefited from large cost reductions.

Indeed, during the period 1982–2005, the area-based tariff increased from EUR 20/ha up to EUR 155/ha (in 2005) for all farmers in the area. It followed that, for most of the district members, water tariffs were considered "wrong" because the cost allocation was not related to actual use. For this reason, the introduction of the EPI in Tarabina was easily justified. The actual implementation took place by way of the use of water meters by those farmers who planned to irrigate in the future and consequently in the shift to the volumetric water pricing system. The volumetric water pricing system was tested in the first year (2006) and improved in the following year (2007). As for the previous tariff system, the payment that each farmer bears in year t is calculated on the basis of the cost (for the flat tariff) and the actual

Table 9.1 Volumetric tariff system adopted in 2006

Trinomial tariff	Non-irrigators	Irrigators
Fixed component	EUR 29/ha	EUR 29/ha
Volumetric component	Not paid	EUR 0.15/m^3
Variable component	Not paid	Paid (EUR/ha)

Source: LRIBRO data

use (for the volumetric tariff) of water in the year *t-1*. Table 9.1 shows the amount of the three components of the volumetric water pricing system related to usage in 2007.

A fixed component (EUR/ha) is paid by both irrigators and non-irrigators and represents the payment component for M&O costs. The volumetric component (EUR/m^3) represents the real water used in year *t* and is controlled by water meters. The variable component (EUR/ha) is computed in year *t+1* and is paid by irrigators only. The latter component (variable each year) is introduced to recover all the remaining costs (not covered by the previous two components). This part could include additional costs beyond ordinary interventions, such as unmetered water and M&O costs.

After only few years from the implementation of the EPI, a very first assessment of the impact of the volumetric pricing can be made. At the global level (whole area), the general efficiency of the system increased because the reduced water use resulted in an abatement of the cost of water provision (as commented by the LRIBRO technical staff). In fact, the total amount of water used decreased and consequently the M&O costs also decreased. The LRIBRO evidence shows that non-irrigators benefited from a cost reduction of about 70 % in 2006 (from 155 to 29 EUR/ha), whereas irrigators experienced a reduction of around 50 % due mainly to a water use reduction induced by the volumetric pricing system. Based on this information it is likely that the shift to the EPI translated into a prevailing reduction of revenues for the farmers. However, at this stage it is not possible to estimate the overall effect on profits given the short time elapsed since the implementation of the EPI and the relative non-availability of data at farm level. Nonetheless, some information were made available from the technical staff of the LRIBRO. In particular, for non-irrigator farmers, it seems likely that the balance between reduced revenues and costs yields an increase in income. The result is more ambiguous for the other farmers.

9.3.1.3 Distributional Effects and Social Equity

The productive activities in the area have changed due to the introduction of the EPI. At the moment, however, precise data is not available. Hence, the present illustration relies on information reported by LRIBRO technical staff. From February 2011 to October 2011, two LRIBRO technical staff members were interviewed on three separate occasions. The objective of the interviews was to collect information,

Table 9.2 Example of decreasing water costs for a non-irrigator

Year	ha	EUR/ha	Total
2005	1.56	123	192
2007	1.56	29	45

Source: Interview to LRIBRO technical staff

data and opinions about the volumetric water pricing system, the main reasons for shifting to volumetric water tariffs and the main effects observed.

On the basis of the information collected, it is possible to identify three different groups of actors to analyze the change in income distribution due to the implementation of the EPI: (1) the first group includes non-irrigators who decreased their water costs; (2) the second includes those who ceased irrigation after the implementation of the EPI; and (3) the third group includes irrigators. Data is not available for groups 2 and 3 therefore considerations about income changes are not provided. With regard to the first group, farmer income increased because water costs decreased after the implementation of the EPI. In Table 9.2 an example is shown, related to an individual farm that reported a reduction in costs related to water tariffs of more than 70 %.

For those who ceased the irrigation activities after the introduction of the volumetric pricing it can be deduced that some labor savings occurred in the farm. In fact, irrigation activities require time for management and the main consequence of stopping irrigation is likely some savings in terms of labor.

The farmers who saved labor are most likely to re-allocate such time to other farming activities. At this stage, we do not have any direct information about the relevance of this issue, as these considerations came from the qualitative assessments of researchers and LRIBRO technical staff.

For irrigators, however, farm-level data are not available at the moment and a specific analysis of changes in internal organization, costs and profits are not possible to be performed.

9.3.2 The EPI Setting Up

The design and the implementation of the EPI did not encountered particular or specific obstacles, given the appropriate infrastructural predisposition of the irrigation system and the management organization as well. The will of the majority of farmers to abandon the current pricing system pushed the TMC to propose the alternative tariff.

9.3.2.1 Institutions

An important aspect that highly contributed to the realization of the EPI is the governance organization and the good relationship existing between the water authorities at different administrative levels. In order to establish a hierarchy among the water authorities that have been instituted during the years in Italy, different administrative levels can be individuated. In fact, in this case study, the relevant organizations are at upper levels: the first level includes the LRIBRO and the CER, while the second level includes the TMC. These organizations were set up at different times: organizations at second level started at the beginning of the 1900s (1933 and 1939), while the TMC is much more recent (1982).

At the national level, Land Reclamation and Irrigation Boards (LRIBs) were introduced in 1933 and regulated by the Royal Decree (R.D.) 215.[3] The LRIBs are public authorities subject to national laws and, since 1977, to regional laws as well. The functions of the RIBs – the reason of their institutionalization – are mainly related to the reclamation of wetlands and irrigation of agricultural areas. In 1989, the functions of LRIBs have been widened to cover many aspects related to land and subsoil protection, in coordination and subalternity to regional laws. In 1984, the Emilia Romagna Region anticipated such national orientation by emanating the regional law 42/1984 that widened the role of the regional LRIBs with respect to use, monitoring and protection of land and water resources. In 1994, a reform at national level about the management of water resources was realized and the related national laws were joined into a unique law, the Law 36/1994 (called Galli law), that provided the LRIBs with the power of building and managing irrigation networks, plants for the agricultural reuse of wastewater, rural aqueducts and other infrastructures functional to reclamation and irrigation systems. After the introduction of the Water Framework Directive (WFD 60/2000), the Italian legislative decree 152/2006 (named "environmental code") improved the functions of the LRIBs, including in particular the environmental protection intended as the protection and recovery of land and subsoil and the hydrogeological restoration of the territory, in concurrence with national, regional, provincial and municipal institutions (Ferrara 2009).

The relationship between LRIBRO, CER and TMC is considered to be quite good and this facilitated the EPI implementation. In fact, the long-term contracts between LRIBRO and CER guarantee the water supply in the area and this avoids water scarcity problems. The water management activities proposed by the LRIBRO can be supported and shared by farmers through the TMC. The sharing of water pricing amongst farmers represents one of the main points in the EPI implementation process in order to guarantee its acceptance. In addition, TMC can propose changes in the water management on the basis of farmers' needs.

With regard to culture and attitudes, the case study area is characterized by the presence of several cooperatives (lower level) that link farmers through shared preservation, processing and selling of their products. Another aspect that highlights the

[3] In 1933, the name of the boards was Land Reclamation Boards, without any mention to irrigation even though such function was provided by the R.D. 215.

level of entrepreneurship of the farmers in the area is the specificity of the crop cultivated, as industrial seeds require good relationships with market buyers and professional ability for cultivation. The strong presence of national farmers associations (lower level) helps avoiding, or mediates in the case of, conflicts among farmers.

All the cited organizations (upper and lower levels) have been involved in the design, implementation and operations of the EPI through a bottom-up approach, from complainer farmers to LRIBRO administration. The TMC, as representative of farmers' needs, submitted a proposal to the technical staff of the LRIBRO. LRIBRO was in favor of change and suggested shifting to a volumetric water pricing system through water metering installation. The move to a new water pricing system benefited from the definition of the area as a "cost center". The tariff was designed to recover all costs from farmers in the area. In addition, given the small area involved the identification of irrigators and non-irrigators was accomplished by way of a direct verification process (farm by farm).

9.3.2.2 Transaction Costs and Design

On the basis of our knowledge, there are no existing studies in this area that analyze transaction costs. However, it is possible to hypothesize that transaction costs are highly correlated with: (a) the purchase of water meters, (b) a system to control and identify non-irrigator farmers; and (c) data collection related to water use.

The cost of water meters was equal to EUR 193 + VAT and is covered by irrigators. In addition, the infrastructure was not modified, so this did not imply any transaction costs related to the irrigation network. Another point to consider is that transaction costs are correlated to the ability of institutions to deal with administrative and negotiation matters. In this case the good collaboration between the TMC and the LRIBRO likely kept transaction costs low. The only transaction costs that administratively represents an increase in total costs is attributable to the monitoring of the water use and the related reports. In fact, data collection concerning use is undertaken directly on the farm by the LRIBRO technical staff who downloads water meter information. In addition, the time spent in the calculation of water tariffs increased and so did the related costs.

9.3.2.3 Policy Implementability

The flexibility of the EPI is particularly connected with some characteristics of the specific case study. In the Tarabina area, the EPI implementation can be considered simple by virtue of its nature and the existing governance system. The simplicity of the implementation depended on the small size of the area, which enabled tailoring the EPI to the aforementioned local particularities: the existence of a pressure pipe system, a "cost center" definition, and the existence of a management committee (TMC). In addition to these characteristics, the fact that the EPI implementation was voluntarily chosen by the farming community positively impacted on the EPI's implementability.

The authorities that managed the implementation of the EPI were highly able to strengthen the synergies between the volumetric pricing and some sectorial policies. In particular, it is possible to identify two main aspects:

- The volumetric pricing is coherent with the needs of the farmers who claim the need for cost reductions in general and, specifically, related to water use;
- The decoupling of payments introduced in 2005 by the Common Agriculture Policy reform (CAP) likely helped in the reduction of the quantity of water used (at the least the CAP reform was not in conflict with it).

The adoption of the incentive pricing system in Tarabina did not found any legislative or bureaucratic obstacle, because the aims underlying the introduction of such instrument are in line with auspices of the mentioned national and regional laws. Moreover, the indications about incentive pricing and cost recovery provided by the WFD were actually important in facilitating the transition from the design to the process of the EPI implementation.

9.4 Conclusions

The Tarabina case study investigates the adoption of a volumetric water pricing system in the agricultural sector. Even though the area examined is quite small, the EPI application can be considered significant within the Italian context.

Some specific conditions have had a crucial role in the implementation of the EPI. Firstly, a pressure pipe system had already been used in the Tarabina area; in addition, the identification of Tarabina as a "cost center" allows for measuring (and hence potentially recovering) all costs related to it, as they are already separately identified in the LRIBRO accounting system. Moreover, the presence of a Management Committee – who actually decided for the adoption of a new tariff – avoided transaction costs related to the administrative and bureaucratic process of changing the tariff system. Finally, contract between LRIBRO and CER has guaranteed, since the outset of the irrigation district, the supply of water even in periods of scarcity, hence allowing EPI to focus only on economic aspects (as compared to EPIs mainly driven by water savings concerns).

The main reason for the introduction of the EPI was the increase in water tariffs during the period 1983–2005 caused by increases in M&O costs. Such increases also caused high inequalities between users (irrigators and non-irrigators). Accordingly, farmers representatives elected to the TMC, with the assistance of the LRIBRO, sought a solution to reduce inequality and overall costs. The solution identified was the implementation of water metering and the shift to a volumetric water pricing system.

The EPI provided multiple impacts related to economic, environment and social aspects. The economic impacts are most evident, in particular those related to the decrease in water delivery costs and the change in the distribution of contribution costs among farmers. In particular, a noteworthy cost reduction for non-irrigators occurred, due to a more efficient cost distribution based on quantity used. With regard to the environment, due to a decrease in water used, the amount of water remaining in the environment increased. Finally, regarding social aspects, the EPI increased the level of 'social agreement' within the group of non-irrigators.

References

Cross, G. J. (1970). Incentive pricing and utility regulation. *The Quarterly Journal of Economics, 84*(2), 236–253.

European Commission [EC]. (2012). A blueprint to safeguard Europe's water resources (COM673), Brussels.

Ferrara, G. (2009). L'ordinamento dei Consorzi di bonifica nell'evoluzione della disciplina nazionale e regionale. *Economia e Diritto Agroalimentare, 13*(2), 63–85.

Raggi, M., & Viaggi, D. (2009). Valutazione degli effetti delle politiche di tariffazione e di scenari mercato nel Consorzio di Bonifica della Romagna Occidentale in Acqua e Agricoltura in Italia Valutazione di scenari e strumenti di supporto alle decisioni a cura di V. Gallerani, G La Via and G- Zanni. Franco Angeli editore.

Rogers, P., Silva, R. D., & Bhatia, R. (2002). Water is an economic good: How to use prices to promote equity, efficiency, and sustainability. *Water Policy, 4*(1), 1–17.

Ward, F. A., & Pulido-Velazquez, M. (2009). Incentive pricing and cost recovery at the basin scale. *Journal of Environmental Management, 90*(1), 293–313.

Chapter 10
Corporatization and Price Setting in the Urban Water Sector Under Statewide Central Administration: The Israeli Experience

Iddo Kan and Yoav Kislev

Abstract As in many European countries, all water sources in Israel are public property, and are centrally managed by the government. This is to facilitate correction of market failures associated with externalities, natural monopolies and equity considerations. The economic policy instrument (EPI) considered here comprises two aspects of the centralized approach: (1) an institutional reform: local services that were formerly provided by municipal water departments became the responsibility of corporations; (2) a price-scheme reform: urban water prices are set by the regulator subject to the constraint of overall cost-recovery at the national and municipal levels, combined with an egalitarian policy; the latter is realized in identical municipal end-users tariffs. We evaluate the environmental, economic and institutional aspects of these reforms, and point out two main conclusions. First, with respect to EPI implementation from the regulator perspective, the lesson learned can be summarized by the phrase "grasp all, lose all." EPI reformation, in this case the establishment of regional corporations, should take account of unattainable objectives: "sanitizing" the political factors from involvement. The second lesson is associated with the challenge of designing a pricing mechanism that simultaneously achieves several potentially contradicting targets: costs recovery, creation of incentives for efficiency, and equality. Also here the mechanism was distorted by political pressures. According to the social norms as they are reflected by the resultant policy, equality overwhelms efficiency.

Keywords Institutional reform • Price-scheme reform • Urban water sector • Regulator

---

I. Kan (✉) • Y. Kislev
Department of Environmental Economics and Management and The Center for Agricultural Economics Research; The Robert H. Smith Faculty of Agriculture, Food and Environment, The Hebrew University of Jerusalem, P.O.B. 12, Rehovot 76100, Israel
e-mail: iddo.kan@mail.huji.ac.il; kislev@agri.huji.ac.il

M. Lago et al. (eds.), *Use of Economic Instruments in Water Policy*,
Global Issues in Water Policy 14, DOI 10.1007/978-3-319-18287-2_10

10.1 Introduction

By law (MNI 1959), all water sources in Israel are owned by the public and managed by the government. The objective of this legal structure is to enable the government to correct market failures related to water management; particularly, to internalize externalities associated with water pollution and extraction from common water resources, to control supply by natural monopolies, and to design long run nationwide investments in infrastructure and extraction from water resources under the scarcity and the uncertain natural enrichment characterizing the climate in Israel.

The Israeli centralized management framework and its legal base are of interest for comparison with more decentralized structures that are based on private property rights and motivation of efficiency by free trade (e.g., in the USA). Such systems differ with respect to many aspects, including the implications of property right assignment, transactions costs and independency of local communities. Our EPI case study focuses on two aspects of the centralized management approach prevailing in Israel: the first is associated with institutions and organization of decision making and with allocation of responsibilities in the water economy; the second aspect is related to the pricing scheme, according to which urban water prices are set by the regulator subject to the constraints of overall cost-recovery at the national and municipal levels, combined with an egalitarian policy; the latter is expressed in identical municipal tariffs. This pricing technique replaced the previous method under which costs were partly covered by the government's and municipalities' budgets, and prices were only partially identical – sewage treatment tariffs and connection fees were not uniform.

The original Israeli "water law" was changed twice. First, starting at 2001, the government, by offering subsidies, encouraged municipalities to establish regional water corporations as substitutes to the traditional municipal water departments. The rationale was to improve efficiency of the urban water systems by ensuring that municipalities do not use water revenue for other purposes, and utilizing economies of scale by merging water services of adjacent localities. The second reform was enacted in 2006 in order to improve management efficiency at the national level; hence, most of the regulations related to water, which were previously spread among a number of ministries, were concentrated at the hand of a new regulatory entity, the Water Authority. The Water Authority has also been made the price setter of all types of water, including the prices of both waters at the municipalities' gates and at the final urban consumers. The corporatization and the establishment of the Water Authority constitute the organizational aspect of our EPI case study.

The pricing element of the EPI encompasses prices set by the regulator for urban use at both the municipality and the household levels. The regulator (the Water Authority) is responsible for setting the prices paid by municipalities and municipal water corporations, either directly to the government as a pumping tax in case of self-extraction from rivers, lakes and aquifers, or to the government-owned national supply company, Mekorót. The prices paid for Mekorót's deliveries are set such that the total payments of the intermediate consumers – the municipalities (and by agricultural consumers) – cover Mekorót's costs. The regulator also monitors and authorizes operational activities and investments by municipal corporations, and sets the prices paid to Mekorót by each corporation while accounting for the corporation's supply efficiency and costs, such that inefficient corporations pay lower prices. This creates cross subsidization across municipalities. At the same time, the regulator is responsible for setting the price paid by households and other urban users to the municipal water corporations, while accounting for social considerations; equity in particular. Hence, while prices at the city gate (wholesale prices) may differ across municipalities, households everywhere pay identical (retail) prices.

10.2 Setting the Scene: Challenges, Opportunities and EPIs

Water management in a country facing water scarcity such as Israel is challenging, and failures to achieve efficiency can have severe economic consequences. Israel is located on the boundary of a desert, its south is dry and its north is semi-arid. Rain falls only in the winter, yet water consumption is highest in summer. The natural sources receive the water stored therein from precipitation, which in recent years has recorded a marked reduction. The traditional role of the water system is to collect the winter rainfall from rainy years and store it for use in dry years; and to deliver water from the north southward to Israel's populated center, and to irrigated agriculture in the south. The main use of water in the urban sector is not drinking, but rather landscaping, bathing, cleaning and other household chores, and removal of contaminants. Thus, the volume of wastewater in the urban sector now exceeds half the potable water used. Almost all the sewage is collected and transferred to wastewater purification and treatment plants. The treated wastewater is taken into recycling facilities, where it is stored and transferred in season to agriculture and natural habitats.

The Israeli experience with respect to the EPI under consideration is of interest particularly in light of its unique elements and the tight association between economic objectives and political interests. Opportunities for reforms in the water economy emerged and were driven by both economic and political trends. Although the share of water's contribution to Israel's GDP is only 0.5 % (Kislev 2011), it is consistently a subject of public debate. The corporatization process and the development of the mechanism of urban water price setting considered in this EPI case study cannot be disentangled from the evolution of Israel's economic conditions and its governmental economic policy. There has been a long run process of reduction in the share of the government in the economic activity of the country; this is reflected by steady reductions in the public expenditures and transformation of responsibilities for the finance of services from the public to the private sector. This process has been accelerated in the early 2000s due to the Second Palestinian Intifada (uprising) and the global recession at that time, that have resulted in a dramatic decline in Israel's GDP.

Following the political and economic changes described above, the government has established a mechanism according to which Mekorót's costs are no longer supported by state budget, but rather covered completely by revenue collected from users. Water prices are adjusted to meet this goal. The recent period has also experienced increased income inequality and this change was among the drivers of political pressures. The centralized structure exposes the government to political pressures by two major interest groups. The first are political parties representing low-income sectors; by waving the "equity principle" flag, they opposed the original plan to increase efficiency by differentiating water prices at the household level in order to reflect spatial variation of costs, and successfully managed to make the Water Authority to set identical retail prices. The second group includes city mayors who resisted the water corporations program that would result in reduction in the cities' revenue and flexibility in financing various municipal activities. They also rejected the plan of establishing regional water corporations, probably because of their concern of losing their independence as separated municipal entities in the long run. As a result, instead of the original plan of establishing only 15 corporations that would serve all the 251 municipalities throughout Israel, there are today as many as 52 corporations serving only 132 municipalities. The city-mayors lobby persistently struggles to reduce the independence of the water corporations, with some recent success in reforming the regulations by increasing their representation in the corporations' directorates.

10.3 The Corporatization and Price Setting in Action

In order to enable evaluation based on historical data and projected future trends, we concentrate on the assessment of the EPI in comparison to the one it has replaced, which therefore constitutes our baseline scenario. The main differences between the two are:

(a) Institutional arrangements: under the current EPI, (1) regulation and management of the water economy is at the hand of the Water Authority in contrast to the spread of authorities across ministries and institutions in the baseline scenario; (2) there is a corporatization process in the municipal sector, which replaces the traditional municipal water departments.
(b) Price settings: under the current EPI, (1) wholesale prices are set such that Mekorót's costs are fully covered by its water sales, whereas costs were partly covered by the government under the baseline scenario; (2) prices paid by the final water consumers (retail prices) are identical, and those paid to Mekorót (wholesale) may differ across municipalities, whereas under the baseline scenario also the wholesale prices were generally identical.

10.3.1 The EPI Contribution

We analyze the EPI contributions with respect to environmental and economic outcomes, and distribution effects.

10.3.1.1 Environmental Outcomes

The EPI's direct environmental impact is associated with the creation of incentives for improving municipal water infrastructures. This implies fewer events of pipes collapsing, water eruptions and sewage discharges to reservoirs, waterways and the sea. Such events may end up with environmental damage, health hazards, leisure constraints (particularly prohibition of sea swimming) and nuisances. Another environmental effect is related to the reduction in water consumption and water losses. Due to the implementation of the policy of full-costs recovery, the EPI has increased prices compared to the baseline scenario, and thereby reduced urban water consumption. The overall savings in freshwater enable reduction in the pressure exercised on natural water resources. The main impact is on the Sea of Galilee – the single large lake in Israel. The basin of the lake is the source of nearly 25 % of Israel's freshwater provision, and the lake's water level is heavily dependent on pumping rates to the National Carrier, which delivers water from north to south. The water level, in turn, affects the lake's ecosystem, its water quality, the basin's natural environment and tourism. Since 2004 the lake's water level has steadily declined, until stabilization in recent years, partly thanks to the reduction in domestic water consumption. Larger water stocks also allow higher provision of ecosystem services through allocation of more freshwater to nature. According to a governmental decision, 50 million cubic meters per year are to be allocated to the nature (MoEP 2011).

An additional environmental implication is associated with the impact on gardening. Quotas of water for watering private gardens, which were previously sold at a lower price, were cancelled, and thereby led to a reduction in watering of private gardens. Irrigation of public gardens exhibits a similar tendency. Once the EPI came into power, the municipalities, in addition to the loss of income that have been previously derived through the supply of water to their residents, are now facing higher expenses, since they are charged the full price by the water corporations. An evident of the welfare implications is the willingness to pay for installation of water-related items in public urban gardens in Tel Aviv, as estimated by Ben Shlomo (2010) to amount to EUR 4.0 per household a month. In addition, following the reduction in freshwater consumption for domestic purposes, the amount of treated wastewater also decreased. This implies lower allotments of recycled water for agricultural irrigation, which in turn leads to changes in the landscape services provided by rural areas (see Fleischer and Tsur 2009; Kan et al. 2009).

10.3.1.2 Economic Outcomes

Our economic assessment focuses on two elements of the EPI: (a) the corporatization process in the municipal sector and (b) the countrywide regime of cost-recovery prices; we commence with the corporatization process.

For years the municipalities have been responsible for water and sewage services in their jurisdictions. Because water is an essential commodity, the fact that one's water bill was attached to one's municipal tax charge helped, in many peoples' opinions, to expedite collection of this tax, thereby constituting a stable cash flow into the municipalities' coffers. Yet, this arrangement did present difficulties. Water services were provided as part of the overall activity of the municipalities, i.e., there was no separate, full accounting of the water supply on its own, such that it was impossible to know its proportion in the total municipal budget; neither was it possible to evaluate its efficiency. Political and other considerations made it easy for some municipal leaders to postpone costly works needed on their water and sewage systems, and instead, divert the accumulated funds to other tasks, particularly to the more visible ones (public buildings, pavements, etc.). Also, there were local authorities that failed to run a proper payment regime, water loss was high, and wastewater was not properly collected and treated. In light of incomplete information on what was occurring in the urban water sector, assessments by professionals invariably resulted in conclusions that the system was not efficient and was exhausting its own capital.

Today, the water and sewage corporations gradually replace urban water departments; they are operating under a business-economic model and under the professional supervision of the Water Authority. Each corporation is required to follow a set of rules for operation and maintenance expenses, as well as targets for gradual reduction in water losses; attaining such reductions requires investments, which affect the cumulative assets value owned by the corporation. In turn, the assets value are factors considered by the Water Authority when setting the prices paid for the water purchased from Mekorót – higher values may reduce this price; by this means the incentive to invest is formed.

Indeed, significant improvements in some aspects of the water services can already be observed. The corporations can now recruit additional workers (particularly in the managerial level) from outside the rigid employment constraints of the municipal sector; i.e., at lower salaries. All incomes and costs are earmarked and transparent. Monetary reports of the corporations are standardized, and are available to the public through the internet. Operation and maintenance of the municipal water system is not conditional on the municipality's financial situation anymore, and the corporations are able to approach the capital market for financing their activities; consequently, investments in infrastructures and in advanced technologies for metering consumption and monitoring water and sewage flows have sharply increased. These investments also encourage the Israeli water-related industry.

Yet, the formation of the corporations also raises certain problems, because of which more than a few municipalities have avoided or postponed joining the corporatization process. The inflow of payments for water and sewage services helps the budgetary management of the municipalities, even those wherein these incomes remain within the domain of water services; this cash source will now dry up. Moreover, a municipality that establishes a water corporation loses the ability to prevent water supply as an enforcement tool for municipal tax collection. Removal of the responsibility for water and sewage services from the municipalities may weaken local democracy; municipalities lose flexibility in defining preferences regarding the allocation of resources among services, and it is uncertain that the marginal benefits of water-leakage prevention equal the marginal cost when opportunity costs (e.g., those associated with education and culture) are taken into account. The outsourcing of services and the resultant distancing of accountability for the services raises difficulties for residents; the corporations, particularly those serving a few municipalities, are likely to become "foreign entities" in the communities, and run into problems in gaining the cooperation of residents and their representatives; the fact that corporations operate on local infrastructures such as roads and parks raises the likelihood of disputes between them and the local authorities over domains of responsibility, thereby rising costs to the community as a whole.

We turn now to discuss cost-recovery pricing. In order to combine the two principles of the water management at the national level – cost-recovery and uniform consumers' rates – the prices paid by the corporations for Mekorót water are not identical: corporations whose approved internal cost is high pay Mekorót a low price, and vice versa. In this way, low cost corporations indirectly support the others and a uniform tariff structure is maintained for the end-users level.

The Water Authority sets prices based on approved cost per cubic meter of water. The approved internal cost for corporations contains several components, such as labor, interest, and return on equity. Three items form most of the differences between corporations in their approved costs: one is the capital invested in the local water system (assessed in a property survey conducted when a corporation is established). The capital-rich corporations have a higher approved cost per cubic meter in this item. The other two items are "normative": the first is loss rates, including both physical water loss and incomplete charge of water bills. High losses are approved for "weak" corporations; i.e., those operating in low socio-economic localities (this group mainly includes municipalities located at the periphery and those populated with minorities). This means that the approved cost on this item per cubic meter sold is higher in the weak corporations than in the stronger ones. Another cost factor with a normative component is wastewater treatment, for which the cost per cubic meter is calculated by formulae dictated by regulations and based on the size of the facility and the quality of the effluent; these differ between the corporations.

The Water Authority expects that the corporations will all converge in a few years to the same normative loss values. Consequently the approved costs in the corporations should converge to similar levels, and the support of weak corporations by the strong ones will be eliminated. Yet this expectation is only a hope, not to say an illusion. The differences between the corporations are large, and the reported gaps between municipalities with low and high socio-economic levels were growing along time. Some corporations will succeed in streamlining operations, while others may not. The result may be that some corporations will be profitable, while others will suffer growing financial losses. The Water Authority will face difficulties in the future in using norms as the basis for approved costs, rather than using actual performances. A structural incentive problem is added to this issue: under the adopted tariffs structure, it doesn't make much sense for a corporation to increase its efficiency; those that show low costs and high profits will see their payments to Mekorót increased. The management of every corporation will attempt to convince the Water Authority that its costs are especially high. The government for its part will not be able to allow the corporations to accumulate profits, and even less to let them accumulate losses and go under, particularly given the governmental extensive support of the corporatization process; in other words, a regulatory capture may emerge.

10.3.1.3 Distributional Effects and Social Equity

The corporatization and the setting of cost-recovering Mekorót prices have distributional impacts on income, authority and political power at the national scale, as well as between and within municipalities. Most of the governmental income is derived from the high-income sector, which pays the lion share of taxes. Hence, by abolishing the financial support to Mekorót from the government's budget, and instead setting higher water prices so as to cover Mekorót's costs only through its water sales, the EPI imposes larger burden on the low-income sector. This policy increases inequity since, as water is an essential commodity, the share of expense on water consumption in low-income households' total expenses is larger than that of the high-income ones (CBS 2011). On the other hand, the reduction in budgetary expenses on water supply enables allocating more governmental resources to other public services that may mostly support weak populations.

The Mekorót-water pricing scheme implies cross subsidization between corporations in weak and strong municipalities. By this means, strong communities support the weak ones. However, at least according to the Water Authority's expectations, these income transfers will be gradually reduced, as the differences in approved costs between corporations will be eliminated. Similarly, prices of water supplied by Mekorót for agriculture are subsidized by urban water consumers; today, nearly EUR 0.18/m^3 of the water price for domestic use is allocated to this purpose. This subsidization would gradually vanish as agricultural water prices are planned to rise.

10.3.2 The EPI Setting Up

10.3.2.1 Institutional Set-up

At the upper level of institutions affecting the EPI stands the "water law," which assigns the property rights over all water sources to the public, and nominates the government to manage and control water. An additional law enacted in 2001 has launched the corporatization process in the municipal sector. The other element of the EPI, the cost-recovering prices, is associated with the reform that has established the Water Authority in 2006. By law, the Water Authority is responsible both for management of water resources and for economic regulation. The combining of these two spheres under the aegis of a single regulating agency is unique to Israel; in other countries on which we have information, regulation is separated by sphere. The pricing mechanism set by the Water Authority constitutes a secondary level of legislative; it defines the rules of the game, and therefore can be considered as an institution level second to that of the water law. The prices themselves lay in the third institutional level.

Two types of institutions have influenced the shape of the EPI and its success. The first are the municipalities, that so far successfully blocked the formation of many regional corporations, and, according to a recent governmental decision (Globs 2011), will even increase their hold and impact on the corporations. The second are political parties representing low-income sectors, who prevented the original intention of the law to set different prices in different urban corporations, in each case to cover locally specific cost, in order to enhance water-supply efficiency.

10.3.2.2 Transaction Costs and Design

The EPI has two major aspects of transaction costs. The first is related to management and control: the establishment of the Water Authority was particularly aiming at concentrating the data collection, decision making and control of the water economy in one institution, and thereby reducing the transaction costs associated with coordination among multiple ministries and institutions. The second aspect is associated with asymmetric information: the information on water management in the urban sector was vague and incomplete as long as the intra-municipality water delivery was managed by the municipalities' water departments. The EPI, by establishing the corporations and setting strict reporting and monitoring standards, has reduced asymmetric information. Yet asymmetric information still exist; for example, the corporations now have the incentive to present exaggerated costs figures, particularly those associated with investments, in order to signal the Water Authority to reduce the prices they pay for water they receive from Mekorót.

10.3.2.3 Policy Implementability

The implementability of the EPI is associated with public debates over distributions of political power and incomes. The institutional component of the EPI (i.e., the corporatization process) has targeted the allocation of responsibilities, authorities and incomes within the municipal sector. The objectives of this policy were only partly achieved: not all the municipalities made the transition, and many of those who did, particularly the large ones, established a single-municipality corporation rather than a regional entity. Moreover, as will be discussed later, the corporations themselves are now most likely going to lose much of their independence.

The EPI's pricing component has also been negotiated and changed along the way. The design of the pricing mechanism has encountered two fundamental problems associated with pricing principles. The first is the question of fairness versus efficiency. Setting a price equal to the marginal cost, which is the cost of desalination, signals the consumers about the costs that they are causing to the economy, and thereby brings about efficient consumption. Yet, because there are reservoirs from which the pumping cost is lower than desalination (e.g., the Sea of Galilee and aquifers), if the consumer pays for all water a price equal to the marginal cost, her total payment will be higher than the total supply costs. On the other hand, if the price of water equals the average of supply cost, total consumer payments will equal the total costs. Here the question arises: is it fair to set prices higher than the average cost? This problem has been partly solved by setting block rate prices at the end-user (retail), while the level of prices are set so as to cover Mecorot's costs, as well as the intra-municipal water delivery and sewage service costs.

Another matter is the question of equity in sharing the water cost burden; for instance, the cost of supply to Tel Aviv is lower than that to Jerusalem. Although it's original intention to set end-user prices which vary between municipalities, ultimately the Water Authority have discriminated only the wholesale prices paid to Mekorót, while maintaining parity in fees to urban consumers. In setting identical prices despite varying costs, the Water Authority sacrifices economic efficacy on the altar of equality in sharing the burden. Equality has actually become one of the objectives that justify state intervention in regulating the water supply.

10.4 Conclusions

The effects of institutional and economic changes are recognized in the long run; it may now be too early to identify and assess the full range of aspects associated with the EPI. We do believe, however, that two lessons can already be learned.

The first lesson drown is associated with the way a reform in EPIs is implemented, and can be summarized by the phrase "grasp all, lose all." Suppose that a local council could freely set the prices for the services it provides as a monopoly in its municipality. According to the well-known Ramsey-Boiteux pricing principle, a welfare maximizing not-for-profit monopoly should assign relatively higher price mark-ups to relatively inelastic price-demand commodities. This argument supports high urban water prices compared to prices charged to the farm sector, since water distribution is characterized as a natural monopoly service, and the demand for urban water is relatively inelastic. However, due to social and equity (and therefore political) considerations, the municipal water supplier is not free to set water prices; the latter are regulated at the national level. Therefore, higher incomes to a water providing monopoly – the municipality – can be derived only through cost reductions. As municipal mayors may be short-sighted politicians, they may favour reducing costs by postponing the expensive investments in replacement of non-visible water-supply infrastructures; thus allowing increased water losses. This strategy is particularly prevalent in municipal systems acting in a non-transparent financial and budgetary environment. A preventive measure may be to separate financially the water and sewage services from the other municipal functions. This separation requires only reorganization within the municipal financial and management administration, and it doesn't necessitate the establishment of separated entities such as new urban corporations. However, in some municipalities in Israel, particularly in those populated by minorities, water and monetary loses are particularly high, due to failure of enforcement resulting in water theft and incomplete collection of taxes and fees. These failures are attributed to cultural and political customs that limit the power of local authorities, and even that of the central government. In such municipalities, corporatization can augment the separation between local politics and water services, and thereby improve the performance of the services. However, the government did not distinguish between municipalities and has been trying to establish water and sewage corporations in all of them. The plan was to exploit economies of scale and establish several regional corporations each servicing 20–30 municipalities. But hastily, the government permitted, and sometimes forced, the creation of many single locality corporations. It will now be difficult to merge them into regional entities. Moreover, the mayors of the affected municipalities, who feared losing power, succeeded in forcing upon the government changes that may eventually make the corporation again subject to local political control. They will lose their independence. Thus, EPI reform should take account of unattainable objectives; in this case, "sanitizing" the political factors from involvement.

The second lesson learned is associated with the challenge of designing a pricing mechanism that simultaneously achieves several potentially contradicting targets: costs recovery, creation of incentives for efficiency, and equality. Replacing government support with uniform end-user cost-recovery prices may increase the burden on low-income families whose share in taxes to meet the state's budget is minimal. This observation is one reason for the criticism of the prices set for the water and sewage corporations. This criticism is particularly strong when costs increase, and therefore prices have to be increased as well. Indeed, trying to avoid criticism, the Water Authority recently refrained from increasing prices. It succeeded in getting the government to cover part of Mekorót's costs. Apparently, according to the social norms as they are reflected by this policy, equality overwhelms efficiency. This time, since it was done, not by direct subsidy, but rather by the government freeing Mekorót of rents it was supposed to pay, the principle of cost-recovery tariffs was maintained, at least from a public relationships perspective. This avenue for mitigating political opposition will not be open for ever and as costs rise (increased share of desalinated seawater is expected to increase costs) the public will have to accept higher rates.

Another problem that the new tariff structure raises is the use of Mekorót's wholesale prices to cross-subsidize weak municipalities and the prices of water supplied to agriculture. This pricing regime does not encourage the management of corporations to improve the efficiency of the services.

Yet, all of these obstacles can be viewed as a reflection of dynamic struggles between public institutions on the allocation of power and authority, and between societal norms on the preferred dominance of contradicting economic effects such as equity and efficiency. The EPI has shed light on these dilemmas, and brought them to a public discussion, while feeding the disputes with more reliable and consistent data; this is by itself a contribution: a problem well defined is a problem half solved.

References

Ben-Shlomo, V. (2010). The economic value of aesthetic aspects in urban gardens (Hebrew). Ph.D. thesis, The Department of Geography, Tel-Aviv University.

Central Bureau of Statistics (CBS). (2011). The state of society in Israel; environment, Report # 4 (Hebrew). http://www.cbs.gov.il/webpub/pub/text_page.html?publ=54&CYear=2009&CMonth=1#8. Accessed 22 Nov 2011.

Fleischer, A., & Tsur, Y. (2009). The amenity value of agricultural landscape and rural–urban land allocation. *Journal of Agricultural Economics, 60*, 132–153.

Globs. (2011). Nisan-committee's recommendations approved: Will the water corporations return to the municipal authorities? (Hebrew). http://www.globes.co.il/news/article.aspx?did=1000695209. Accessed 7 Nov 2011.

Kan, I., Haim, D., Rapaport-Rom, M., & Shechter, M. (2009). Environmental amenities and optimal agricultural land use: The case of Israel. *Ecological Economics, 68*(6), 1893–1898.

Kislev, Y. (2011). Water pricing in Israel in theory and practice. http://departments.agri.huji.ac.il/economics/teachers/kislev_yoav/Kislev-Acc-6-11-10.pdf. Accessed 15 Nov 2011.

Ministry of National Infrastructures (MNI). (1959). Water law. http://www.nevo.co.il/law_html/Law01/235_001.htm. Accessed 30 Nov 2011.

Ministry of Environmental Protection (MoEP). (2011). Governmental decisions (Hebrew). http://www.sviva.gov.il/bin/en.jsp?enPage=HomePage Accessed 25 Nov 2011.

Chapter 11
Water Budget Rate Structure: Experiences from Several Urban Utilities in Southern California

Ariel Dinar and Tom Ash

Abstract Being a semi-arid state, California faces frequent and prolonged droughts. In a typical policy intervention to deal with less water, state agencies and water utilities responded (in the urban sector) by either cutting water allocations to users or by dramatically increasing water tariffs, or both. Drought returned to California in 2007, and lasted, with various levels of severity until 2014. Starting 2008, with the slowdown in economic activity in the state, water rates that were adjusted, led to reduction in water consumption and decline in revenues of water utilities; customers that saved water have faced increased rates again and again, much to their dissatisfaction. The Water Budget Rate Structure (WBRS) (called also by some analysts sustainable rate design, since it seeks to stabilize revenues and drive conservation at the same time) has emerged as a practice that allows water utilities obtain a high level of conservation without jeopardizing the financial and political stability of the water utility. This chapter reviews the legal, economic and political aspects of the design and implementation of the WBRS in southern California in three water utilities, starting in 1990 until a recent implementation by the Western Municipal Water District. The chapter draws lessons and suggestions regarding the possible implementation of the WBRS by other utilities.

Keywords Water pricing • Drought • Southern California • Water budget rate structure • Efficiency • Financial stability • Equity • Political acceptability

---

A. Dinar (✉)
School of Public Policy, University of California Riverside, Riverside, CA, USA
e-mail: ariel.dinar@ucr.edu

T. Ash
Western Municipal Water District/Inland Empire Utilities Agency, Riverside, CA, USA
e-mail: t-ash@sbcglobal.net

M. Lago et al. (eds.), *Use of Economic Instruments in Water Policy*,
Global Issues in Water Policy 14, DOI 10.1007/978-3-319-18287-2_11

11.1 Background

Like many other countries/states, most of California's precipitation falls and is stored in its northern regions (Fig. 11.1) while most of the population and the economic activity are concentrated in the south. To close this gap the state of California and the federal government developed sophisticated water delivery systems that move water across the state, from north to south. However, population growth rates in Southern California, with the relatively high rate of water scarcity necessitate some demand management efforts.

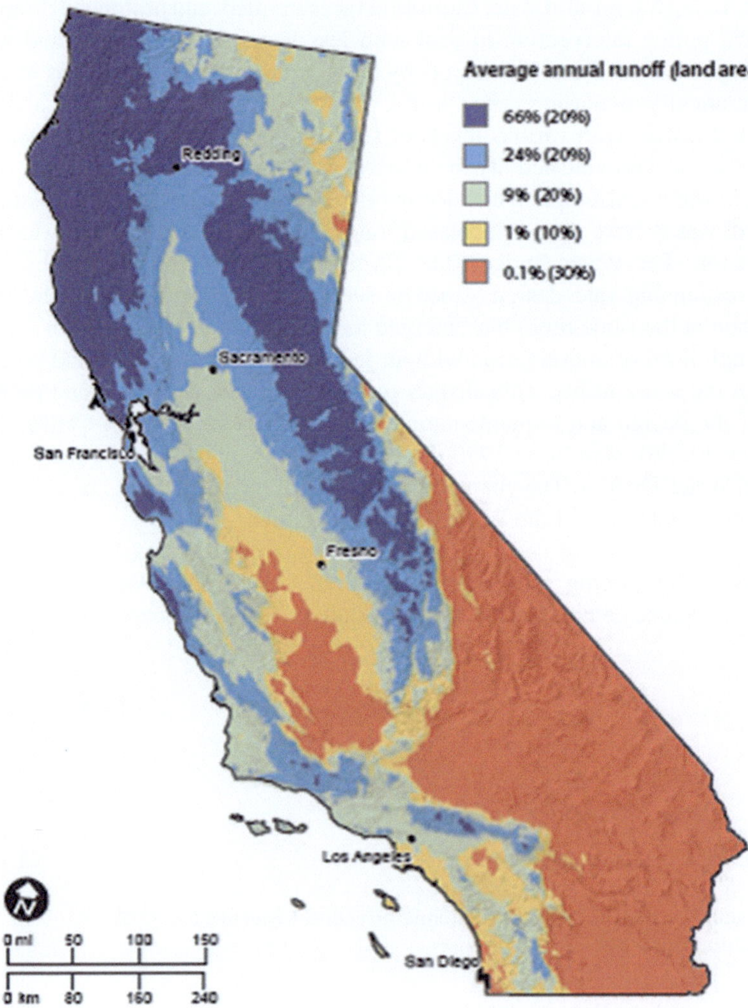

Fig. 11.1 Precipitation in California (Source: Hanak et al. 2011)

In an effort to cope with water scarcity, California introduced various mechanisms of pricing of water to induce water conservation. This has long been a challenge to water utilities and regulatory agencies in the urban sector (Hewitt 2000; Hall 2000), especially in the Western US where water supply is subject to major variation due to prolonged droughts and the semi-arid climate in that region. Traditional volumetric water pricing methods such as the uniform volumetric rate, the increasing block rate, and the decreasing block rate tariffs have had difficulties in addressing efficiency (conservation), financial stability of the water utility, and at the same time to provide fairness and equity issues across customer groups. These issues became the trigger for the dissatisfaction from the existing marginal cost rate structures in Tucson Arizona and Los Angeles, following the 1976–1977 and the 1986–1991 droughts they faced, respectively. Having one rate structure that has to fit all customers may not allow the water utility to reach highest possible efficiency, without jeopardizing several of the fundamental conditions for stable social optimum. They include financial (revenue) and political stability for the water utility, reasonable cost of service prices, satisfaction and fairness as perceived by customers with respect to water rates and conservation (Maria-Saleth and Dinar 2001). Indeed volumetric pricing methods have achieved a great deal of increased efficiency and conservation, but because they were designed based on an 'average household', their ability to achieve highest efficiency, customer acceptance and revenue stability under extreme water supply conditions are questionable.[1] Under prolonged drought conditions in California, water utilities faced continued water supply cuts that, given the 'traditional' marginal cost pricing instruments in use, reduced water sales could be met only by increased rates to all customers, even efficient users. Higher flat rates and tiered rates have produced some conservation, albeit inequitably across customers. But they have mostly created financial instability at the water utilities. What agencies missed in the rate design is how to achieve revenue stability and customer equity. "Raising rates" were the only tool they believed they had to drive the necessary conservation. This narrow view has created significant political/social conflict for the simple reason: customers who use water efficiently see their rates go up as the penalty for using water efficiently. Therefore, it is not surprising that what is known as a Water Budget Rate Structure (WBRS) has been adopted and attracting water utilities in regions facing high water scarcity such as the Western US.[2] However, the economic and public relations fundamentals of WBRS have the ability to assist any agency in any type of climate to price water accurately, recover costs accurately and to incentivize water use efficiency. The locations of the various water

[1] There is also the philosophy, often sought by the environmental community, that most of the fixed costs of water be moved to the variable costs side, making the cost of water high to encourage conservation, but putting the agency at great financial risk should users reduce demand. The initial concept appears correct, making the cost of water high to encourage conservation, yet denying that the actual costs of water delivered are mostly fixed. Activist groups are not responsible to the local voter, yet have amassed significant political power, causing this philosophical dilemma.

[2] It is only adopted in a small group of innovator agencies, but is generating discussions, mainly without the fundamental details understood and the typical conservation pricing still in the minds of rate consultants, environmentalists and most public agencies.

Fig. 11.2 Three Southern California water utilities that implemented the WBRS (Source: http://www.mwdh2o.com/mwdh2o/pages/memberag/member03.html)

utilities in Southern California that have been involved with the WBRS and are part of the analysis in this paper are depicted in Fig. 11.2.

The WBRS,[3] which will be explained later in details, allows the water utility to tailor the rate structure essentially to each household served. This flexibility could be enhanced, as we will see below, by use of the advancement in the information technology field (such as remote sensing, finer Evapotranspiration—ET—estimates, Geographic Information Systems, Automated/remote Meter Reading, etc…), although the main technology needed is an adequate billing system software that allows customer-specific variables and adjustments.

In the past quarter of the century, there has been an increase in the number of water utilities in Western US (Fig. 11.3; Table 11.1), and in particular in Southern California that have implemented WRBS. This case study will focus on three water utilities in Southern California that have implemented WBRS between early the 1990s and late 2011 with various levels of sophistication. While the number of implementing agencies was stable between 1990 and 2007, WBRS attracted water utilities in Southern California, starting 2008 as a result of a combination of

[3] "Water budget-based water rates—also known as individualized, goal-based, and customer specific rates—are block rates, where the block is defined by using one or more customer characteristics. Water budget-based rate structures can be thought of as an increasing block rate structure where the block definition is different for each customer, based on an efficient level of water use for that customer" (Mayer 2009: 4).

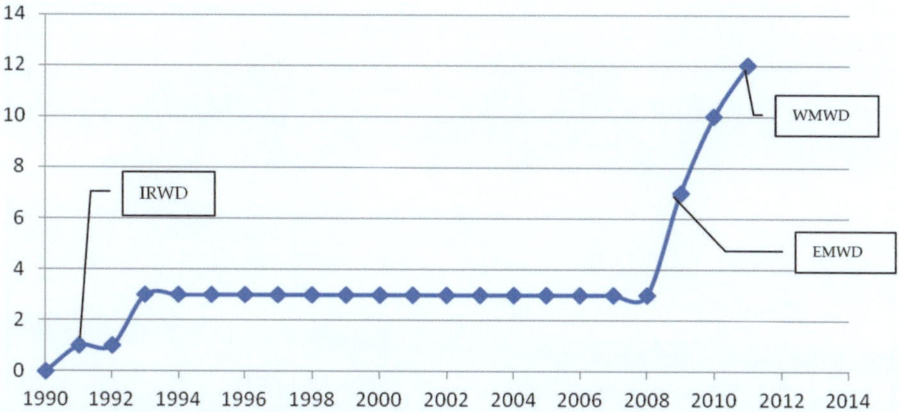

Fig. 11.3 Diffusion of WBRS in California between 1990 and 2011 (Source: Authors. Note: East Valley WD in San Bernardino county will implement in 2015; Las Virgenes WD will implement in 2016)

Table 11.1 Water utilities in Southern California that adopted WBRS and years of adoption

Utility	Year of adoption
Irvine Ranch Water District	1991
San Juan Capistrano Water District	1993
Otay Water District	1993
Eastern Municipal Water District	2009
Palmdale WD Water District	2009
Coachella Valley Water District	2009
Elsinore Valley Water District	2010
City of Corona	2010
Rancho California Water District	2010
El Toro Water District	2010
Moulton Niguel Water District	2011
Western Municipal Water District	2011

Source: Ash, T. (2011, November 28). Personal communication

economic slowdown and prolonged drought, both of which lead to reduction in demand for water and a direct impact on the revenue stability of the water utilities.

11.2 Characterisation of the Region

The three water utilities that comprise our case study are located in the Santa Ana river basin (Fig. 11.4).

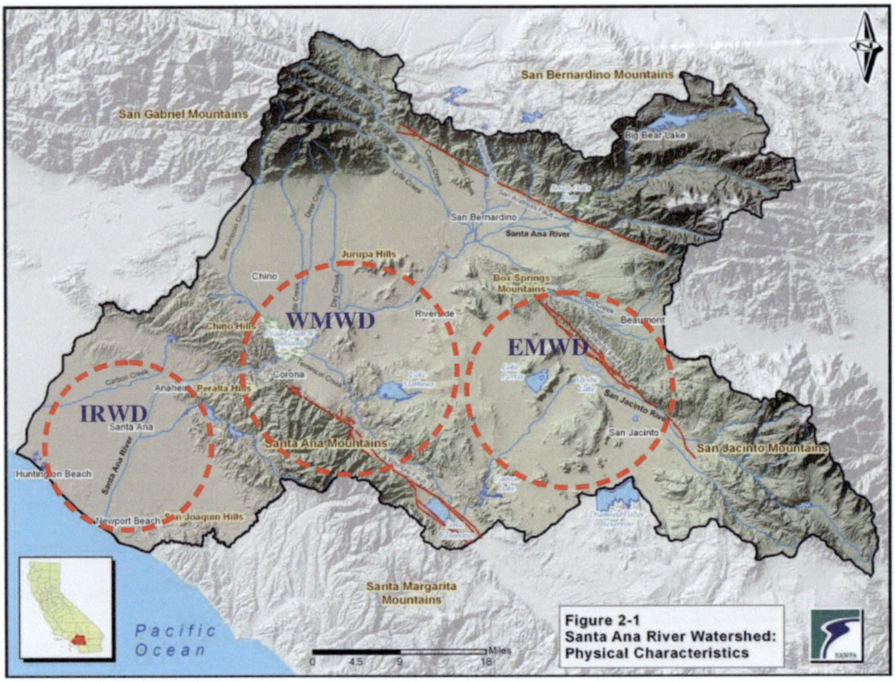

Fig. 11.4 Map of the Santa Ana Watershed (Source: SAWPA 2010). *Red circles* do not represent service area boundaries

The Santa Ana River Watershed drains a 2,650 square-mile area. The watershed is home to over six million people and includes the major population centers of parts of Orange, Riverside, and San Bernardino Counties, as well as a sliver of Los Angeles County. The Santa Ana River flows over 100 miles and drains the largest coastal stream system in southern California. It discharges into the Pacific Ocean at the City of Huntington Beach. The total length of the Santa Ana River and its major tributaries is about 700 miles (SAWPA 2010).

The Irvine Ranch Water District (IWRD) is an independent special district serving Central Orange County, California. It provides high-quality drinking water, reliable wastewater collection and treatment, ground-breaking recycled water programs, and environmentally sound urban runoff treatment to more than 330,000 residents. IRWD encompasses approximately 181 square-miles extending from the Pacific Coast to the foothills and serves the City of Irvine and portions of Costa Mesa, Lake Forest, Newport Beach, Orange, Tustin and unincorporated areas of Orange County. Approximately 65 % of the drinking water supply comes from local groundwater sources. The remaining 35 % of IRWD's drinking water comes from the Colorado River (Colorado River through the Colorado River Aqueduct) and the State Water Project (the Sacramento-San Joaquin Delta in Northern California) and is imported by the Metropolitan Water District of Southern California (MWD) (IRWD Water Facts 2011).

Eastern Municipal Water District (EMWD) services an area of 555 square miles and population of about 700,000 people. The major water sources are imported water from the Colorado River and the state water project (66 %), local groundwater and desalinization (16 %), and recycled wastewater (18 %) (EMWD 2011).

Western Municipal Water District (WMWD) serves a region of 527 square-miles with a population of about 850,000. The water sources are from the Colorado River (about 20 %, purchasing from MWD), the state water project and groundwater. This district operates and maintains domestic and industrial wastewater collection, treatment, and conveyance systems. Annual water deliveries are 125,000 acre-feet (1.05 billion cubic meters). About two-thirds of the water that Western sells is treated; the remaining is untreated or raw water. About 25 % of the water sales are for agricultural uses, and 75 % is for domestic purposes (WMWD 2011).

11.3 The WBRS Methodology

WBRS is a tiered pricing system, but it differs from the traditional inclining tier pricing design in that it is designed to provide revenue security to the water utility and at the same time guarantee fairness to the customers.

Fixed costs of service are handled, mainly by political considerations and compromise. Of the amount calculated as fixed cost of service, utilities distribute a certain percentage as a fixed (irrespective of water use by the customer) charge on the water bill and the remaining percentage as part of the variable charge on the water bill. Therefore some of the fixed costs are assigned to the variable amounts of water used. Utilities are aware of the trade-off between risk of low cost recovery of the fixed share and customer dissatisfaction from higher rates. Common practice among water utilities is to set the ratio off fixed cost distribution between the fixed and the variable portion of the bill to 20–30 % and 80–70 % respectively.

The WBRS is comprised of fixed costs and variable cost components. The fixed cost portion is kept at a both a reasonable level for the customers and the water utility. The variable costs are comprised of several increasing tiers (between 3 and 5), depending upon the number of sources of water, as per State legislation, Proposition 218, where the price of the tier must be linked to a water source and a relative cost. The first and second tiers (efficiency tiers) represent reasonable use of water by customers, as recommended by State legislation and empirical studies (State does set a standard for indoor and outdoor efficiency. Agencies use or adapt the standards based on political and regional needs). The first tier in each WBRS refers to indoor water need and the second tier refers to outdoor water need. Both of these two tiers are anchored to legal and scientific parameters, expressed in ccf/billing period, as follows:

$$\mathrm{IDU} = (\mathrm{R}) \cdots (\mathrm{IS}) \cdots (\mathrm{D}) \tag{11.1}$$

$$\text{ODU} = (\text{ET}) \cdots (\text{LF}) \cdots (\text{SF}) \cdots (\text{DF}) \tag{11.2}$$

$$\text{MWA} = [\text{IDU} + \text{ODU}] \tag{11.3}$$

where IDU is indoor water use allocation to the residency; R is the number of residents in the household; IS is the indoor water use standard per capita, recommended at 55 gal per capita per day (gpd/d)[4]; D is the number of days in the billing cycle; ODU is outdoor water allocation to the residency (ccf); ET is the evapotranspiration value (inches)[5] during the billing period; LF is the periodical landscape factor of a representative fescue grass (fraction).[6] SF is the irrigated area in the lot (square foot); DF is a drought factor (fraction), representing the water reduction the retail agency may face in an emergency[7]; MWA is billing period indoor and outdoor water allotment (ccf).[8]

Customers that exceed the first two tiers are considered not-efficient and face significantly higher prices per unit of water consumed in the over-allocation tiers. Many water utilities compute the prices of the tiers following the second tier, by using the next alternative for water (the opportunity cost approach), such as imported water or water that are associated with much higher cost of provision. The WBRS is applied to the service area of the utility, using normative parameters. Customers are given the ability to adjust the individual allocation/efficiency tiers (Variance) to their own unique parameters. A simple example of the WBRS with two customers, A and B (where customer B requested to adjust tier 1 to her specific conditions, is provided in Fig. 11.5. Customers can request variance for tier 1 and/or 2 only, or the variables for indoor and/or outdoor water need, based on changing site conditions (i.e. more family, added irrigation area, medical need, pool or large animals.

The three water utilities comprising the case study use an allocation-based conservation rate structure, described in general terms above, which offers property specific water budgets and tiered pricing to provide each of its customers with economic incentives for efficient water use. In addition to providing incentives for saving water to the customers, the WBRS provides incentives to the water utilities to set the fixed costs and the tier levels in such a way to transparently price the cost of water and water services. It also educates the consumer to what the water agency actually does…providing reliable water, and changes the relationship between the water user and the agency. All together it increases the confidence and satisfaction of the customers and thus, the long-term stability of the water utility.

It should be pointed out that a WBRS is modelled to be revenue neutral or to recover only the cost of service if, as intended, every customer is efficient. Only when

[4] 1 gal ≅ 4 l.

[5] 1 in. ≅ 2.54 cm.

[6] The annual LF for fescue grass is 0.8 of ET. Monthly values may exceed or be below 0.8, depending on the month.

[7] Some water utilities use the DF to adjust both the ODU and the IDU.

[8] 1 ccf ≅ 100 ft^3 or 748 gal; 1 in. = 0.083333 ft^3.

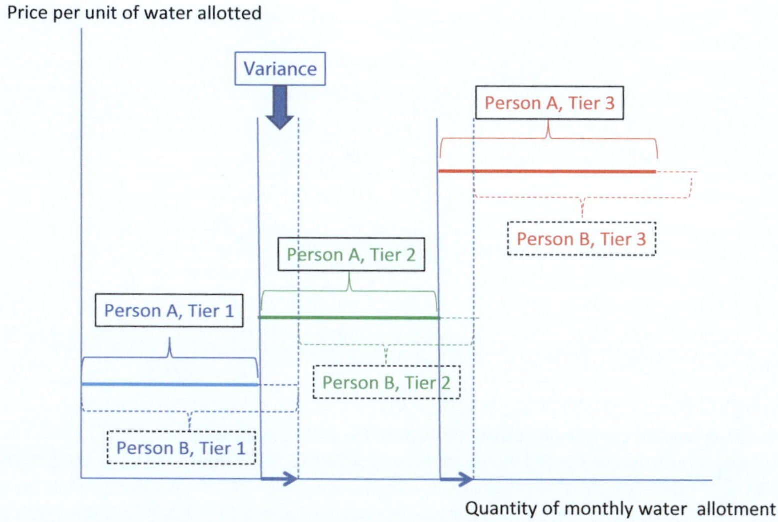

Fig. 11.5 Scheme of the water budget rate structure (Source: Authors)

consumers exceed their individualized water budget is there excess revenue collected. The excess revenue, as part of the rate design, becomes another pillar of the WBRS achievement, where the revenue collected from higher tier water use is reinvested in promoting long-term improvement programs in water use efficiency and support the water utility urban runoff programs that reduce pollution of aquifers, beaches and wetlands.

The three water utilities established customized and equitable water for each customer by allowing a 'variance' program—an increase in the normalized amounts of indoor and outdoor allocations—such as: updated number of people in the household; people with special needs, irrigated area, livestock on premise, or business type. The rate structures of the three subject agencies as of July 2011 is presented in Table 11.2.

11.4 Performance of the WBRS

The WRBS will be assessed, using several criteria, including environmental outcomes, economic and financial aspects, water savings, reduction in runoff pollution, and distributional effects and social equity.

11.4.1 Environmental Outcomes

While the WBRS's declared motivation is for the water utility financial stability, for water conservation, and for customer satisfaction, environmental benefits are an integral outcome of WBRS and can be estimated from the performance of the water utility before and after the implementation of the WBRS.

Table 11.2 Residential rates (US$/ccf) in IRWD (effective July 1, 2011), EMWD and WMWD (effective October 1, 2011)

IRWD[a]			EMWD[b]			WMWD		
Tier	Rate (US$/ ccf)	% of allocation	Tier	Rate (US$/ ccf)	% of allocation	Tier	Rate (US$/ ccf)	% of allocation
Low volume	0.91	0–40	Indoor	1.483	0–50	Efficient indoor	1.77	
Base rate	1.22	41–100	Outdoor	2.714	50–100	Efficient outdoor	1.87	
Inefficient	2.50	101–150	Excessive	4.864	100–150	Inefficient	2.41[c]	100–125
Excessive	4.32	151–200	Wasteful	8.898	150+	Excessive	3.78[d]	125–150
Wasteful	9.48	200+	N/A	N/A	N/A	Unsustainable	4.67[e]	150+

Sources: IRWD 2011; EMWD 2011; WMWD 2011
Note: First two tiers of each water utility constitute the total allocation
[a]The original Rate structure set in 1991 were more restrictive, as follows: (1) Low volume 0–40 % of allocation at three-fourth of the base rate; Conservation 41–100 % of allocation at base rate; Penalty 101–110 % of allocation at twice the base rate; Excessive 111–120 % of allocation at four times the base rate; and Abusive +120 % of allocation at eight times the base rate. This rate has evolved over time and went through several modifications
[b]EMWD initiated a WBRS in 1992 for new customers only and then adopted a tiered rate structure for all its service area in 1993. Due to economic recession and drought EMWD increased tariffs by 34 % in the summer of 1993 and Faced angry protests from customers that led to retrieval from the tiered pricing to increased fixed rates (Pekelney and Chessnut 1977:2-1–2-14). The IRWD structure described above was for irrigation accounts, not residential or commercial. The EMWD attempt in 1992 was very short-term and was ill-conceived at best, and should have been disbanded as it was. In 2009 it was restarted (See Baerenklau et al. (2014) for many more details its performance)
[c]Including US$0.30 to fund efficiency and environmentally-related programs
[d]Including US$0.60 to fund efficiency and environmentally-related programs
[e]Including US$1.49 to fund efficiency and environmentally-related programs

At this point several environmental outcomes are identifiable, which are quantifiable and will be estimated and presented at the next version of the report:

1. Reduction of pollution of water bodies (aquifers, wetlands) from pesticides, nitrates in outdoor irrigation runoff;
2. Reduction in import of lower quality (higher salinity content) water from the Colorado River resulting in (a) need for less energy for water treatment and (b) less contamination of aquifers and soils from use of water with higher levels of salinity;
3. Reduction of negative environmental impact in the source (Colorado River Basin) from transporting water out of basin;
4. Establishment of stable urban carbon sequestration patterns by allowing sustainably growing trees in a reasonable cost of water.

11.5 Economic Assessment Criteria

IRWD, facing an extended drought (1987–1993), reduction in regional allocations set by MWD, wholesale price increases, and revenue loss from lower water sales, set out to re-design water rates that would meet all of the needs of the agency. IRWD requested the University of California to place a water conservation advisor (Tom Ash) at the district in January 1991 to assist with conservation programs and water rates, then referred to as a "water budget rate structure".

With internal agency staff including finance, customer service and public affairs, the design of a new conservation rate structure was delineated to address the following fundamental questions: (1) How can a rate structure recover costs accurately (reducing revenue risk when demand is reduced)? (2) How can a rate structure identify water wasters and send a consistent economic signal to use water efficiently? And (3) Achieve stable revenue recovery, establish an efficiency ethic and be fair and equitable to end-users?

IRWD arrived at a water budget tiered rate structure that includes (1) recovery of 75 % of fixed costs on a fixed "service" charge (a change from 25 % of fixed cost recovery in its pre-existing rate structure); (2) individualized customer allocations (based on per resident gallons per day (gpd) for indoor use, local evapotranspiration and size of irrigated area for outside use). To achieve the desired results of revenue stability, conservation and consumer equity required, daily downloads of three microclimate evapotranspiration zone data into the billing system; low variable base price for efficient users; steep inclining tiered prices for water wasters; and a variance system to adapt individual customer allocation variables as necessary.

IRWD implemented the new rate structure in June of 1991. The drought and regional restrictions lasted another 2 years until March 1993 when heavy rains ended the 6-year drought.

The impact of the IRWD water budget rate structure was immediate and documented by the agency and reviewed in an independent study by MWD, the regional wholesale agency (Pekelney and Chestnut 1997). Overall the first water budget rate structure accomplished the following within the first 5 years of implementation, (1) 58 % reduction in landscape irrigation water use (dedicated irrigation meters); (2) 19 % residential water use reduction; (3) Stable fixed revenue recovery; (4) Reduced water runoff (water quality improvement) (MWDOC-IRWD 2004); (5) Fully funded conservation programs (paid only by water wasters); (6) 85 % customer satisfaction (independent customer surveys); and (7) re-election of all water board members since 1991 (continuing for 22 years up through 2013). The rate structure has operated as designed and envisioned for 23 years, during drought and rain, good and poor economic years. The service area of the Irvine Ranch Water District is considered one of the most water efficiency in the State of California and has continued to recover appropriate revenues for the water agency.

The EMWD service area is located in the hot inland of southern California, where customers have a wide range of lot sizes, pools, equestrian properties and residents per household. In 2008 the EMWD was facing a significant drought, State and regional water restrictions and declining revenues as customers cut water use due to the declining economy and water restrictions. The board of directors agreed with the goals of a classic water budget rate design especially in terms of customer equity, and directed staff to create a WBRS implementation plan. In 2009 EMWD implemented the WBRS.

Features of the EMWD water budget rate structure include (1) individualized allocations for all residential, commercial and irrigation accounts; (2) daily ET for 50 microclimates in the service area; (3) indoor and outdoor allocations that is modeled on the State legislation for indoor and outdoor water efficiency standards; and (4) a variance program to insure accurate allocations for individual customer accounts. The impacts to date include (1) water use reductions of 13 % (with drought and recession factored into the findings); (2) revenue increase of 6 %; (3) accumulation of capital for funding for conservation programs paid only by water wasters.

In 2008 the WMWD was also facing drought restrictions and declining revenues with their traditional low fixed service charge and flat variable cost rate structure. With an educational workshop for elected officials, the agency decided to adopt the WBRS and directed staff to develop an implementation plan.

The WMWD billing system was antiquated and was scheduled for a full software and hardware upgrade by 2011. The agency carefully re-built its billing system, navigated through elections and was mindful of the impact of recession and water rates on customers in the service area as they moved toward WBRS. The features of the WMWD water budget rate structure include the successful elements used in the WBRS deployments of other agencies, including (1) individual allocations for residential, commercial and irrigation accounts; (2) a drought factor built into the allocation equation if needed to meet local and/or regional supply limitations; (3) a variance program for individual customer allocation adjustments; (4) fully funded conservation programs paid only by water wasters (tiers 3–5); (5) increased emphasis on customer services; and (6) purchase of private sector provided daily ET for 450 microclimates in the service area.

WMWD implemented WBRS in November of 2011. The WMWD implementation represents the most advanced WBRS design and may serve as a model of how an agency can carefully study, consider and coordinate an a deployment plan, including billing system upgrades, public outreach, politics, staff training and total costs to change a water rate structure and reform how agencies meet the cost of service and reduce demand in an equitable and defensible manner.

In the State of California required public hearing process, under Proposition 218, WMWD received 98 % customer approval of the new rate structure by customers. WMWD has met cost of service budgets since implementing water budget rates, has increased the availability and funding of conservation programs to customers, and has seen a 17 % decrease in water use since 2010, despite hotter weather and drier winters.

11.5.1 Water Savings Potential Seen with Water Budget Rate Structures

Landscape irrigation accounts for at least 50 % of urban water use in Southern California (Hanak et al. 2011:97, Fig. 2.12). An analysis of water usage in outdoor landscape irrigation by urban customers in IRWD between 1988 and 1995 suggests savings from 34 % to 41 % between pre WBRS implementation (1988–1990) and post WRBS implementation (1991–1995). The results are summarized in Figs. 11.6 and 11.7 below.

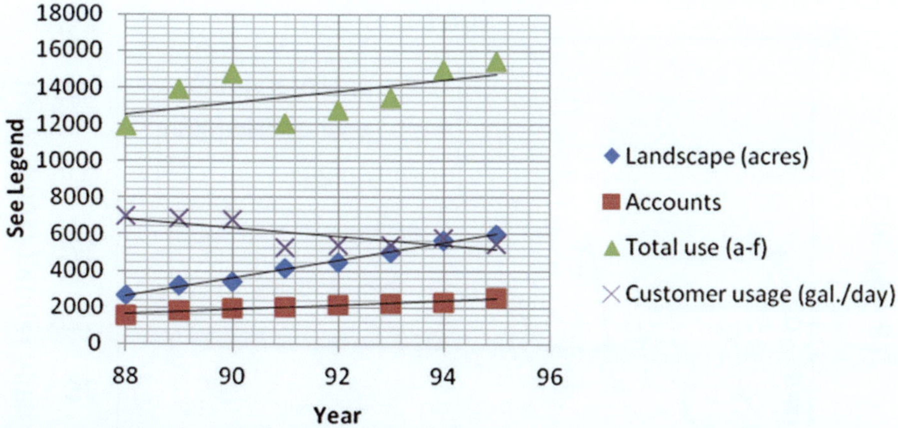

Fig. 11.6 Physical parameters of water use in IRWD during 1988–1995 (Note: Based on data in Pekelney and Chessnut (1997: Table 4.3))

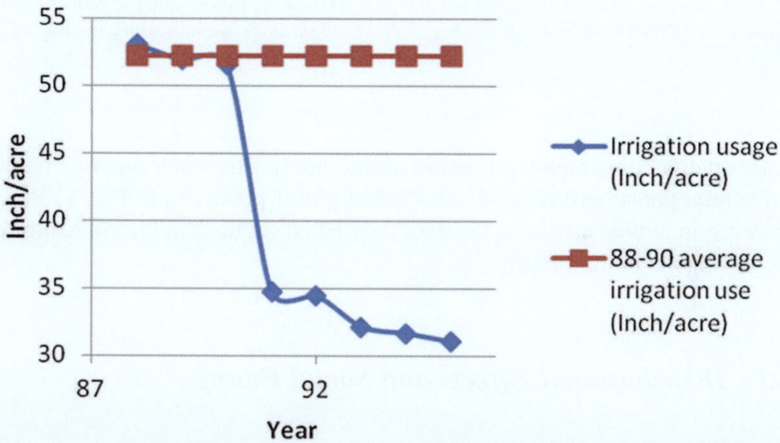

Fig. 11.7 Actual reduction in landcape water use by IRWD customers between 1988 and 1995 (Note: Based on data in Pekelney and Chessnut (1997: Table 4.3))

11.5.2 Reduction in Runoff Pollution

With the significant percentage of residential water demand used for outdoor purposes excess landscape irrigation results also in increased runoff that is the transport mechanism of pollutants that enter natural waterways and, ultimately, the Pacific Ocean.

A study focusing on estimation of runoff from residential plots and the level of pollutants transported was conducted between 2000 and 2002 in a small residential area of IRWD (MWDOC-IRWD 2004) comparing runoff and concentration of pollutants in the runoff during the dry season of the year. The study collected data on

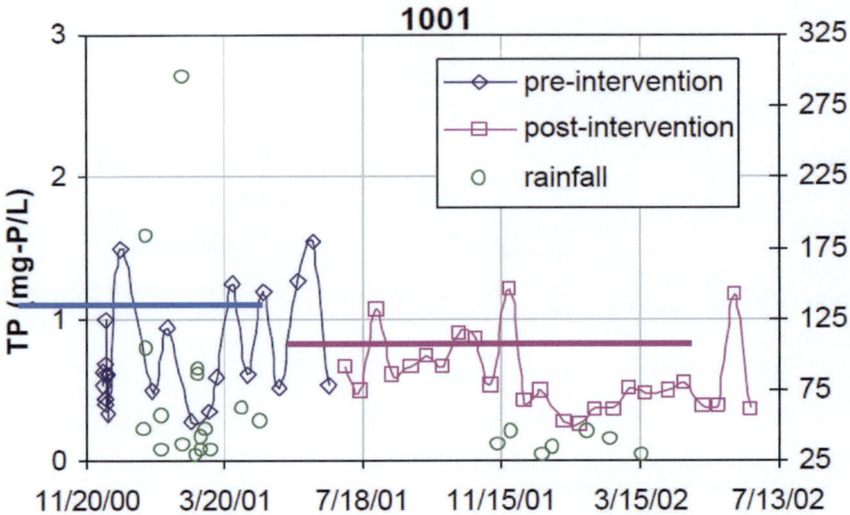

Fig. 11.8 Time-series of total phosphorus from plot 1001 of the runoff study at San Diego Creak, IRWD (Source: MWDOC-IRWD (2004: Fig. 5.3)). *Straight lines* are indicatory means

the water quality constituents present in urban runoff. The water quality component related to total phosphorous in one residential plot is presented in Fig. 11.8.

However, in almost all cases, the data showed no changes in the concentration of these constituents in the runoff.

11.5.3 Distributional Effects and Social Equity

Although the objectives of WBRS are to conserve water while recovering the cost of service, there is still a very significant component of improved distributional effects and social justice. The suggested procedures for distributional effects and social justice can be easily estimated for each water utility.

The WBRS provides for what is called a 'variance', which is a request for a change of the individual variables that either increase or decrease the water budget either in tier 1 or 2 for each customer. We will use the allowed increase in indoor and outdoor water allocation (that is associated with tier 1 and 2—the "budget") following a variance request process as the indicator for the distributional effects and social equity derived from the WBRS.

The water districts allocate to the household, under WBRS, a given quantity of water in tier 1 and another allocation as tier 2 based on State guidelines and the individual household situation. The sum of the tier 1 and tier 2 allocations are the water budget to that household. These allocations are based on normative coefficients and may or may not be representative of the household conditions. Following the variance process (appeal by the household that corrects either tier 1, or tier 2, or both, subject to updated parameters), the water district re-calculates the allocation with the new parameters. Usually, the variance process culminates in a higher allocation for the tiers under consideration.

The variance process allows for an accurate water budget to the household, mainly the household gets more water compared to the 'before variance'. One indicator of the household benefits after the variance process is the sum of income saved (gained) compared with the pre-variance payment.

With information about the household accounts in each of the water districts and the variance levels requested and approved in the service area of the water utility for representative households, it is possible to estimate the total welfare transfers in each water utility and the distribution of such welfare.

WMWD data is used here to demonstrate the impact of the WBRS on the water consumption and the cost of water to several arbitrarily selected households (with the intent to provide a range over lot sizes and persons per household). Data and analysis was provided by WMWD staff in January 2012 for this report.

Data of ten households, ranging in their family size and lot size was selected from the billing accounts of WMWD. The benefit calculation refers to the months of November and December of 2011, following the implementation of the WBRS and the initiation of a possible variance process. Each household was given the option to appeal their normative parameters (used by WMWD for setting the household budget) by submitting Request for Water Budget Adjustment (Annex II). Of about 25,000 accounts, 6,000 households used the appeal process by November 2011 and 2000 more households submitted their Request for Water Budget Adjustment in the month of December 2011. The original normative factors used for water budget allocation as well as the adjusted factors, the revised tier 1 and tier 2, and the actual consumption in the months of November and December 2011, are presented in Table 11.3.

A comparative analysis of the impact of WBRS on the cost of water for each household with the pre-WBRS rate compared with the WBRS rate is presented in Fig. 11.9.

The comparative analysis spans over January 2009 and December 2011, where between January 2009 and October 2011, the previous charging system, which is based on a flat rate of 1.87 per CCF, was in effect. The new WBRS was imposed on the existing consumption to demonstrate what would have happened if the WBRS was in place. While this exercise does not introduce any behavioural responses, it does suggest the following observations (Please refer also to figures in Annex I in Dinar (2011) for the graphical analysis of the ten selected households): (1) There is quite a wide range of the household parameters (persons per household and irrigated area that affects the retroactive performance of the WBRS across the analysed years. Each customer has different situations and the bills are only comparing their use to their specific standard; (2) Some households have not been affected by the implementation of WBRS, some households were already efficient, so they slightly gained by lowering their water cost, and some households were already abusive in

Table 11.3 Incremental gains for various households from submitting Request for Water Budget Adjustment as reflected in the month of November and December of 2011

Household #	Consumption Nov 2011 (CCF)	Consumption Dec 2011 (CCF)	Budget Nov 2011 In+out before variance (CCF)	Budget Dec 2011 In+out before variance (CCF)	Budget Nov 2011 In+out after variance (CCF)	Budget Dec 2011 In+out after variance (CCF)	Incremental gain Nov 2011 (US$/month)	Incremental gain Dec 2011 (US$/month)
232	21	10	10.456	10.899	10.509	10.955	0.128	0.136
218	13	5	10.861	11.329	10.909	11.380	0.115	0.122
368	24	15	10.861	11.329	13.176	13.718	5.579	5.758
833	13	15	10.861	11.329	16.611	17.298	13.857	14.385
984	69	27	26.873	28.342	68.495	72.494	100.308	106.406
003	52	26	12.058	12.601	15.438	16.192	8.146	8.655
696	36	32	11.341	11.839	27.063	28.263	37.891	39.582
368	20	6	10.983	11.459	16.071	16.865	12.262	13.028
941	11	5	10.861	11.329	13.629	14.270	6.672	7.088
342	47	27	26.873	28.342	28.346	29.907	3.550	3.772

Calculations of tier 1 and 2 for the months of November and December and conversion factors are provided below

$$\text{Indoor budget for Nov} = (\text{R}\cdot 60\text{gpd}\cdot 30\text{days})/748(\text{gallons per CCF}) = \text{Tier1}(\text{CCF})$$

Outdoor budget for

$$\text{Nov} = \left[\text{SF}(\text{square feet})\cdot 0.65(\text{LF Nov})\cdot 2.45(\text{ET Nov, inch})\cdot 0.083333(\text{conversion from inch to feet})\right]/100(\text{conversion from cubic feet to CCF}) = \text{Tier2}(\text{CCF})$$

$$\text{Indoor budget for Dec} = (\text{R}\cdot 60\text{gpd}\cdot 31\text{days})/748(\text{gallons per CCF}) = \text{Tier1}(\text{CCF})$$

Outdoor budget for

$$\text{Dec} = \left[\text{SF}(\text{square feet})\cdot 0.60(\text{LF Dec})\cdot 2.82(\text{ET Dec, inch})\cdot 0.083333(\text{conversion from inch to feet})\right]/100(\text{conversion from cubic feet to CCF}) = \text{Tier2}(\text{CCF})$$

Table 11.3 (continued)

Household #	Adj. form status	Persons per household (R) default	Persons per household (R) adjusted	Default irrigation area (SF) (square foot)	Irrigation area adjusted (SF) (square foot)
232	Used default figures	3	3	2,439	2,479[a]
218	Used default figures	3	3	2,744	2,780[a]
368	Sent in adj. forms	3	4	2,744	2,675
833	Sent in adj. forms	3	5	2,744	3,450
984	Sent in adj. forms	3	4	14,810	4,4360
003	Sent in adj. forms	3	3	3,646	6,193
696	Sent in adj. forms	3	7	3,106	7,700
368	No adj. form recvd	3	3	2,836	6,670[a]
941	No adj. form recvd	3	3	2,744	4,830[a]
342	No adj. form recvd	3	3	14,810	15,920[a]

Source: WMWD Staff (January 2012)

[a]Adjustment made by WMWD

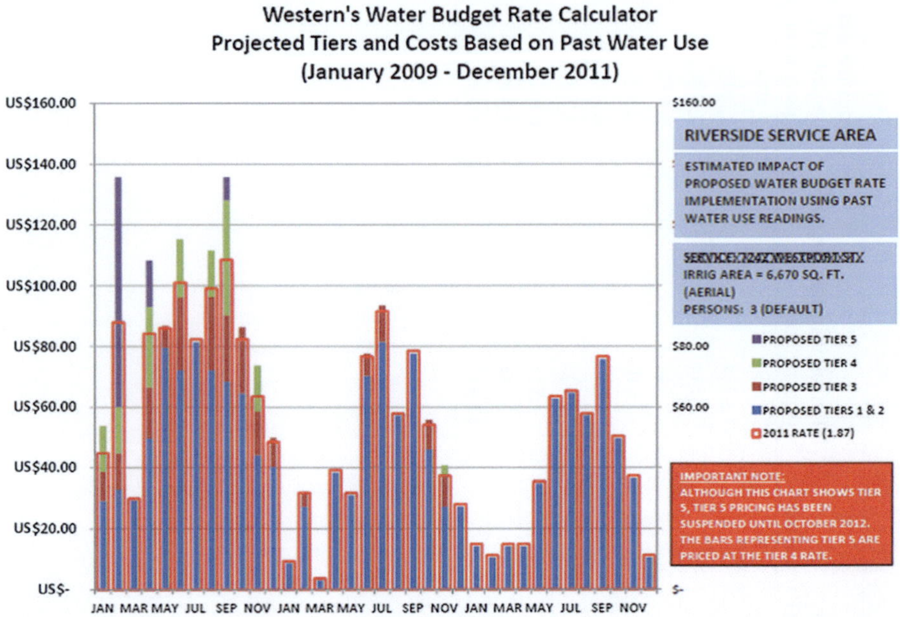

Fig. 11.9 Comparison of water consumption and cost of water for one household in the WMWD service area under the pre WNRS rate and under the WBRS during 2009–2011 (Source: WMWD Compliments (January 2012))

water use so that their cost from the WBRS has increased; (3) Such analysis can be used as a wakeup call for the 'wasters' households, demonstrating how the WBRS can drive up their water cost if they continue to use at such high rates, suggesting that they adjust their behaviour accordingly.

Figure 11.9 presents the results for one household with three persons and an irrigated area of 6,670 square feet. The red framed bars indicate the monthly cost of water using the previous flat rate of US$1.87 per CCF. The imposition of the new rate on past and present consumption suggests that under 2009 weather-consumption that household would have exceeded significantly its water budget (the blue filling of the red framed bars); under 2010 weather-consumption the same household had minor excess of its budget (small glitches to tier 3 and 4); and under 2011 weather-consumption the same household consumed exactly according to the budgeted allocation of water.

11.6 Institutions for Implementing WBRS

While the state provided legal standing for the design and implementation of the WBRS, there are also local institutions following the individual water utility bylaws. Both will be discussed below.

WBRS is supported by various state legislations, and follows various bills since 1990s.[9] In 2004, (Assembly Bill) AB 2717 was passed, which requested the California Urban Water Conservation Council (CUWCC) to convene a stakeholder task force, composed of public and private agencies, in order to evaluate and recommend proposals for improving the efficiency of water use in new and existing urban irrigated landscapes in California. Based on this charge, the Task Force adopted a comprehensive set of 43 recommendations, essentially making changes to the AB 325 of 1990 and updating the Model Local Water Efficient Landscape Ordinance. The recommendation of the bill charges (the State Department of Water Resources) DWR in updating the Model Efficient Landscape Ordinance and to upgrade (California Irrigation Management Information System) CIMIS.

The Water Conservation in Landscaping Act of 2006 (AB 1881) enacts many, but not all of the recommendations reported to the Governor and Legislature in December 2005 by the CUWCC Landscape Task Force (Task Force). AB 1881 requires DWR, not later than January 1, 2009, by regulation, to update the model ordinance in accordance with specified requirements, reflecting the provisions of AB 2717. AB 1881 requires local agencies, not later than January 1, 2010, to adopt the updated model ordinance or equivalent or it will be automatically adopted by statute. Also, the bill requires the Energy Commission, in consultation with DWR, to adopt, by regulation, performance standards and labelling requirements for landscape irrigation equipment, including irrigation controllers, moisture sensors, emission devices, and valves to reduce the wasteful, uneconomic, inefficient, or unnecessary consumption of energy or water. Senate Bill (SB) 7 (approved on 12/2009) requires the state to achieve a 20 % reduction in urban per capita water use in California by December 31, 2020.

11.7 Policy for Implementation

The following outlines the ideal steps for designing a Water Budget Rate Structure, based on experiences from water utilities which have implemented WBRS.[10]

1. Determine the agency costs for service, both fixed and variable:
   - Determine revenue requirements for the agency, parameters for a revenue neutral cost recovery, etc.
2. Accurately identify customer issues and expectations:
   - Conduct customer surveys to understand user perceptions of water use and the water agency

[9] http://www.water.ca.gov/wateruseefficiency/landscapeordinance/updatedOrd_history.cfm#summary

[10] Based on Ash, T. (2011, November 28). Personal communication.

3. Determine the allocations and variables affecting demand for each customer group:
    - Residential allocation
    - Irrigation/Landscape allocation
    - Multi-family allocation
    - Agriculture allocation
    - Commercial allocation
4. Accumulate customer data:
    - Residents per household
    - Square footage of outdoor irrigated area
5. Identify accurate ET data for daily downloading into billing system:
    - Based on service area microclimates, availability of ET weather stations and/or private sector ET data, etc…
6. Test (simulate) customer use in the WBRS:
    - How many customers would meet allocations at current use patterns
7. Test financial requirements in the WBRS:
    - Model different fixed/variable recovery scenarios
8. Finalize policies on rates with elected water board officials:
    - Allocations
    - Tiers (number and width)
    - Prices per tier
    - Excess funding apportionment to go for conservation and environmental programs (see item 12 below)
    - Adjustments and credits
9. Identify billing system requirements/upgrades
10. Identify implementation timeline:
    - Billing system upgrade completed
    - Board election schedule
    - Prop 218 process (California only)
    - Public outreach/education campaign
11. Staffing needs (if any)
12. Efficiency programs design and implementation:
    - Programs to assist customers to reduce water waste
        - Residential programs
        - Landscape efficiency programs
        - Agricultural efficiency programs
        - Commercial efficiency programs

13. Website upgrade:
    - Customer education of WBRS details
    - Water budget estimator tool (estimates of future billing period allocations)
    - Efficiency program workshops
14. Internal staff training:
    - Customer service, conservation, board, general employee
15. Internal tracking tools
16. Implementation
17. Continuing customer education
18. Excess revenue/conservation fund monitoring
19. Board and public education/reporting

Examples of ways that WMWD followed such implementation guidelines can be seen in Annex III in Dinar (2011).

These suggested steps are associated with several difficulties and risks that have to be addressed.

11.7.1 Transaction Costs

The main transaction costs associated with the implementation of WBRS are associated with the Proposition 218, which requires meeting the cost of service standards, including a process of hearing and approval of changes in water rates by customers. Water utilities are therefore obliged to submit themselves to a serious and long process of customer education. Following a necessary educational process, the agency interacts with customers via a public hearing, where customers can make their opinions heard. There are several examples where the public opinion of frustrated customers derailed the process of tariff change (such as the case of EMWD in 1992 (Pekelney and Chessnut 1997).

The second type of transaction cost is the process of adjustment (variance), which necessitate validation by the water agency of appeals on the part of the households. While a quantitative assessment of the processing of the thousands of Requests for Water Budget Adjustment forms (Annex II, Figure II.1 in Dinar 2011) are not available, in retrospect, the WMWD is satisfied by this investment of time of its staff in light of the gain in customer confidence and support. The use of GIS-based techniques to verify irrigated areas of the household (Annex II, Figure II.2 in Dinar 2011) simplified the verification process.

11.7.2 Uncertainty

The current rate structures are very uncertain in terms of revenue generation, thus they inflict on the ability of the water utilities to sustain their services. That is due to the design of a collection of only a small portion of the fixed costs in the structure and linking the remaining share of the fixed cost recovery to water sales, while at the same time working to get customers to use less water. The reason for having a small share of the fixed cost recovered independently of water use is certainly politically driven. Therefore, with improved saving, namely, with reduction in water sales, the part of the fixed cost that is linked to water sale will be jeopardized and may lead to a change in the rates.

A safer water budget rate structure suggests that the majority of fixed costs are recovered independent of water sales. When that is done the agency is free to pursue conservation at the rate they need, and eliminates the negative political and socially unjust action of raising rates if not enough water is sold. The agencies with WBRS experience more stable revenue recovery (reduced uncertainty).

Since the WBRS is dependent on ET, uncertainty in finding reliable ET values that may be over or under determine the monthly tiers. Depending on the climatic conditions in the service area of the water utility, it has to 'optimize' the number of micro-climatic zones to be used. To remind the reader, IRWD uses 3 ET zones, EMWD uses 50 ET zones, and WMWD uses 450 ET zones. The trade-off between more reliable (and representative) information and the cost of information is an important aspect in deciding on the level of precision. This is a subject for a separate study. The reader can find a map with the ET zones used by WMWD in Annex IV in Dinar 2011.

11.8 Conclusions

Water is delivered in California by wholesale and retail agencies. WBRS are typically used at present by retail agencies as a means to establish efficiency standards for end-users. Legislation in California has set efficiency standards and allocations, such as per capita per day indoor use (SB 7–7) and 80 % of local ET for outdoor use, as current and reasonable allocations (AB 1881). Wholesale agencies in California also operate under State law in terms of water efficiency goals; however the wholesale rate structures do not incorporate water budget methodology to set standards for retail agencies and pricing triggers for excessive water purchases. With State of California efficiency guidelines now set, it could be useful to align the entire chain so that wholesale agencies and retail agencies apply water budget rates. The benefits to wholesale agencies would be very similar as those for retail agencies, specifically a wholesale agency would:

1. Recover fixed costs separately from water sales;
2. Establish agency by agency water budgets (as per SBX7-7 guidelines);
3. Charge increasing tier prices for water used above the agency allocation;
4. Align wholesale rate structure with State legislation and retail agency practices for a more consistent public message and education.

Agencies with water budget rates have succeeded in stabilizing revenues, reducing risk of revenue loss when customers use less water, increasing water efficiency, improving customer services and even reducing urban runoff. Many agencies are unaware or apprehensive about making a rate structure change, particularly to a more sophisticated structure that would require technical upgrades, public education and staff training. However, current rate structure designs are the cause for agencies losing necessary revenues, angering customers who save water or have large families or large properties. Currently agencies have only one method to recover revenue lost if customers use less water, and that is to raise water rates. A properly designed water budget rate structure, that reflects the actual costs of water and water service, can permanently fix the structural problem of current rate structures, drive more water conservation and appease customers with individualized allocations.

The experiences of the various water utilities (not only those included in the case study) suggest the following aspects as enabling/disabling factors in the implementation of WBRS:

- Appropriate billing system to allow addressing all the aspects of WBRS and provide needed flexibility in the adjustment (variance) process;
- Access to appropriate climate data to allow proper calculations of ET per unit of consumption and prevent using averages;
- Technological advancements to verify claims by households and to record usage and wastage in order to help the utility address disputes by customers.

References

Baerenklau, K., Schwabe, K., & Dinar, A. (2014). The residential water demand effect of increasing block rate water budgets. *Land Economics, 90*(4), 683–699.

Dinar, A. (2011). Water budget rate structure: Experiences from urban utilities in California. Working paper of the EU project Evaluating Economic Policy Instruments for Sustainable Water Management in Europe, WP6 IBE EX-POST Case studies. Accessed on 7 Aug 2014 at http://www.feem-project.net/epiwater/docs/d32-d6-1/CS27_California.pdf

Eastern Municipal Water District (EMWD). (2011). http://www.emwd.org/, visited on 2 Dec 2011.

Hall, D. C. (2000). Public choice and water rate design. In A. Dinar (Ed.), *The political economy of water pricing reforms*. Oxford: Oxford University Press.

Hanak, E., Lund, J., Dinar, A., Gray, B., Howitt, R., Mount, J., Moyle, P., & Thompson, B. (2011). *Managing California's water: From conflict to reconciliation*. San Francisco: PPIC.

Hewitt, J. A. (2000). An investigation into the reasons why water utilities choose particular residential rate structure. In A. Dinar (Ed.), *The political economy of water pricing reforms*. Oxford: Oxford University Press.

IRWD water facts. (2011). http://www.irwd.com/assets/files/About%20Us/IRWD-Fact-Sheet.11.11.pdf. Accessed on 30 Nov 2011.

Maria-Saleth, R., & Dinar, A. (2001). Preconditions for market solutions to urban water scarcity: Empirical results from Hyderabad city, India. *Water Resources Research, 37*(1), 119–132.

Mayer. (2009). *Published in metering and smart energy international*. http://www.metering.com/water-budget-based-rate-structures-powerful-demand-management-tools/. Accessed 30 Apr 2015.

Municipal Water District of Orange County, Irvine Ranch Water District (MWDOC-IRWD) 2004. The residential runoff reduction study, July. Found on: http://www.irwd.com/assets/files/Conservation%20Research/R3-Study-Revised11-5-04.pdf, visited on 4 Dec 2011.

Santa Ana Watershed Project Authority (SAWPA). (2010).One water, one watershed plan adopted 16th Nov 2010 (11615 Sterling Avenue, Riverside CA 9250, (951) 354–4220.

Pekelney, D. M., & Chessnut T. W. (1997). Landscape water conservation programs: Evaluation of water budget based rate structures. Report submitted to the Metropolitan Water District of Southern California, September.

Peter, M., DeOreo, W., Chessnut, T., Pekelney, D., & Summers, L. (2008). *Water budgets and rate structures: Innovative management tools*. Denver: AWWA Research Foundation.

Western Municipal Water District (WMWD). (2011). http://www.wmwd.com/, visited on 2 Dec 2011.

WMWD Staff (January 2012). Interactions and data provision by Staff of the Water Conservation unit headed by Mr. Tim Barr.

Chapter 12
Green Energy Certificates and Compliance Market

Jaroslav Mysiak, Fabio Farinosi, Lorenzo Carrera, Francesca Testella, Margaretha Breil, and Antonio Massaruto

Abstract In the economies striving for low-carbon footprint hydropower plays an important role, as one of few sources of renewable energy for which the technology is available, affordable, and reliable. Hydropower is an important source in the mix of renewable energy sources (RES) on the pathway to meet the ambitious targets set in the EU Directive 2009/28/EC and the Europe 2020 strategy. However, hydropower development may impair the integrity of water courses and river health, in contrast to the objectives of the Water Framework Directive (2000/60/EC). In this chapter we review a mix of economic policy instruments, designed separately and at least partly for different purposes, but all acting together in a way hydropower potential was exploited in Italy. *Feed-in tariffs* (FIT) and especially tradable *green energy certificates* (GEC) had been introduced to build supply-side competition among the RES and to curtail the costs of renewables. The actionable concession award or operating large hydropower plants are an opportunity to coerce environmental improvement. Yet these opportunities have not been used so far.

Keywords Hydropower • Feed-in tariffs • Renewable energy sources • Green energy certificates • Italy

J. Mysiak (✉) • F. Farinosi • L. Carrera • M. Breil
Fondazione Eni Enrico Mattei (FEEM) and Euro-Mediterranean Centre on Climate Change (CMCC), Isola di San Giorgio Maggiore, 30124 Venice, Italy
e-mail: jaroslav.mysiak@feem.it; fabio.farinosi@feem.it; lorenzo.carrera@feem.it; margaretha.breil@feem.it

F. Testella
Department of Political Sciences, Communication and International Relations, University of Macerata, Piazza Strambi n.1, Macerata, Italy
e-mail: francesca.testella@unimc.it

A. Massaruto
Dipartimento di Sscienze Economiche e Statistiche, Università di Udine, Via Tomadini 30a, 33100 Udine, Italy
e-mail: antonio.massarutto@uniud.it

M. Lago et al. (eds.), *Use of Economic Instruments in Water Policy*, Global Issues in Water Policy 14, DOI 10.1007/978-3-319-18287-2_12

12.1 Introduction

Energy and water security are arguably among the most important, and interconnected, present-day societal (and environmental) challenges (WEF 2014). Water is one of the most important factors of energy production. It is used in power generation, in fossil fuel extraction, in fuel transportation and process, and for the production of biofuels. Attending to the 2012 World Energy Outlook, some 580 billion cubic metres of water are withdrawn every year for energy production (IEA 2012). Water is also used for cooling thermal power plants (TPPs) and as a source of kinetic energy in the hydropower plants (HPPs). On the other hand, energy, and in particular electricity, is important for water transportation, treatment and distribution. In 2011, some 8 % of the Italian electricity demand was represented by the requirements of the water treatment and distribution sector.

Amidst the early signs of human induced climate change, both energy and water management systems are undergoing sizeable transformations. The EU Directive 2009/28/EC (and before in the Directive 2001/77/EC) and the Europe 2020 Strategy (EC 2010) set ambitious energy goals and renewable energy targets. In Italy these targets entail increasing the share of renewable energy sources (RES) in the gross energy consumption to 17.0 %, and 26.4 % in terms of electricity generation by 2020.

In 2010, thanks to the sizeable incentives described in this chapter and as a result of demand decline amidst economic crisis, Italy was still far away from both targets but exceeded the 2010 milestone and came close to the 2015 milestone. Since then, the exceptional grow of RES in electricity generation set forth and in some months during 2013, with RES replaced surpassing the thermoelectric power generation.

The kinetic energy contained in natural water flow is a *renewable*, *carbon dioxide emission-free* and *easily exploitable* source of energy. In the modern carbon-free economies hydropower plays an important role, as one of few sources of renewable energy for which the technology is available, affordable, and reliable. Hydropower is an important source in the mix of renewable energy sources (RES) on the pathway to meet the ambitious targets set in the EU Directive 2009/28/EC and the Europe 2020 strategy.

The hydroelectricity generation however requires structural modification of water courses and, in the case of larger plants, construction of water impoundments. Hence, hydropower development may impair the integrity of water courses and river health. The clash of the two objectives – renewable energy development and river restoration – caused that hydroelectricity generation grew into a controversy. On the one side, hydroelectricity generation is relatively safe and flexible technology enabling water flow regulation and flood risk management. On the other hand the hydropower development may cause significant negative environmental impacts. The *Hydropower Sustainability Assessment Protocol* (Tollefson 2011) assessing the impacts of dams in all phases, from development to operation, is one of the recent initiative to reconcile the positive and negative environmental effects of hydroelectricity generation.

In this chapter we explore a mix of economic policy instruments, designed separately and partly for different purposes, but all acting together to in a way hydropower potential is exploited in Italy. *Feed-in tariffs* (FIT), and later *tradable green energy certificates* (GEC), had been introduced in Italy in 1990s in order, among others, to reduce the country's carbon dioxide emissions and dependency on energy imports. Both FIT and GEC contributed to increasing the production of renewable energy (Ringel 2006). The latter, more sophisticated among the both, introduce a competition among the RES that should under favourable market conditions curtail the generation costs of renewables (Bertoldi and Huld 2006).

Neither FIT nor GEC as implemented in Italy take into account the environmental impacts of hydropower generation and both treat all renewable energy sources (RES) in the same way. The concessions to build a new HPP are in principle granted upon the results of *environmental impact assessment* (EIA) but this instrument did not prevent excessively concentration of HPP in some places. Besides, to limit the development of hydropower in less or not suitable places, the water abstraction fees and charges can be designed in a way sensible to the environmental impacts. In Italy this has not been done yet but is being discussed. Finally, the government-auctioned concession for operating the state-owned hydropower reservoirs provide another opportunity to control the hydropower operations in a sustainable way and taking into account the costs of decommissioning and removing the dams. Yet Italy extended the concessions in place and postponed the auctions, a move that has been contested both by the European Commission and the Italian Constitutional Court.

This chapter sets to explore the environmental impacts of economic policy instruments targeted at hydropower generation. Differently than the other chapters in this book, the chapter reviews EPIs set to exploit kinetic energy contained in water, which is a conventional water use, analogous to irrigation or cooling, expect for it does not 'consume' water. Nor does it significantly alter water's physical or chemical properties. But it does impact the morphological conditions and flow regimes of water bodies. Although the incentives are pursued in order to developed (renewable) energy sources, the instruments reviewed stimulate thoughts about water-energy interconnection and the extent to which other policy instruments, both regulatory and economic ones, are suitable to counteract or counterbalance the spillover effects.

12.2 Setting the Scene: Challenges, Opportunities and EPIs

In Italy, the number of hydroelectric power plants grew between 2000 and 2010 at an annual average rate of 1.3 % but the installed capacity increased only by 0.7 % per year. Large hydropower facilities (>10 MW) account for around 86 % of the total installed hydropower capacity. Most of the hydropower plants (HPP) are located in the north of the country, comprised in the Po-River Basin District (P-RBD); one of the eight river basin districts (RBDs) established under the EU Water Framework Directive (2000/60/EC) (AdBPo 2006). Four administrative regions comprised in the P-RBD (Piedmont, Lombardy, Vale D'Aosta, Emilia Romagna) (ISTAT 2011) account for more than a half of the installed hydropower capacity and hydroelectricity production in Italy. The installed gross capacity in the P-RBD has increased steadily from 10,210 MW in 2000 to 11,285 MW in 2012. In 2012, Lombardy alone produced 10,646 GWh and Piedmont 7,113 GWh, respectively 49 % and 33 % of the total hydropower production in the PRBD regions. Yet although the capacity in the district grew on average by 1 % point per year, the net hydroelectricity produced remained below the 2000 level for all years except 2011. The number of HPP increased from 839 in 2000 to 1,273 in 2012 (Terna 2000–2012; APER 2011) (Table 12.1).

There are different types of hydropower plants (HPP). Conventional hydroelectric plant exploits the gravitational force of falling water stored in a reservoir. Run-of-the-river hydroelectric plants do not require a reservoir as they exploit the power of flowing water. Pumped-storage hydroelectric plant is a semi-closed circuit consisting of two reservoirs between which the water conveyed and electricity produced on-demand, helping so to 'store' energy and make it available at times of peak demand. In terms of capacity, the HPP are usually classified into small capacity (<1 MW), medium capacity (1–10 MW) and large (>10 MW). Small and medium size HPPs have higher rate of expansion (<1 MW and 1–10 MW), while the number of larger HPP remained constant. The data highlights a strong increment of small HPP from 2009, due to the connection of small capacity plants to the grid. The first peak of new plants was observed in 2002 and then again in 2008–2010.

In order to boost the development of renewable energy sources (RES), in late 1990s the Italian government introduced compliance market, first specified by the decree 79/1999. The compliance market is based on mandatory targets from renewable energy to be supplied by each energy provider every year, and a scheme of renewable energy certificates (GEC). The mandatory target for renewable energy share was first set to 2 % of the previous year's production or import of electrical energy. The target applies to the importers and producers of electricity from non-renewable sources. The rule exempts the first 100 GWh of yearly production/import.

Table 12.1 Hydropower and hydroelectricity production in 2012 in the regions of the PRBD

Region	Nr of plants	Change to 2000 (%)	Gross capacity [MW]	Change to 2000 (%)	Gross product [GWh]	Change to 2000 (%)
Piedmont	635	50	3.681	17	7.113	–9
Valle d'Aosta	97	80	921	11	3.063	8
Lombardy	428	43	6.039	7	10.646	–19
Emilia Romagna	113	82	645	6	895	–27
Total PRBD regions	1.273	52	11.285	11	21.716	–13
Italy	2.977	51,5	22.249	7,7	43.854	–14
PRDB as a % of Italy	43	0,3	51	1,0	50	1,0

Based on data Terna (2013, 2011, 2010)

The companies falling short of meeting the target are obliged to purchase the GEC for the equivalent of the underperformed renewable energy. For HPP, a tradable certificate was issued for each 1 MWh of renewable energy produced in the previous year by plants with installed capacity exceeding 1 MW. The HPP built after December 31st, 2007 with installed capacity smaller than 1 MW were excluded from the GEC scheme but remunerated with an **feed-in tariff** (FIT). For the most part, these HPP are of the run-of-the-river type. In 2010, the volume of GEC traded, under this scheme, amounted to EUR 301 million (GSE 2010b).

The quota were first set to 2 % and later increased by an annual rate of 0.35 % (from 2004 to 2006) and by 0.75 % (from 2007 to 2011). The producers of energy from renewable sources benefit from a double source of income, from both the sale of electrical energy and the sale of green certificates. The compliance market was first set for 8 years, then extended to 12 years by the decree 152/2006, and 15 years by the law 244/2007 for power plants built or restored after 2007. The legislative decree 28 of March 3rd, 2011 (the so-called Romano decree) marks the end of the GEC system in Italy. It gradually phases out the compulsory quota between 2012 and 2015. Green certificates exceeding the demand will be withdrawn from the market at a price corresponding to 78 % of the previously determined level. The incentives introduced in favour of small renewable energy plants will remain in place for the whole envisaged incentive period.

The environmental impacts of hydroelectricity development can be controlled by the mandatory environmental impact assessment (EIA), and the fees for water concession fees (WCF) that are based on the installed capacity of the HPP. The WCF, introduced in 1930s and in the 1990s delegated from the central government to the administrative regions, may in principle, but is not, be differentiated according to the environmental pressures on water bodies, and hence prevent overexploitation of some basins with high HE potential. Supplementary fees introduced to compensate riverine and mountain communities are discussed further down in the chapter. Besides, the renewal of the concession for large water abstraction and operation of the HPP, pursued by auctions and rewarding the efforts to reduce the impact of water flow modification can be but have not yet been used.

12.3 The Green Energy Certificates and Feed-in Tariffs in Action

12.3.1 The EPI Contribution

12.3.1.1 Environmental Outcomes

The persistent and contentious debate about the benefits and costs (in the largest sense) of hydroelectricity is triggered by the environmental and social effects of hydropower (Schiermeier et al. 2008; Kramer and Haigh 2009). The HPP disrupt river habitats (Vannote et al. 1980) and fish migration routes. The alterations of river flow patterns influence river stages and temperature; both have an effect on riverine and riparian flora and fauna (Nilsson and Berggren 2000). Alternation of sedimentation processes lead to lesser sediment supply downstream, amplifying so coastal subsidence and erosion. Reduced downstream river flow creates condition for saltwater intrusion (Milligan et al. 2006; Vorosmarty et al. 2003; Walter and Merritts 2008). Processes of coastal erosion and subsidence represent a serious concern for the low lying Adriatic coasts at the mouth of the river basin which are reducing their potential of natural adaptation processes to sea level rise.

Hydropower reservoirs are also a potential source of greenhouse gas (GHG) (Giles 2006), as a result of bacterial decomposition of organic material (see for instance Rosenberg et al. 1997) According to Barros et al. (2011), hydroelectric reservoirs worldwide emit about 4 % of global carbon emissions from inland waters, with varying contributions from the single reservoirs according to their age (higher emissions in the first years after flooding due to decomposition of previous vegetation) and climate zone (highest contributions from reservoirs in tropical climates). Rosenberg et al. (1997) expect these impacts to last for even 100 years after the first flooding of the reservoir, whereas the statistical analysis of different measurements on GHG emissions made by Barros et al. (2011) indicates of 20 years as the critical period after flooding.

Not all environmental effects are negative. Hydropower reservoirs help to regulate river flows and cushion against too high or low river stages (Verbunt et al. 2005; Dugan and Allison 2010).

12.3.1.2 Economic Outcomes

The market of the green certificates has been subject to different shocks. The main problem materialised through a large increase of supply and general stagnation of the demand for green certificates (GSE 2011). The general surplus of supply registered from the end of 2007 determined a collapse of the GEC price that reached its minimum value of EUR 58/Mwh in August 2008 (GME). The fall of the demand has been provoked by the exemption of some operators from the quota system (Barbetti 2009). The exemptions were introduced since 1999 for cogeneration, energy produced for self-consumption, energy produced using coal coming from national mines, and for the first 100 Gwh yearly produced/imported by each operator. It has been estimated that, on 2008, due to the exemptions, demand for GEC has been reduced by the half (ibid).

The compliance market was reserved only through the intervention of the of the regulatory agency (GSE) (Poletti 2009). The excess of supply has been controlled by the introduction of the Ministerial Decree 18/12/2008 (Ministry of Economic Development) obliging the GSE in purchasing the unsold GEC at the average price of the 3 years before till 2010. This intervention artificially stimulated the demand side and consequently the rise of the GEC price from 2009 avoiding the market failure. This reached values substantially high during the period 2007–2008, with the excess of supply of GEC, and fell down in the first trimester of 2009 after the introduction of the Ministerial Decree 18/12/2008.

The compliance market has been designed to promote the exploitation of renewable energy sources, otherwise not able to compete with fossil fuel. The EPI triggered investments with positive ripple effects on the sub-suppliers and technological innovation.

The costs borne by the operators are passed on to final electricity consumers. The costs of incentives sustained by the operators, in relation to the GEC purchased to satisfy the compulsory quota, converge into the final price for energy that consumer has to pay. Moreover the final consumers are charged of the costs of the GSE through a section of the electricity bill. It has been estimated that the cost of CIP6/92 for final consumers for the year 2009 was EUR 1.8 billion; for 2010 was EUR 800 million (AEEG 2010). At the same time, the compliance market weighted final consumers with indirect costs for EUR 600 million and direct costs for EUR 1 billion (AEEG 2010; Capicotto 2011).

12.3.1.3 Distributional Effects and Social Equity

Hydropower development has been met with increasing social resistance fuelled by perceptions of social and geographic injustice. Concentrated in less developed, mountainous areas, the hydroelectricity generation is associated with negative externalities (negative environmental impacts, modification of water courses and landscape) in proximity of the plants, whereas the downstream communities take most benefits. The history of hydropower exploitation in Italy is punctuated by incidents among which the most prominent one is the Vajont disaster in 1963. At the time of the completion the tallest dam in the world (262 m), the reservoir built on the Vajont river became centre stage of a tragedy claiming the life of some two thousand people. A landslide with speed of 110 km/h hit the reservoir, causing a seiche that overtopped the dam and destroyed the villages downstream. Another major disaster occurred in Val di Stava in 1985, claiming a death toll of some three hundreds.

The Italian legislation introduced compensation for the local communities in hydropower project's influence areas. Supplementary water abstraction fees and charges have been introduced to benefit local communities. Supplementary fee benefiting riverine communities is split between the municipalities in the territory of which the water is derived, and the higher order administrative units – districts, usually by three-quarter to one-quarter ration (Regione Piemonte 2003). Supplementary fee for mountainous basins is distributed too, but according to different patterns. Usually, the local communities constitute a consortium and distribute the collected fees according to an agreement (Regione Piemonte 2003). For other cases the central government offers an equitable scheme for dividing the collected fees: 10 % is equally distributed among the communities; 20 % is distributed in relation to the municipal territory; 30 % in relation to the number of inhabitants; 40 % in relation to size and impact of the plants installed in the municipal territory (Regione Piemonte 2003). The wealth from the supplementary fees is used to finance local infrastructures and economic development of the local communities.

The high incidence of existing hydropower plants in the territory of the Province of Sondrio is fuelling resistance of inhabitants in the valley, opposing any project for new concessions for plants regarding the area. Further to the high percentage of exploitation of water flow in the area (some 90 % of the rivers in the province are already exploited), the fact that the area provides almost half of the hydro power generated in the entire Lombardy region, but only 20 % of this production is consumed within the province. Small dams are opposed because of the environmental impacts, landscape alteration, and impacts on the existing water uses (including sport and leisure fisheries) judged disproportionate in relation to the increase in electric capacity generated (IAPS 2010). Since 2006, a number of civil society initiatives have been launched to oppose any new project for water abstractions.

Triggered by the local resistance, and upon invitation of almost all political parties and civil society organisations, the 13ª permanent commission (Territory, environment and environmental goods) of the Senate held hearings about the water crisis in Sondrio district, and asked the government to limit the hydropower concessions in the district for 2 years. Successively, the 2007 Financial Law (law 296/06 article 1, 1106 commas) established that new concessions for both large and small hydropower plants, exclusively for the Province of Sondrio, from 1st January 2007 to the 31st December 2008, were granted only after the binding advice of the Ministry of Environment. This moratorium was due to the critical situation of the hydrographical basin of Province of Sondrio caused by the extraordinary weather conditions of July and August 1987.

12.3.2 The EPI Setting Up

12.3.2.1 Institutions

The system of green energy certificates (GEC) had been introduced by the Bersani Decree (79/1999) and later modified by laws 244/07 and 239/04, and the Legislative Decree 387/03. The Bersani Decree (law 79/99) transposed the provisions of the Directive 96/92/CE. The Decree set off the process of energy liberalisation. Whereas the import, export and production of electricity was privatised; the transmission, dispatching and management of electricity lines remained under state control. Regulation of the free energy market was entrusted to the Energy Service Authority (*Gestore dei Servizi Energetici*, GSE). GSE certifies the renewable energy plants and oversees the market with green energy certificates. The energy sector regulator (*Autorità per l'Energia Elettrica ed il Gas, AEEG*), constituted in 1995 as part of the liberalization process. The AEEG defines the rules – on equitable and neutral basis – for of transmission and distribution of energy. The Authority also regulates the feed-in tariffs applicable to small renewable energy plants (<1 MW) and the modalities of financing the GEC.

The Bersani Decee introduced the scheme of green energy certificates (GEC). The law obliges the electricity companies to supply a certain share of their production by energy from renewable sources, including hydropower. The companies that fail short of meeting the target may purchase tradable Green Energy Certificates (GEC) for the equivalent of the underperformed renewable energy. Initially, the mandatory quotas for renewable energy sources (RES) were set to 2 % and the period of the incentive scheme was set to 8 years. The nominal value of the green certificates was set to 100 MW. The renewable energy plants, in order to be admitted into the system, had to be certified. The tradability of the certificates was limited to 1 year.

The law specified a number of exemptions reducing the overall volume of the RES to be supplied. Most importantly, the obligation applies to energy production or import exceeding 100 GWh. Exempt is also electricity produced from coal from national mines and cogeneration; water pumping, and electricity for self-consumption.

The Bersani decree was modified by the decree 387/2003 (so-called Marzano decree) transposing into Italian legislation the EU Directive 2001/77/CE. The main changes of the GEC system included: (i) increase of the compulsory quota by 0.35 % every year for the period 2004/2006; (ii) extension of the tradability of the certificates from one to three consecutive 3 years; (iii) reduction of the nominal size of the certificates from 100 to 50 MWh. Further modification to the GEC regime was introduced in the law 152/2006. In order to increase the profitability of the energy production from RES and to favour the flow of private investments into the sector, the duration of the incentives was increase from 8 to 12 years.

The law 244/2007 (financial bill for the year 2008) partially overhauled the GEC system (Repubblica Italiana 2007). First, it introduced a new feed-in tariff for certified small renewable energy plants certified with capacity <1 MW (200 KW for wind power). Second, the compulsory quotas were increase annually by 0.75 % for the period 2007/2012. Third, the nominal size of the green certificates was further reduced from 50 to 1 MWh. Fourth, the number of certificates issued for a given volume of renewable energy was made dependent on the type of energy. This has not affected hydroelectricity. Fifth, the incentive period was extended from 12 to 15 years.

The Decree of the Minister for Economic Development 18/12/2008 compelled the authority (GSE) to stimulate the market with green certificates by purchasing the certificates in excess until the end of 2010. The fixed price at which the GSE was to bay the certificates was set to the average price over the precedent 3 years. Subsequently, the obligation to purchase the certificates in excess was extended until 2011. In 2009, the legislators shifted the obligation to supply renewable energy from the producers and importers of energy to the companies dispatching energy to the final consumers (law 99/2009). Only a year after this provision was withdrawn by the law 72/2010. The legislative decree 28 of March 3rd, 2011 (the so-called Romano decree) marks the end of the GEC system in Italy. It gradually phases out the compulsory quota between 2012 and 2015. Green certificates exceeding the demand will be withdrawn from the market at a price corresponding to 78 % of the previously determined level. The incentives introduced in favour of small renewable energy plants will remain in place for the whole envisaged incentive period.

12.3.2.2 Transaction Costs and Design

To be admitted to the GEC incentive system, the renewable energy plants are certified by the authority (GSE *Gestore dei servizi Energetici*). The applicant is required to register and submit detailed technical and administrative information relative to the plant. With respect to hydropower, the investors are requested to submit a detailed report about the technical and hydrological information from the area the HPP is situated. The authorisation for building a new renewable energy plant is issued by regional or provincial authorities. Concession for water derivations for hydropower purpose is a separate and cumbersome legal procedure. The water concessions are issued by regional authorities. In some cases the environmental impact assessment is required. The competent authority attests the availability of water resource and impact on the minimum environmental flow based on the River Basin Plan. Subsequent to the release of the concession, the applicant is to submit the executive project relative to the concession. The project is assessed and approved based on the criteria specified in the legislative decree 387/2003. The application for water derivation is aggravated if territorial development plan for hydropower sector is not in place, and by the lack of centrally managed water information systems. Between 2005 and 2011, the Sondrio district authority received some 68 applications for new concessions, out which only 22 have been authorised so far.

The Constitutional court intervened several times on the matter related to hydropower in Italy. The latest sentence n. 205 of July 13th, 2011 the Court found unconstitutional the extension of the water concessions for hydropower generation introduced in the law decree 78/2010 (see Sect. 3.5). In 2008 the Court intervened on the matter of tendering procedures to renew expired concessions for large water derivations, declaring the provisions of the law n. 266/2005 in parts unconstitutional. The European Commission started in 2004 the infringement procedure against Italy for similar reasons and drop the case in 2006, after the publication of the above Court's decision.

12.3.2.3 Implementation

Sondrio district situated in Lombardy is an emblematic case for overexploitation of the hydropower potential and social uproar. Given the abundant water endowment and topography favourable for hydropower generation (Provincia di Sondrio 2008), the Sondrio district became one of the most hydropower-developed areas in Italy. Some 12.45 % of the national and about 40 % of the Lombardy's hydroelectric production is generated here (GSE 2010a). The further hydropower development was suspended several times, most recently in the late 2000s. Triggered by the local resistance and upon invitation of almost all political parties and civil society organisations, the Italian Senate asked the government to limit the hydropower concessions in the district for 2 years. Successively, the 2007 Financial Law (law 296/06 article 1, 1106 commas) established that new concessions for both large and small hydropower plants, exclusively for the Province of Sondrio, from 1st January 2007 to the 31st December 2008, were granted only after the binding advice of the Ministry of Environment.

In 2010–2011, the "Industry, Commerce and Tourism Parliamentarian Committee" of the Italian Senate held hearings related to national energy strategy (ENEA 2011; GSE 2010c). The experts witnessing in Senate include representatives of public authorities, energetic companies, research organisations, professional associations, electric network operators, and energy providers. Hydropower, the most important renewable energy source in Italy, is captured by a technology that is widely believed efficient, advanced and technically mature (Markandya et al. 2010). It is hard to believe that the plans to construct new large (>10 MW) hydropower plants in Italy would obtain the necessary political support, local acceptance, and financial backing. Even small (<1 MW) and medium-sized (1–10 MW) HPP are occasionally opposed because of the implied environmental impacts and social effects. What is left is (i) increase of efficiency and/or capacity of existing plants, and (ii) development small and medium-sized HPP.

The economic incentives for renewable energy sources (RES) made the further expansion of hydropower profitable. In order to increase the participation of local communities on the profits, the government proposed to extend the large hydropower water concessions by 5 years, or 7 if the public municipal or district authorities were engaged in running the business. In July 2011, the Italian Constitutional Court declared unconstitutional the article 15, commas 6-ter and 6-quarter of the Law 122/2010. The Court recognized that the article infringed the regional competence and represented an obstacle for the market. Before the Court sentence, the European Commission expressed the intent to open infringement procedure.

12.4 Conclusions

The ambitious goals set in the Directive 2009/28/EC (and before in the Directive 2001/77/EC) can be achieved if available renewable energy sources (RES) are efficiently exploited. By 2020, Italy has to increase the share of RES in the gross energy consumption from 5.2 % to 17.0 %. Electricity from renewable sources has to be increased from 14.5 % to 26.6 %. The transition to less carbon-intensive economies should be pursued at lowest possible costs, to reduce overall economic costs of emissions reductions. Green energy certificates (GEC) schemes are among the means to this end, in synergy with other economic policy instruments incentivising production of RES and greater energy efficiency.

The GEC system as introduced in Italy is comparable with similar schemes introduced in other counties. Under market conditions, the producers of RES bear the price uncertainty and the competition between the different sources of renewables ensures that the policy targets are achieved at lower costs. In Italy, the market became soon saturated with the excessive certificates and the price of GEC started to decline. Partly, this is a result of the (many) exemptions from the initial obligation to supply energy form renewable sources granted to the producers or importers by the initial design of the scheme. The government intervened by guaranteeing a fixed price of the certificates, and by doing so removed the price uncertainty and competition between the different renewables. In principle, through this intervention the initial tradable incentive scheme had been turned into indirect subventions. Overall the costs of RES were borne by final consumers, contributing so to making the electricity price for consumers one of the highest in Italy.

The market with tradable CGE has not been insulated from political inference. The design of the GEC has been adapted more to changing political mood than to the requirement of the renewable energy sector. The regulatory mistakes in managing the market with tradable GEC have been remedied by overhauling the whole incentive system, phasing out the CGE and introducing a new system of auctions.

Hydropower development can only be reconciled with environmental concerns and social responsibility if planned in a holistic way, within a well-articulated river basin management plan. A precondition for the latter are clearly defined competences and authority over water resources within hydrographic boundaries. The existing water abstraction charges can be integrated with the GEC to control the environmental impacts particularly of the small HPP. To this end the abstraction charges can be differentiated according to the marginal environmental impacts of a new plant. In order to guarantee sustainable and socially beneficial hydropower exploitation, the whole system of concession and certification has to be embedded within a well-developed river basin plan that identifies and priorities the sites suitable for hydropower development.

The hydroelectricity production are susceptible to production breaks due to low river flows. This is manifested by the declining trend in hydroelectricity production, despite increased installed generation capacity. The climate projections for PRBD provide a doom prospect to what used to be and partly still is water-abundant river basin district. If the decline of annual water endowment of the P-RBD continues, Italy may face an additional burden to meet its renewable energy goals.

Hydropower energy differs from other renewable energy sources (RES) in two important aspects: First, as a mature technology it offers relatively little room for improvement in the efficiency of generation (Schiermeier et al. 2008). The existing and easy-to-tap potential has been already exploited. In 1999 when the GEC system was introduced, the already installed gross capacity exceeded 10,036 MW. Reclamation of existing, mostly large hydropower power plants (HPP) could increase the operating efficiency and the environmental performance of hydropower facilities. Alternatively, the deployment of small (>10 MW) 'run-of-river' HPP that produces power from the natural flow of water provide potential for greater hydropower exploitation, with lesser environmental impacts but at much higher costs.

Second, impact assessment and certification of HPP require different, more comprehensive and meticulous procedures than in the case of other RES. The assessment should not only address the marginal effect of a single HPP, but the cumulative impacts of hydropower exploitation across the entire river system, identifying the best sites and coordinating energy production between the up- and downstream plants.

Furthermore, the reclamation of existing, and construction of new HPP, may require different incentive schemes. Recall that the law 79/1999 had extended the concessions to operate large HPP that would have otherwise expired between 2004 and 2010, up to 2029. This is because the reclamation of large HPP requires investments that are likely not paid back within the 8 years of incentivised RES. In addition, the law put the incumbent – outgoing concession-holder in a favourable condition when tendering the renewal of the concession. The concession tendering would have been a more suitable economic policy instrument to address the specificities of the large HPP.

References

AdBPo. (2006). Caratteristiche del bacino del fiume Po e primo esame dell'impatto ambientale delle attività umane sulle risorse idriche del Fiume Po. Parma: Autorità di Bacino del fiume Po (AdBPo).

AEEG. (2010). Indagine Conoscitiva Sulla Strategia Energetica Nazionale (Italian). Rome, Autorità per l'energia elettrica e il gas.

APER. (2011). Rapporto Rinnovabili 2010–2011. Situazione e prospettiva delle fonti rinnovabili in Italia. Associazione Produttori Energia da Fonti Rinnovabili.

Barbetti, T. (2009). Certificati Verdi. Cosa accadrà? (Italian). Milano: Rete Ambiente, Edizioni Ambiente S.r.l. Milano*.

Barros, N., et al. (2011). Carbon emission from hydroelectric reservoirs linked to reservoir age and latitude. *Nature Geosciences, 4*, 593–596.

Bertoldi, P., & Huld, T. (2006). Tradable certificates for renewable electricity and energy savings. *Energy Policy, 34*(2), 212.

Capicotto, L. (2011). "Nuove politiche legislative in materia di energia rinnovabile: c.d. "Decreto Romani" e "Quarto Conto Energia" (Italian)." Diritto.it: -*Diritto & Diritti*.

Dugan, P., & Allison, E. H. (2010). Water: Act now to restore river health. *Nature, 468*(7321), 173.

EC. (2010). Communication from the Commission: EUROPE 2020 A strategy for smart, sustainable and inclusive growth. COM(2010) 2020 final.

ENEA. (2011). Indagine conoscitiva sulla strategia energetica nazionale. Audizione al Senato della Repubblica Italiana. (Italian), Agenzia Nazionale per le Nuove Tecnologie, L'energia e Lo Sviluppo Economico Sostenibile.

Giles, J. (2006). Methane quashes green credentials of hydropower. *Nature, 444*(7119), 524.

GSE. (2010a). Idroelettrico 2009, Rapporto statistico 2009, Gestore dei Servizi Energetici.

GSE. (2010b). Guida agli incentivi per la produzione di energia elettrica da fonti rinnovabili, Gestore dei Servizi Energetici.

GSE. (2010c). Indagine conoscitiva sulla strategia energetica nazionale. Audizione al Senato della Repubblica Italiana (Italian), Gestore Servizi Energetici.

GSE. (2011). Incentivazione delle fonti rinnovabili: Certificati Verdi e Tariffe Onnicomprensive. Bollettino aggiornato al 31 dicembre 2010, Gestore dei Servizi Energetici.

IAPS. (2010). Un movimento dal basso vicnente per la vita delle nostre acque. Sondrio: Intergruppo Acque Provincia di Sondrio.

IEA. (2012). *World energy outlook 2012*. Paris: International Energy Agency.

ISTAT. (2011). Noi Italia. 100 statistiche per capire il paese in cui viviamo, ISTAT (Istituto Nazionale di Statistica).

Kramer, G. J., & Haigh, M. (2009). No quick switch to low-carbon energy. *Nature, 462*(7273), 568.

Markandya, A., Bigano, A., & Porchia, R. (2010). The social cost of electricity. Scenarios and policy implication.
Milligan, J., O'Riordan, T., & Watkinson, A. (2006). *Designing coastlines fit for the future*. English Nature Research Reports, No. 702.
Nilsson, C., & Berggren, K. (2000). Alterations of riparian ecosystems caused by river regulation. *BioScience, 50*(9), 783–792.
Poletti, C. (2009). Le prospettive del mercato dei certificati verdi. Gestore Mercato Elettrico (GME).
Provincia di Sondrio. (2008). Relazione tecnica, Bilancio Idrico della Provincia di Sondrio.
Regione Piemonte. (2003). Bacini Imbriferi Montani: Gestione dei sovracanoni provenientiagli impianti idroelettrici.
Repubblica Italiana. (2007). Legge n. 244 del 24/12/2007: Disposizioni per la formazione del bilancio annuale e pluriennale dello Stato (Legge finanziaria 2008).
Ringel, M. (2006). Fostering the use of renewable energies in the European Union: The race between feed-in tariffs and green certificates. *Renewable Energy, 31*(1), 1–17.
Rosenberg, D. M., Berkes, F., Bodaly, R. A., Hecky, R. E., Kelly, C. A., & Rudd, J. W. M. (1997). Large-scale impacts of hydroelectric development. *Environmental Review, 5*, 27–54.
Schiermeier, Q., Tollefson, J., Scully, T., Witze, A., & Morton, O. (2008). Electricity without carbon. *Nature, 454*(7206), 816–823.
Terna. (2010). Dati Statistici sull'energia elettrica in Italia 1997–2010.
Terna. (2011). *Dati statistici.* http://www.terna.it/default/Home/SISTEMA_ELETTRICO/statistiche/dati_statistici.aspx
Terna. (2013). *Dati statistici.* http://www.terna.it/default/Home/SISTEMA_ELETTRICO/statistiche/dati_statistici.aspx
Tollefson, J. (2011). Worth a dam? *Nature, 474*(7352), 430–430.
Vannote, R. L., Minshall, G. W., Cummins, K. W., Sedell, J. R., & Cushing, C. E. (1980). The river continuum concept. *Canadian Journal of Fisheries and Aquatic Sciences, 37*(1), 130–137.
Verbunt, M., Zwaaftink, M. G., & Gurtz, J. (2005). The hydrologic impact of land cover changes and hydropower stations in the Alpine Rhine basin. *Ecological Modelling, 187*(1), 71–84.
Vorosmarty, C. J., Meybeckc, M., Feketea, B., Sharmad, K., Greena, P., & Syvitski, J. P. M. (2003). Anthropogenic sediment retention: Major global impact from registered river impoundments. *Global and Planetary Change, 39*, 169–190.
Walter, R. C., & Merritts, D. J. (2008). Natural streams and the legacy of water-powered mills. *Science, 319*(5861), 299–304.
WEF. (2014). *Global risks 2014* (9th ed.). Geneva: World Economic Forum.

Chapter 13
Subsidies for Ecologically Friendly Hydropower Plants Through Favourable Electricity Remuneration in Germany

Verena Mattheiß

Abstract Whereas hydromorphological alterations represent one of the major ecological challenges for European river systems, very few economic instruments exist to mitigate their impacts. The German Renewable Energy Sources Act has established an innovative instrument for the hydropower sector. By guaranteeing higher remuneration for electricity produced by hydropower installations that comply with selected ecological requirements, it provides incentives for improving the morphological situation next to the plants.

The present case study describes the most important aspects of this economic policy instrument (EPI) and provides a critical evaluation, taking in particular environmental outcomes, economic effects and institutional aspects into account. It aims at being a useful source of information on this EPI which is so far not much discussed at international level but which constitutes nevertheless a very interesting example of how the promotion of renewable energy sources can be reconciled with nature conservation objectives as well as the requirements of the EU Water Framework Directive.

Keywords Hydropower • Hydromorphology • Renewable energy • Water Framework Directive

13.1 Introduction

In Germany, the Renewable Energy Sources Act (EEG) is since the year 2000 the main instrument to promote the use of renewable energy sources. It guarantees for electricity production a defined remuneration per kWh which is above free market prices. The present case study looks at the environmental preconditions for the eligibility of hydropower plants to increased tariffs which form part of the EEG since its amendment in 2004. The environmental measures required aim at substantially

V. Mattheiß (✉)
ACTeon, 5, Place Sainte Catherine, 68000 Colmar, France
e-mail: v.mattheiss@acteon-environment.eu

M. Lago et al. (eds.), *Use of Economic Instruments in Water Policy*,
Global Issues in Water Policy 14, DOI 10.1007/978-3-319-18287-2_13

improving the ecological status of water bodies next to hydropower plants, if not at reaching good ecological status, as asked by the Water Framework Directive (WFD). However, neither quantitative targets nor a time span for reaching these objectives have been set at the introduction of the instrument.

The case study has been chosen for several reasons. Hydromorphological pressures – including those originating from hydropower use – are an important barrier for reaching the good ecological status (or the good ecological potential) of water bodies in European countries. The favourable EEG remuneration for ecologically friendly hydropower plants in Germany is one of the rare economic policy instruments (EPIs) that have been developed to target those pressures, and not much documentation at international level is available so far. The case study provides a good example of how the promotion of renewable energy produced by hydropower can take nature conservation issues into account. Existing since 2004, the ecological requirements have been further specified in the EEG amendments which entered into force in 2009 (and 2012), following the regular reports of experiences on the implementation of the law. They concern, among others, the biological passability of the weirs and the provision of minimum water flow. The present analysis focuses on the EEG amendments from 2004 to 2009.

13.2 Setting the Scene: Challenges, Opportunities and EPIs

Germany's river water bodies are to a large extent subject to hydromorphological degradation. At present, only 10 % of the watercourses have a high or good ecological status (UBA 2010b). Next to uses like agriculture, navigation and flood protection this is due to hydropower use. Taking into account Germany's ambition to significantly increase the share of renewable energy in the future electricity production, an important challenge consisted and still consists in reconciling the extension of hydropower use and its impact on nature conservation needs (BGBI 2004; Naumann and Igel 2005; BMU 2010).

Water management in Germany today has to be seen against the background of the European Water Framework Directive (WFD). Reaching a good ecological potential (GEP) or a good ecological status (GES) includes both the need for structural changes (installation of fish ladders, smaller grill sizes) and modifications to operation (e.g., guaranteed flow rates during fish migration periods). Those changes are linked to profit losses for the operators of hydropower plants. As the plants are provided with very long concession periods of several decades (or even unlimited rights), reaching the GES soon will depend on the voluntary participation of operators as well as on effective incentives (UBA 2010a).

In this context, the EEG has been amended in July 2004 in order to provide economic incentives for hydropower plant operators to take ecological considerations into account. The conditions which have to be fulfilled include location bound requirements (e.g., the construction must take place next to already existing barrage weirs or dams). In its § 6, the EEG (2004) requires furthermore that either the water body which is affected by hydropower use must reach GES, or it has to be substantially improved compared to the previous status.

The latter can – for existing hydropower plants – be reached through a modernisation of the plant. As the decisive aspect lies in the improvement of the state of the water ecology and of the accompanying floodplain, also measures which are only targeting the ecology can be seen as a modernisation in the sense of the EEG (Naumann and Igel 2005). In the EEG amendment of 2009, the terms 'substantial improvement of the ecological status' are further defined by indicating that they need to refer to the following criteria (EEG 2009):

- Storage capacity and management,
- Biological passability,
- Minimum water flow,
- Solids management, or
- Bank structure,
- Or shallow water zones have to be established or abandoned channels or branches have to be connected, in so far as the measures in question are necessary individually or in combination, taking into account the relevant management goals, in order to achieve good ecological status.

The requirements depend on the capacity of the hydropower plants as well as on the year in which the permission to construct or to operate the plant has been obtained.

13.3 The Subsidies for Ecologically Friendly Hydropower Plants Through Favourable Electricity Remuneration in Action

The EEG and its ecological conditions for hydropower plants are applied all over Germany and are in theory relevant for all of the existing 7,500 hydroelectric power stations. Since its introduction in the year 2000, the EEG constitutes an important instrument for maintaining and extending hydropower production. This effect is untouched by the ecological provisions, which do not preclude the plants to be remunerated according to the EEG 2000 conditions. The focus of this case study lies, however, on the increased remuneration proposed after the establishment of ecological improvements on the plants.

13.3.1 The EPI Contribution

13.3.1.1 Environmental Outcomes

When looking at the environmental impact of the EEG favouring ecologically friendly hydropower plants, two different aspects are worth considering. In the first place, with a remuneration paid per kWh, the EEG provides incentives for investments in an extended production capacity or the construction of new hydropower plants. Thereby, the EPI has an effect in terms of reducing the emission of greenhouse gases by promoting the use of renewable energy sources. At the same time – and this will be the focus of the following considerations – the conditions defined by the EEG aim at improving the hydromorphological situation of water bodies next to existing hydropower plants by providing incentives for voluntary or early adaptation of the plant structure and/or operation.

The legislation on the sale of electricity to the grid (StrEG) from 1990, which has been replaced by the EEG in the year 2000, stimulated the operation of small hydropower plants (SHPs) and has prevented its impending decline (BMU 2010). In the year 2007, the predominant part of the electricity generated stemmed from big plants which were not remunerated according to the EEG (BMU 2008a). In terms of numbers of plants, from the 7,500 existing ones 6,925 have been remunerated according to the EEG in 2009.[1]

The amendments of the EEG in 2004 and 2009 which introduced the ecologically bound fees for hydropower plants have also successfully provided incentives for the construction or extension of plants with a capacity above 5 MW. This concerns for example the extension of the hydropower station in Albbruck-Dogern in 2009 (BMU 2010), or the new construction of the power station Rheinfelden (Energie-Chronik 2011).

Whereas in 2008 only about 100 plants have been modernised or new constructed, this was the case for more than 600 in 2009 (Dumont and Keuneke 2011), as operators waited for the more attractive remuneration conditions of the EEG 2009 to come into effect. It can be expected that the majority of new constructions of hydropower plants in 2009 took place on already existing hydropower sites. According to Dumont and Keuneke (2011), most of those works are probably modernisations, which have been classified as new constructions due to the high investments. In those cases, the EPI gave an incentive to accelerate the adaptation of the plants to recent regulations – which ask to comply with the WFD requirements – by making new approvals necessary.[2]

[1] Please note that this number includes both plants which fulfilled the ecological requirements, and plants which do not.

[2] Please note that a difference can be made regarding the incentive effect of the EPI. In cases where new hydropower plants are built to replace stations which were at the end of their concession period or at the end of their economic lifetime, the incentive effect of the EPI is primarily leading to increased electricity production, with environmental standards being fulfilled according to regulation. The same applies to the rare case of constructions on new sites.

The main pressure exerted by hydropower use is linked to its dam constructions (see for example Dumont 2005). According to Naumann (2011), the most relevant criteria for the removal of ecological deficits on hydropower plants are: establishing the biological passability upstream, ensuring a sufficient fish protection downstream, and providing the ecological minimum flow. Those measures form therefore rightly part of the EEG conditions foreseen for hydropower plants (see above). Umweltbundesamt (2012) tries to provide a comprehensive indication on the number and type of ecological measures applied to hydropower plants which have been induced by the EEG amendments. They indicate that about 10 % of the existing hydropower plants possess equipment which assists the upstream migration of fishes and/or provide minimum water flow conditions. Figure 13.1 shows the relative importance of the different measures, and indicates that a great part of them go back to the increased remuneration of the EEG. Examples of concrete improvements on existing hydropower plants are illustrated for example in UBA (2008).

Although improving the status of water bodies is a precondition for receiving the increased EEG remuneration, no study is available which investigates comprehensively

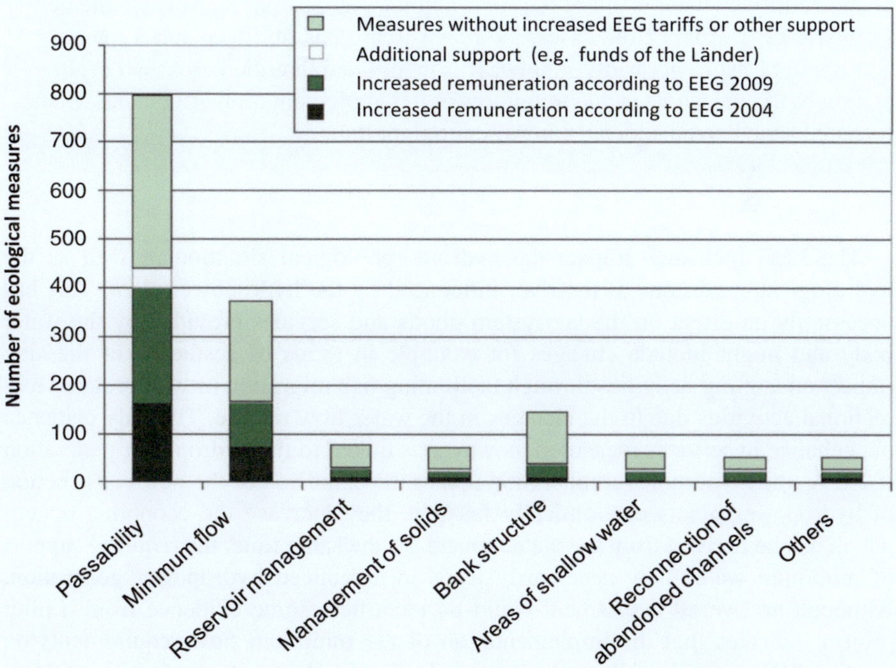

Fig. 13.1 Support to ecological measures on hydropower plants according to the seven measures of the EEG 2009 (Source: Umweltbundesamt (ed.) 2012, translated by the author. Note: Data is coming from a survey targeted to all German hydropower plant operators. The figures summarise the returned answers of 859 plants (15 % of the total))

the real ecological functionality of the measures applied (Naumann 2011).[3] Some reports indicate, however, that the actual status improvement is questionable in a significant number of cases (see illustration box below).

Evaluation of the Effectiveness of Ecological Measures for Selected Hydropower Plant Installations

Anderer et al. (2012) examined exemplarily 16 hydropower plants for the criteria upstream passability and minimum flow which were admitted to the increased remuneration according to the EEG. Only four of the investigated plants had actually reached good status, with regards to the two criteria examined.

Another study (BfN 2009 *unpublished*, cited in Dumont and Keuneke 2011) selected ten hydropower installations which received a higher remuneration according to the EEG 2004. They found out that in four cases the biological upstream passability was not given, although the establishment of fish passes has been the modernisation measure which led to the increased remuneration in two of the cases. In the remaining six cases, the upstream passability was either moderately or considerably limited. The downstream passability is given in six of the sites, amongst others due to the implemented measures. In two of the cases, the downstream passability is interrupted, or migrating fishes get badly injured. It is mentioned that the partly bad evaluation of the measures is due to failures in the implementation of details, which could have been avoided through better planning.

The EEG measures impact the hydromorphological situation as well as the hydrological conditions of the river influenced by the hydropower plant. This has necessarily an effect on the ecosystem goods and services provided by the water body and might include changes for example in terms of aesthetics of the site, impact on angling activities through facilitating fish migration or water related recreational activities due to the changes in the water flow regime. The only evidence on a change in services measured, however, is linked to the hydropower generation itself. As the favourable remuneration makes the extension or the new construction of hydropower plants economically feasible, they increase the economic benefit which can be derived from the water course. At the same time, the required support of minimum water flow necessarily leads to a reduced hydropower generation. Although no overall assessment could be identified, some evidence from a pilot project indicates that the implementation of the minimum flow requirements on existing SHPs would lead to an average reduction of electricity production of 25 % (Knödler and Wotke 2009).

---

[3] Naumann, S. German Federal Environment Agency, author of the operational guideline for the EPI implementation; telephone interview in November 2011.

13.3.1.2 Economic Outcomes

The present EPI has the advantage of favouring both ecologically friendly practices and of providing incentives to increase hydropower production activities. However, no cost-effectiveness-analysis has been carried out to compare the chosen EPI to alternatives. It is in the responsibility of the local authorities to ensure that the investments corresponding to the ecological improvement measures are reasonable with regards to the additional receipts provided by the EEG (Naumann and Igel 2005; Dumont 2005). The amendments made on the EEG are furthermore not linked to negative effects for hydropower operators compared to the previous situation, as the status quo is conserved for plants which received remuneration from the EEG before. The different remuneration rates according to the size of the hydropower plants take into account that smaller plants have to deal with proportionally higher costs for their efforts to comply with ecological minimum standards. This has an impact on the economic efficiency of the SHPs, which is often characterized by higher electricity production costs (Nitsch et al. 2004).

Once the ecological improvement of the plants approved and the higher remuneration accorded, it is guaranteed for a period of 30 years for plants up to 5 MW according to the EEG amendment of 2004 (20 years since the EEG 2009). The EPI provides hence for investment security, taking the importance of the investments and the long depreciation periods of the hydropower stations into account. It would have been difficult to provide this risk reduction for the operators through another instrument, like e.g., certification schemes.

As the higher electricity remuneration is paid by the electricity consumers, costs are recovered from the users. In fact, the financing of the ecological measures for hydropower plants can be seen as a way of internalizing the external costs of the plants. Cost recovery is possible in a much more direct way through the electricity tariffs as it would have been the case for example through state subsidies to investments – which are indirectly paid by the tax payer.

An essential factor for the success of the instrument is given by the comparison between the surplus provided by the EEG and the corresponding investments (Naumann and Igel 2005). Table 13.1 shows the electricity remuneration for hydropower plants given in the EEG amendments of 2004 and 2009. In 2004, operators which were already remunerated according to the EEG 2000 with 7.67 ct/kWh had the possibility to receive 9.67 ct/kWh if they fulfilled the ecological criteria. In order to be economically viable, necessary investments needed hence to be refundable by the difference of 2 ct/kWh. The scope for investments is all the more restricted the smaller the hydropower plant capacity is, as less electricity is generated and hence less remuneration received (Dumont 2005).

The comparison of the remuneration levels given in the table above shows that the amendment of the EEG from 2009 significantly raised the tariff rates for hydropower plants up to 5 MW. This provides higher incentives for financing ecological

Table 13.1 Remuneration rates of hydropower plants according to the EEG in 2004 and its amendment in 2009 in EUR cents/kWh

Plants up to and including 5 MW – new plants; share in production capacity	EEG 2004	EEG 2009
Up to 500 kW	9.67	12.67
500 kW to 2 MW	6.65	8.65
2–5 MW	6.65	7.65
Plants up to and including 5 MW – modernised, revitalised plants; share in production capacity	EEG 2004	EEG 2009
Up to 500 kW	9.67	11.67
500 kW to 2 MW	6.65	8.65
2–5 MW	6.65	8.65
Modernisation of plants over 5 MW – increase of capacity	EEG 2004[a]	EEG 2009
Up to 500 kW	7.29	7.29
Up to 10 MW	6.32	6.32
Up to 20 MW	5.80	5.80
Up to 50 MW	4.34	4.34
Over 50 MW	3.50	4.34

Source: Knödler and Wotke 2009; BMU 2008c, adapted

[a]The remuneration for plants with a production capacity of over 5 MW depends on the year in which it started operation. The tariffs given here are applicable for plants which started operation in 2009 (BMU 2004)

Table 13.2 Specific costs for the modernisation of hydropower plants up to and including 5 MW

Installed capacity	Specific measure costs ct/kWh	Average increase in remuneration ct/kWh
100 kW	3.94–5.73	4.00
500 kW	2.06–2.57	4.00
1 MW	1.59–2.23	3.75–4.00
2 MW	1.15–1.58	2.83–3.02
5 MW	0.83–1.13	2.32–2.39

Source: Dumont and Keuneke 2011, translated by the author

improvement measures (BMU 2008b), but has also to be seen against the background that the guaranteed remuneration period has been reduced at the same time from 30 to 20 years (Knödler and Wotke 2009). Dumont and Keuneke (2011) calculated specific ecological modernisation costs and compared them to the average increase in remuneration according to the (EEG 2009) (as compared to the remuneration of the EEG in 2000). As shown in Table 13.2, the remuneration level is in particular not high enough to cover investments for SHPs up to an installed production of 100 kW. However, existing SHPs, which are not yet modernised in accordance with the provisions of the EEG, have often significant ecological deficits (Knödler and Wotke 2009; Deutsche Umwelthilfe 2006).

13.3.1.3 Distributional Effects and Social Equity

When looking at the wider impacts of the EPI, the main stakeholder groups affected are electricity consumers and hydropower plant operators.

As electricity produced with renewable sources and remunerated according to the EEG is in average more costly than electricity stemming from fossil or nuclear sources, electricity consumers see their material living standard affected by the EPI through the higher prices they have to pay per kWh (Bundeskabinett 2002; BMU 2011). The EEG apportionment in the electricity price increased from 1.1 ct/kWh in 2008 (and representing about 5 % of the total price per kWh) (Kluge 2009) to 3.5 ct/kWh in 2011 (EEG/KWK-G 2011). As a general rule, entities which are high energy consumers are more affected by the higher prices. Several energy intensive industries, however, are exempted and allowed for lower prices (BMU 2007, 2011). The EEG related extra costs are going back to expenses for all types of renewable energy sources, the higher costs due to the ecological requirements for hydropower plants only represent a very small share.

With regards to the hydropower plant operators, they will only choose to meet the ecological requirements when the EEG remuneration is linked to financial gains in the long-term – given that the EEG is based on voluntary participation. An increase in material living standards due to increased revenues can be expected. Uphoff (2011)[4] notes, however, that this is hardly the case for SHPs, as the surplus provided by the EEG is offset completely by the investments.

The local community living next to the installations is potentially also concerned, but more investigations are necessary. Bouscasse et al. (2010) shows that the wider population of a hydrographic basin can obtain environmental benefits from hydromorphological improvements on rivers which enhance the development of fish populations.

13.3.2 The EPI Setting Up

13.3.2.1 Institutions

The most embedded institutions relevant for the EPI are given by the existing hydropower plants (Lehr et al. 2011). The plants are endowed with very long concession periods (up to 100 years or unlimited in case of "old rights"; Naumann 2011) or even unconditioned user rights (Bunge et al. 2001).[5] Concessions for about half of the installed capacity are expiring in the next twenty years (Umweltbundesamt 2012). Also the high share of hydropower plants which are considered as small (7,100 out of 7,500) form a relevant part of the embedded insti-

[4] Uphoff, H. Leading office manager of the German Federal Association of Hydropower Plants (BDW); telephone interview in November 2011.

[5] The current legislation foresees concession periods which are in general not longer than 30 years (Anderer et al. 2012).

tutions. SHPs are particularly relevant from an ecological point of view as they are often situated on less modified rivers (Clearingstelle 2011).

The very important basis of the EPI is given by the EEG, which has been adopted in the year 2000. It is an important piece of the German strategy to expand the use of renewable energy sources (Bundeskabinett 2002; BMU 2007).

The second important policy which triggered essentially the elaboration of the ecologically bound remuneration system for hydropower plants is the WFD, to which the requirements set for hydropower plants are directly referring. The Directive has been translated into the German legislation through the Federal Water Resources Law (WHG, last complete amendment in 2009). The latter provides also – together with the water laws of the German Länder – the basis for the approval procedure for hydropower plants – which include the approval of the ecological measures according to the EEG (Umweltbundesamt 2012).

In the design phase of the EPI, existing institutions played a preeminent role. While little evidence is available in the literature, interviews revealed that the essential initiative for the instrument in its current design is going back to discussions between different parts of the Bundestag at that time which were either supporting the extension of hydropower use for the production of renewable energies or advocating nature conservation (Uphoff 2011; Naumann 2011). The resulting compromise was to provide hydropower operators with higher remuneration rates, while requiring efforts to increase their environmental sustainability, as it can be found in the EEG amendments since 2004. With the long concession periods providing legal security to the hydropower operators, voluntary incentives as given by the EEG seemed most appropriate to change the environmental conditions at a relatively short notice.[6]

The tradition of very long concession periods has a strong influence on the operational phase of the EPI. Once the ecological measures implemented and approved, the eligibility of the hydropower plants to the increased EEG remuneration is guaranteed and no control of their functionality is taking place afterwards. For hydropower plants with unconditioned user rights, some reluctance can be observed to touch upon them, limiting the scope of implementation of the EEG (Clearingstelle 2011).

13.3.2.2 Transaction Costs

Breitschopf et al. (2010) indicate that transaction costs for operators of electricity production plants are not relevant, as those costs are considered by the operators in their reflections on whether they will follow the ecological requirements of the EEG or not. They are hence internalised – because refinanced by the remuneration. The only cost component identified is linked to the proof that the good ecological status has been reached for big hydropower plants according to § 6 EEG, which has to be

[6] Please note that new concessions are obligatorily in line with the WFD as they have to comply with the current legal provisions given by the WHG.

given by the distribution system operators. It is estimated to be EUR 20 per demand (Breitschopf et al. 2010).

The EEG amendment of 2004 was based on the recommendations of the report of experiences elaborated for the former EEG 2000 (Bundeskabinett 2002). Different research projects have also been carried out, dealing with the ecologically optimal extension of the use of renewable energy sources (Nitsch et al. 2004), the area of conflict between biodiversity conservation and climate change in particular for SHPs (Dumont, *unpublished*, mentioned in Ammermann 2011) or the ecological effectiveness of measures induced by the EEG (Umweltbundesamt 2012).

Concerning the implementation process of the EPI, it is crucial to emphasise that it forms part of the provisions for the remuneration of all renewable energy sources in Germany, which are regrouped in one legal text (the EEG). Furthermore, the ecological conditions for hydropower plants have not been introduced together with the remunerations of the plants, but they have been added to the existing system. Hence, not the remuneration per se is of interest (regarding transaction costs), but only its obligatory link to the ecological requirements.

With regards to monitoring and enforcement costs, the control of whether the ecological improvements have been carried out is done by the competent water agency as part of the approval procedure of hydropower plants according to water law (Knödler and Wotke 2009). No additional controls are foreseen afterwards.

13.3.2.3 Implementation

The EPI design provides for some flexibility in its implementation. On the local level, the decision whether the conditions for being eligible to the EEG remuneration system are fulfilled, lies in the responsibility of the competent authority. They can consider local particularities, as well as economic reasonableness of the corresponding investments – compared to the additional receipts through the increased EEG remuneration (Naumann and Igel 2005).

A second level of flexibility is linked to the EPI design itself. The EEG foresees regular reports of experiences every 4 years which include recommendations for further amendments (e.g., Bundesregierung 2011). After the introduction of the ecological requirements for hydropower plants in 2004, an amendment was adopted in 2009, which concerns in particular the amount of remuneration provided and the duration of the guaranteed remuneration. Other amendments entered into force in January 2012 and in August 2014.

During the development process of the EPI, a public hearing took place before each EEG amendment (see for example Deutscher Bundestag 2008). Furthermore, technical experts, including hydropower representatives, had been directly consulted in the forefront (Uphoff 2011). During the implementation process of the EPI, no specific importance of public participation could be noted.

An important element of the EPI which supports its implementation is an operational guideline which has been developed with the support of several stakeholder groups, and which aims at ensuring a nationwide consistent and transparent implementation (Naumann and Igel 2005; see also BMU 2008b).

Table 13.3 Interactions with different EU policies

EPI-objective	Improving the ecological status of water bodies next to hydropower plants by improving the hydromorphological situation	
Other sectoral policies	Objectives of the sectoral policy	Synergies
Water Framework Directive	Reaching good ecological status for all water bodies	+++
		Supported the establishment of the EPI; provides a broader legal background
EU energy policy	Promoting the use of renewable energy sources	++
		Promoting renewable energy sources is an essential reason for the existing design details
EU nature conservation policy	Ensuring biodiversity through the conservation of natural habitats and of wild fauna and flora	+
		No direct interaction. But nature conservation aspects play a role in the selection of the ecological criteria

Source: Elaborated by the author
+ represents a positive synergy between the objectives of the EPI and the other policy; three levels: + (low positive interaction), ++ (medium), +++ (high positive interaction)

When putting the EPI into the context of relevant sectoral policies, important synergies can be identified (see Table 13.3). They are focussing on the impact the policies had on the implementation and operation of the EPI.

13.4 Conclusions

Providing subsidies for ecological improvements on hydropower plants in the form of higher electricity remuneration is an interesting EPI. It represents a smart solution which manages to reconcile the political will to promote renewable energy sources with nature conservation objectives as well as requirements set by the EU Water Framework Directive. The EPI takes furthermore the specificities of the hydropower sector into account: despite their long lasting concession rights which provide them with legal security, the ecological improvement process of the plants gets accelerated through economic incentives, which at the same time provide for planning and investment security. Introducing the EPI through the amendment of an existing law and ensuring its implementation through the existing system – including the remuneration procedure, reporting rules etc. – significantly helped to keep the transaction costs of the EPI low. At the same time, the possibility of continued law amendments allows for flexibility to improve the EPI and to adapt it to the current state of knowledge. The instrument is furthermore designed as a cost recovery mechanism without imposing disproportionally high costs to consumers.

Nevertheless, in particular the environmental evaluation of the EPI is linked to important uncertainty. Neither quantified targets in terms of a number of hydropower plants which should comply with the requirements of the EEG were set at the beginning, nor has there been a time horizon set in which the measures should be applied (Naumann 2011). Measuring its success is furthermore complicated by a lack of knowledge on the number of plants having implemented ecological improvement measures, and in particular also the lack of information about the actual ecological effectiveness of the measures. Controlling this effectiveness would be an important element to be improved.

Another limit is recognised with regards to the economic incentive effect for small hydropower plants. The remuneration level is not high enough to provide sufficient incentives for the ecological modernization of most of them. According to the policy makers, the level that would be required is not justifiable from a political point of view. Other solutions have hence to be found to promote the ecological improvements next to small hydropower plants.

A different potential adaptation of the EPI, which is subject to discussions, is to loosen the direct link between the eligibility to increased remuneration through ecological investments to a specific hydropower plant (Naumann 2011). In its current form, the EEG promotes ecological improvements where they are economically feasible (which concerns mainly bigger plants) and not where they would be most ecologically effective. One idea is to redistribute money by means of a fund (Naumann 2011). This would entail, however, a significant change in the structure of the present EPI.

In summary, the EPI can be considered as being successful. Even if not all measures had a positive effect or even if part of the works would have been done also without the EEG incentives, a positive net effect of the instrument is uncontested.

References

Ammermann, K. (2011). Ökologische Verbesserung bei der Modernisierung von Wasserkraftanlagen. Presentation held the 13 May 2011 on the 8th expert meeting of the Clearingstelle on hydropower. http://www.clearingstelle-eeg.de/files/01-Vortrag_Ammermann.pdf. Accessed 24 Oct 2011.

Anderer, P., Dumont, U., Massmann, E., & Keuneke, R. (2012). Wasserkraftnutzung in Deutschland: Wasserrechtliche Aspekte, ökologisches Modernisierungspotenzial und Fördermöglichkeiten. Umweltbundesamt (ed.), Texte 22/2012, Dessau-Roßlau and Aachen. http://www.umweltbundesamt.de/publikationen/wasserkraftnutzung-in-deutschland. Accessed 8 Sept 2014.

BfN (Bundesamt für Naturschutz). (2009). Aus- und Neubau der kleinen Wasserkraft im Spannungsfeld von Biodiversitätsschutz und Klimawandel. Leipzig (Unpublished).

BGBl. (2004). Konsolidierte Fassung der Begründung zu dem Gesetz für den Vorrang Erneuerbarer Energien (Erneuerbare-Energien-Gesetz – EEG) vom 21. Juli 2004. BGBl. S.1918. http://www.bmu.de/files/erneuerbare_energien/application/pdf/eeg_begruendung.pdf. Accessed 29 Sept 2011.

BMU. (2004). Mindestvergütungssätze nach dem neuen Erneuerbare-Energien-Gesetz (EEG) vom 21. Juli 2004. http://www.erneuerbareenergien.de/files/pdfs/allgemein/application/pdf/verguetungssaetze_nach_eeg.pdf. Accessed 14 Nov 2011.

BMU. (2007). Erfahrungsbericht 2007 zum Erneuerbare-Energien-Gesetz (EEG-Erfahrungsbericht). http://www.bmu.de/files/pdfs/allgemein/application/pdf/erfahrungsbericht_eeg_2007.pdf. Accessed 3 Oct 2011.

BMU. (2008a). Erneuerbare Energien 2007 – Aktueller Sachstand – Juni 2008. Bonn. http://www.bmu.de/files/pdfs/allgemein/application/pdf/sachstand_juni08.pdf. Accessed 26 Sept 2011.

BMU. (2008b). Die Nutzung der kleinen Wasserkraft in Deutschland im Spannungsfeld von Klima-, Natur- und Gewässerschutz. http://www.erneuerbare-energien.de/files/pdfs/allgemein/application/pdf/nutzung_wasserkraft.pdf. Accessed 29 Sept 2011.

BMU. (2008c). Vergleich der EEG-Vergütungsregelungen für 2009. http://www.solarsys.at/web/download/solarsys_Foerderungsrichtlinien_fuer_Deutschland.pdf. Accessed 11 Oct 2011.

BMU. (2010). Erneuerbare Energien in Deutschland – Aktueller Sachstand für das Jahr 2009. Berlin. http://www.erneuerbare-energien.de/files/pdfs/allgemein/application/pdf/ee_sachstand_bf.pdf. Accessed 26 Sept 2011.

BMU (Bundesministerium für Umwelt, Naturschutz und Reaktorsicherheit). (2011). Erneuerbare Energien in Zahlen. Nationale und internationale Entwicklung. Berlin. http://www.erneuerbare-energien.de/files/pdfs/allgemein/application/pdf/broschuere_ee_zahlen_bf.pdf. Accessed 4 Nov 2011.

Bouscasse, H., Defrance, P., Strosser, P., Rozan, A., & Spitzenstetter, F. (2010). Evaluation économique de l'amélioration de la richesse piscicole dans le bassin du Rhin supérieur franco-allemand. Report for EDF elaborated by ACTeon & ENGEES. Unpublished.

Breitschopf, B., Klobasa, M., Sensfuß, F., Steinbach, J., Ragwitz, M., Lehr, U., Horst, J., Leprich, U., Diekmann, J., Braun, F., & Horn, M. (2010). Einzel- und gesamtwirtschaftliche Analyse von Kosten- und Nutzenwirkungen des Ausbaus Erneuerbarer Energien im deutschen Strom- und Wärmemarkt. Bestandsaufnahme und Bewertung vorliegender Ansätze zur Quantifizierung der Kosten-Nutzen-Wirkungen des Ausbaus Erneuerbarer Energien im Strom- und Wärmebereich – Arbeitspaket 1. ISI, GWS, IZES and DIW Berlin, commissioned by the BMU, Berlin. http://www.erneuerbare-energien.de/files/pdfs/allgemein/application/pdf/endbericht_ausbau_ee_2009.pdf. Accessed 8 Oct 2011.

Bundeskabinett. (2002). Bericht über den Stand der Markteinführung und der Kostenentwicklung von Anlagen zur Erzeugung von Strom aus erneuerbaren Energien (Erfahrungsbericht zum EEG). http://www.erneuerbare-energien.de/files/pdfs/allgemein/application/pdf/eeg_erfahrungsbericht.pdf. Accessed 3 Oct 2011.

Bundesregierung. (2011). Erfahrungsbericht 2011 zum Erneuerbaren-Energien-Gesetz (EEG-Erfahrungsbericht). Adopted by the Bundeskabinett the 6 Jun 2011. http://www.erneuerbare-energien.de/files/pdfs/allgemein/application/pdf/eeg_erfahrungsbericht_2011_bf.pdf. Accessed 10 Nov 2011.

Bunge, T., Dirbach, D., Dreher, B., Fritz, K., Lell, O., Rechenberg, B., Schmitz, E., Schwermer, S., Steinhauer, M., Steudte, C., Voigt, T. (2001). Wasserkraftanlagen als erneuerbare Energiequelle – rechtliche und ökologische Aspekte. Umweltbundesamt (ed.), UBA-Texte 01/01, ISSN 0722-186X, Berlin. http://www.umweltbundesamt.de/wasser/veroeffentlich/Wasserkraftanlagen.pdf. Accessed 5 Oct 2011.

Clearingstelle EEG. (2011). Wasserkraft – Tagungsprotokoll. Minutes from the 8th expert meeting of the EEG Clearingstelle in Berlin, 13 of May 2011. http://www.clearingstelle-eeg.de/files/Tagungsbericht_0.pdf. Accessed 11 Oct 2011.

Deutsche Umwelthilfe. (2006). Lebendige Flüsse & Kleine Wasserkraft – Konflikt ohne Lösung?. Berlin. http://www.duh.de/uploads/media/Kleine_Wasserkraft_klein.pdf. Accessed 8 Nov 2011.

Deutscher Bundestag. (2008). 64. Sitzung (05.05.2008) – Entwurf eines Gesetzes zur Neuregelung des Rechts der Erneuerbaren Energien im Strombereich und zur Änderung damit zusammenhängender Vorschriften, Ausschuss für Umwelt, Naturschutz und Reaktorsicherheit. http://webarchiv.bundestag.de/archive/2008/0506/ausschuesse/a16/anhoerungen/64_sitzung_05_05_2008/index.html. Accessed 31 Oct 2011.

Dumont, U. (2005). Entwicklung eines beispielhaften bundeseinheitlichen Genehmigungsverfahrens für den wasserrechtlichen Vollzug mit Anwendungsbeispielen im Hinblick auf die Novellierung des EEG – Gutachten für das Umweltbundesamt. Ingenieursbüro Floecksmühle, Aachen. http://www.floecksmuehle.com/img/c6ea762a3558aa18.pdf. Accessed 11 Oct 2011.

Dumont, U., & Keuneke, R. (2011). Vorbereitung und Begleitung der Erstellung des Erfahrungsberichtes 2011 gemäß § 65 EEG – Vorhaben IId Spartenspezifisches Vorhaben Wasserkraft. Final report, Ingenieursbüro Floecksmühle, FICHTNER, Forschungsstelle

UmweltEnergieRecht e.V. & IHS. http://www.bmu.de/files/pdfs/allgemein/application/pdf/eeg_eb_2011_wasserkraft_bf.pdf. Accessed 28 Nov 2011.

EEG. (2004). Act on granting priority to renewable energy sources (Renewable Energy Sources Act). German version published in the Federal Law Gazette (Bundesgesetzblatt) 2004 I No. 40, 1918. http://www.bmu.de/files/english/pdf/application/pdf/eeg_en.pdf. Accessed 28 Nov 2011.

EEG. (2009). Act on granting priority to renewable energy sources (Renewable Energy Sources Act, EEG), Renewable Energy Sources Act of 25 October 2008 (Federal Law Gazette I p. 2074) as last amended by the Act of 11 August 2010 (Federal Law Gazette I p. 1170). http://www.bmu.de/english/renewable_energy/downloads/doc/42934.php. Accessed 28 Nov 2011.

EEG/KWK-G. (2011). EEG-Umlage 2011. Press release. http://www.eeg-kwk.net/cps/rde/xchg/eeg_kwk/hs.xsl/484.htm. Accessed 31 Oct 2011.

Energie-Chronik. (2011). Neues Wasserkraftwerk Rheinfelden leistet bis zu 100 MW. Reference 110914. http://www.energie-chronik.de/110914.htm. Accessed 7 Nov 2011.

Kluge, D. (2009). EEG-Umlage steigt 2010 an/Strompreisentwicklung zeigt: Erneuerbare Energien sind keine Kostentreiber. BEE Bundesverband Erneuerbare Energie e.V.. http://www.bee-ev.de/3:333/Meldungen/2009/EEG-Umlage-steigt-Erneuerbare-Energien-sind-keine-Preistreiber.html. Accessed 31 Oct 2011.

Knödler, M., & Wotke, A. (2009). Ökologische Verbesserungen an kleinen Wasserkraftanlagen. Erfahrungen aus einem Pilotprojekt im Einzugsgebiet des Neckars, Büro am Fluss e.V. and Deutsche Umwelthilfe e.V. (eds.), Berlin. http://www.lebendiger-neckar.de/fileadmin/lebendiger-neckar.de/download/EEG/Wasserkraft_Broschure_klein.pdf. Accessed 30 Sept 2011.

Lehr, U., Lutz, C., Edler, D., O'Sullivan, M., Nienhaus, K., Nitsch, J., Breitschopf, B., Bickel, P., & Ottmüller, M. (2011). Kurz- und langfristige Auswirkungen des Ausbaus der erneuerbaren Energien auf den deutschen Arbeitsmarkt. http://www.erneuerbare-energien.de/files/pdfs/allgemein/application/pdf/ee_arbeitsmarkt_bf.pdf. Accessed 4 Nov 2011.

Naumann, S. (2011). Vergütungsvoraussetzungen für Wasserkraftanlagen aus wasserwirtschaftlicher Sicht. Presentation held on the 8th expert meeting of the EEG Clearingstelle, Berlin 13th of May 2011. http://www.clearingstelle-eeg.de/files/04-Vortrag_Naumann.pdf. Accessed 2 Nov 2011.

Naumann, S., & Igel, F. (2005). Leitfaden für die Vergütung von Strom aus Wasserkraft. BMU (eds.), Berlin. http://www.thueringen.de/imperia/md/content/tlvwa2/604/wasserkraft.pdf. Accessed 11 Sept 2011.

Nitsch, J., Krewitt, W., Nast, M., Viebahn, P., Gärtner, S., Pehnt, M., Reinhardt, R., Schmidt, R., Uihlein, A., Scheurlen, K., Barthel, C., Fischedick, M., & Merten, F. (2004). Ökologisch optimierter Ausbau der Nutzung erneuerbarer Energien in Deutschland. Forschungsvorhaben im Auftrag des Bundesministeriums für Umwelt, Naturschutz und Reaktorsicherheit, FKZ 901 41 803, Stuttgart, Heidelberg, Wuppertal. http://www.ifeu.de/landwirtschaft/pdf/Oekologisch_optimierter_Ausbau_Langfassung.pdf. Accessed 12 Sept 2011.

UBA (Ed.). (2008). Wasserrahmenrichtlinie und Hydromorphologie. Dokumente der gemeinsamen Umsetzungsstrategie der EU-Staaten (CIS). Texte 17/08, ISSN 1862-4804, Dessau-Roßlau. http://www.umweltdaten.de/publikationen/fpdf-l/3465.pdf. Accessed 10 Oct 2011.

UBA (German Federal Environment Agency). (2010a). Water resource management in Germany – part 1 fundamentals. Dessau-Roßlau. http://www.umweltdaten.de/publikationen/fpdf-l/3770.pdf. Accessed 11 Sept 2011.

UBA. (2010b). Water resource management in Germany – part 2 water quality. Dessau-Roßlau. http://www.umweltdaten.de/publikationen/fpdf-l/3771.pdf. Accessed 11 Sept 2011.

Umweltbundesamt (Ed.) (UBA). (2012). Wasserkraftnutzung in Deutschland: Wasserrechtliche Aspekte, ökologisches Modernisierungspotenzial und Fördermöglichkeiten. In UBA-Texte 22/2012, Dessau-Roßlau. http://www.uba.de/uba-info-medien-e/4287.html. Accessed 17 Nov 2014.

Uphoff, H. Leading office manager of the German Federal Association of Hydropower Plants (BDW); telephone interview in November 2011.

Chapter 14
Water Trading: An Introduction

Gonzalo Delacámara and Carlos M. Gómez

Abstract Rather than setting water prices and leaving quantities to economic agents, water authorities may rather choose to cap water quantity and set the necessary conditions for voluntary trades to happen. From a wider perspective of water use (one not only constrained to water withdrawal and consumption but also to the disposal of polluting substances), water rights or entitlements could also be defined as pollution credits and be traded in water quality trading (WQT) schemes. This chapter presents a wide array of experiences both on water quantity and water quality trading. A successful experience on nutrient credit trading in the Great Miami River (Ohio, USA) is presented along with a non-fully successful one in North Carolina, from which insightful lessons can be drawn in terms of optimising the incentive design. Furthermore, a salinity offsetting scheme in Australia is also analysed. In terms of water quantity trading, incipient experiences in central Spain (Tagus river basin district) are analysed together with mature and dynamic experiences of deep markets in Chile, the Murray-Darling Basin (Australia) and Colorado (USA).

Keywords Water rights • Water markets • Water trading • Water quality trading • Offset schemes

G. Delacámara (✉)
Madrid Institute for Advanced Studies in Water (IMDEA-Water Foundation),
Punto Net 4, floor 1, Edificio ZYE, Parque Científico-Tecnológico de la Universidad de Alcalá, 28805 Alcalá de Henares (Madrid), Spain
e-mail: gonzalo.delacamara@imdea.org

C.M. Gómez
Departamento de Economía, Universidad de Alcalá and IMDEA-Water Foundation,
c/ Plaza de la Victoria s/n, 28803 Alcalá de Henares (Madrid), Spain
e-mail: mario.gomez@uah.es

M. Lago et al. (eds.), *Use of Economic Instruments in Water Policy*,
Global Issues in Water Policy 14, DOI 10.1007/978-3-319-18287-2_14

14.1 The Role of Markets and Trading in Water Policy

As an economic policy instrument, water (use right) trading entails a voluntary transfer of a quantifiable water allocation, either to be withdrawn or polluted, between a buyer and a seller (Hodgson 2006; Hanemann 2014; Shortle 2013). These two parties enter into a transfer agreement only if and when it is in each party's interest. Water trading is an adaptive management instrument in the sense that, unlike regulation and mandates, it is a flexible economic incentive to fit new and emerging water uses over time (Rosegrant et al. 2014). Further, it is a de-centralized mechanism in the sense that users themselves make decisions on water use so that local conditions and ad-hoc needs are accommodated (Garrick et al. 2013; Colby et al. 2014; Young 2014).

Water trading schemes, as a response to water scarcity and drought risk (Debaerea et al. 2014) have been pervasive in the recent economic literature, even if such schemes are not widespread in the world (let alone in Europe).

Major experiences in water quantity trading are necessarily a driver for research in this area. As a result of that, the Murray-Darling Basin in Australia can be said to be a lab for water trading and steadily yields peer-reviewed articles and other academic work in this research area (see, for example, Docker and Robinson 2014; Grafton and Horne 2014; Grafton et al. 2014; Grafton 2010; Kirby et al. 2014; Loch et al. 2014; Wheeler et al. 2014a, b). Something similar happens in Chile (Wagnitz et al. 2014; Hearne and Donoso 2014; Donoso et al. 2014) or the USA western states (Howitt 2014; Ghosh et al. 2014; Goemans and Prichett 2014).

Besides, literature on water trading is quite diverse. Water trading is perceived as a contribution to water security (via supply reliability) (Colby et al. 2014) and a fertile space to reflect on institutional reforms, water policy design, and transaction cost analysis (McCann and Garrick 2014; Erfani et al. 2014), but also as an economic instrument to tackle water quality concerns (Keller et al. 2014). There is also a wealth of references, from a microeconomic perspective, on farmers' decisions and exposure to risk (Loch et al. op. cit.; Wheeler et al. 2014b; Zuo et al. 2015; Lafreniere et al. 2015) or the effects of alternative irrigation institutions (Ghimire and Griffin 2014). Yet, whilst many authors focus on the economic instrument itself (water trading), others rather explore their different delivery mechanisms (types of trades): see, for instance, Howitt (2014) and Broadbent et al. (2014) on lease contracts or Hansen et al. (2014) on valuing options.

Over the last few years, a number of cross-country analyses on water market activities have been published, always biased towards institutional issues (Hadjigeorgalis 2009; Grafton et al. 2010, 2011). As above, most of the relevant experiences are found in mature markets, such as those in the Murray-Darling Basin (Australia), northern Chile, and the semi-arid states of the western US. Just minor experiences can be found in water markets, to a different extent, in countries such as China, India, South Africa or Spain.

In terms of water quality trading, most experiences can still be found in Australia, the USA, Canada, and New Zealand (Shortle 2013; Greenhalgh and Selman 2012; Keudel 2007).

As Delacámara et al. (2015) point out, water quantity trading in Europe is only in its embrionary state, despite the emphasis of the EU Blueprint to Safeguard Europe's Waters [COM (2012) 673] highlighting the policy interest of water trading as a means to tackle water scarcity and drought risk. Experiences are mostly restricted to some Mediterranean catchments in Spain (Kahil et al. 2014; Garrido et al. 2012; Gómez et al. 2013) and also to somewhat bounded upstream markets in England and Wales (OFWAT 2010; Mitchell and McDonald 2015). In France and, to a lesser extent, in Italy – the latter not yet being supported by national legislation – the status could be described as expectant or, at best, as exploratory.

As per water quality trading, Europe offers "much ado about nothing" or, to put it in a different and more positive way: a huge number of opportunities and not too many facts to date. Wind (2012) when developing an overview, found experiences in Sweden (based on Collentine 2006), Finland (Lankoski et al. 2008), the Baltic Sea (Hautakangas and Ollikainen 2011), Belgium (Klooster et al. 2007), or the Netherlands (Oosterhius and Peeters 2014) All those experiences, though, could be arguably said to be at an experimental stage (i.e. simulations, etc.).

14.2 Water Trading Experiences

The reader will find in this part of the book the following experiences both on water quality trading (Ohio and North Carolina, USA), salinity offset schemes (Australia), and water quantity trading in the Tagus watershed (Spain), Chile, the Murray-Darling Basin (Australia), and Colorado (USA).

In Kieser and McCarthy (Chap. 15), a nutrient credit trading scheme is presented. Nutrient credits were traded between five wastewater treatment plants (WWTPs) and hundreds of diffuse pollution sources (farms) in the Great Miami River, a tributary of the Ohio River (USA). An interesting institutional setup, whereby a watershed-based flood control agency managed a water quality trading (WQT) programme, led to a cost-effective option for WWTP compliance. The WQT scheme includes a specific incentive design (i.e. a reverse auction for securing lowest-cost credit contracts for farmers) that partly explains the success of this programme, one of the ambitious ones in the USA.

Yates (Chap. 16) analyses a nitrogen trading scheme in the Neuse River catchment (North Carolina, USA). In this case, the cap-and-trade programme (setting a mandatory threshold and allowing for trade to comply), WWTPs were allowed to sell or temporarily lease their permits to other plants. Whereas the economic policy instrument managed to meet environmental targets (i.e. abating emissions against baseline), the author argues that it failed to meet an economic objective (i.e. reducing emissions in the least-cost way).

Most interestingly, in what could virtually provide insights on the link between water quality and water quantity trading, Ancev and Azad (Chap. 17) analyse a salinity offsetting scheme. Salinity levels, a major concern in water scarce and drought prone areas, are naturally significantly higher in downstream river sections. As water quantity trading results, at least for countries such as Australia, in large movements of water to downstream areas, in-plot water use may increase groundwater seepage to rivers, thus increasing in-stream salinity levels. This is far from being the only reason to explain higher salinity levels; yet, it has a major potential to draw conclusions in some arid and semi-arid regions of the world where water trading might be explored as an option. Ancev and Azad assess the impact of three offsetting programmes designed to mitigate irrigation-induced salinity in Australia. Salinity offsets are designed to compensate for salinity impacts from a given agricultural activity through a commensurate reduction of salinity impacts elsewhere. In other words, it can be seen as a compensation mechanism.

Trading pollution permits thus require the creation of pollution entitlements subject to property rights. They benefit from the existence of drivers inducing action at the local level, such as national legislation, definite pollution standards, and the possibility of external intervention if lacking local action. The existence of a "champion" i.e. of a well-defined institutional focal point promoting, overseeing and facilitating the activity is essential. They also require institutional cooperation and stakeholder participation. Likewise, salinity offsets in Australia can also be seen as an example of burden sharing in the presence of economic incentives.

Within the context of water quantity trading, Delacámara et al. (Chap. 18) analyse two specific, small-scale water trades in the Tagus River watershed in Central Spain. Given the incipient status of water quantity trading in Spain, the main interest of these two trades is that they can be considered as some of the first experiences in the country, always linked to drought events and providing clear economic incentives to involved parties. The Spanish water legislation was amended in 1999 to allow for the transfer of water rights, which in Spain take the form of and administrative license or concession and are mainly traded through lease contracts. The experience analysed in Chap. 18 shows how Greater Madrid metropolitan area managed to overcome structural water constraints during drought events through voluntary agreements to trade water from agriculture to urban uses.

The immature experience in Spain contrasts with deep markets in Chile and, above all, the semi-arid states in the USA and the Murray-Darling Basin in southeastern Australia.

Donoso (Chap. 19) analyses the Chilean water trading experience. Chile, likewise Australia, defined a water right system based on nominal entitlements. As in the Australian case (presented in Chap. 20 by Young) the Chilean water trading model can be said to have succeeded in terms of harnessing the economic potential of water (for instance, with a major expansion of irrigated land for an export-oriented economy) whereas raising doubts in terms of its environmental outcomes. Chile can be said to be an approach to water trading that has taken up to a fever pitch the notion of private water use rights. Markets have driven investment given the high level of legal security attached to right allocation. Yet, concerns remain as to legal security of some rights (i.e. Copiapó Valley) is supported by water availability given the evidence of overexploitation.

Probably the most active water markets in world are located in the Murray-Darling Basin (Australia) where most of the trade occurs between agricultural users. Young (Chap. 20) does not present a comprehensive nationwide overview of the Australian model but rather an analysis of an interesting milestone in water policy reform in the country: the unbundling of the licensing system. Unbundling sheds light on one of the necessary conditions for the development of market-based approaches to sustainable water management: allowing people to hold water licenses without owning any land.

Last but not least, Howe (Chap. 21) assesses the renowned experience of the Northern Colorado Conservancy District (NCCD) in Colorado (USA). This case would be somewhat difficult to transfer to other realities, given the massive support via subsidies for a major diversion project to make water available for a large irrigation district. However, many lessons can be drawn from its analysis. The NCCD market is the most active water market in the USA in terms of number of transactions per year, due to relatively low transaction costs that stimulate frequent small trades.

Overall, the reader of this book will have access to a very wide diversity of water trading schemes. Water trading has proved to be an instrument to re-allocate water from lower- to higher-value economic activities (notably in Chile, the Murray-Darling Basin and Colorado), providing a clear signal, under appropriate conditions, of the value of water but not necessarily encouraging conservation in all cases. As an economic policy instrument, water trading elicits to water users the opportunity cost of their decisions through setting a price and making market incumbents (and others, in some national legislations) aware of the possibility of buying and selling at that price, if so they wish.

As per water quality trading, the experience in North Carolina, for instance, shows where potential for improved design of the instrument may lie: by restricting trading to occur within zones, rather than having only one single zone.

In many of the cases (remarkably Chile and Australia), a crosscutting issue has to do with the fact that individual rational decisions (i.e. the trade should be beneficial both for buyer and seller) may paradoxically lead to inefficient (and unsustainable) outcomes (i.e. mutual benefit for trading parties at the expense of social welfare), unless environmental outcomes (including physical return flows) are duly factored in.

References

Broadbent, C. D., Brookshire, D. S., Coursey, D., & Tidwell, V. (2014). An experimental analysis of water leasing markets focusing on the agricultural sector. *Agricultural Water Management, 142*, 88–98.

Colby, B., Jones, L., & O'Donnell, M. (2014). Supply reliability under climate change: Forbearance agreements and measurement of water conserved. In *Water markets for the 21st century* (pp. 57–82). Netherlands: Springer.

Collentine, D. (2006). Composite market design for a Transferable Discharge Permit (TDP) system. *Journal of Environmental Planning & Management, 49*(6), 929–946.

Debaerea, P., Richterb, B. D., Davisc, K. F., Duvallc, M. S., Gephartc, J. A., O'Bannonc, C., Pelnikd, C., Powelle, E. M., & Smithf, T. W. (2014). Water markets as a response to scarcity. *Water Policy, 16*, 625–649.

Delacámara, G., Gómez, C. M., & Maestu, J. (2015). Water trading opportunities and challenges in Europe. In K. Burnett, R. Howitt, J. A. Roumasset, & C. A. Wada (Eds.), *Routledge handbook of water economics and institutions* (pp. 279–293). Oxford/New York: Routledge.

Docker, B., & Robinson, I. (2014). Environmental water management in Australia: Experience from the Murray-Darling Basin. *International Journal of Water Resources Development, 30*(1), 164–177.

Donoso, G., Melo, O., & Jordán, C. (2014). Estimating water rights demand and supply: Are non-market factors important? *Water Resources Management, 28*(12), 4201–4218.

Erfani, T., Binions, O., & Harou, J. J. (2014). Simulating water markets with transaction costs. *Water Resources Research, 50*(6), 4726–4745.

Garrick, D., Whitten, S. M., & Coggan, A. (2013). Understanding the evolution and performance of water markets and allocation policy: A transaction costs analysis framework. *Ecological Economics, 88*, 195–205.

Garrido, A., Rey, D., & Calatrava, J. (2012). Water trading in Spain. In *Water, agriculture and the environment in Spain: Can we square the circle?* (pp. 205–216). Leiden: CRC Press/Balkema.

Ghimire, N., & Griffin, R. C. (2014). The water transfer effects of alternative irrigation institutions. *American Journal of Agricultural Economics, 96*(4), 970–990.

Ghosh, S., Cobourn, K. M., & Elbakidze, L. (2014). Water banking, conjunctive administration, and drought: The interaction of water markets and prior appropriation in southeastern Idaho. *Water Resources Research, 50*(8), 6927–6949.

Goemans, C., & Pritchett, J. (2014). Western water markets: Effectiveness and efficiency. In *Water markets for the 21st century* (pp. 305–330). Netherlands: Springer.

Gómez, C. M., Delacámara, G., Pérez-Blanco, C. D., Ibáñez, E., & Rodríguez, M. (2013). WP4 (Ex-Ante Case Studies), Deliverable 4.2 (Report for the task 4.2). Case study final report: Droughts and water scarcity – Tagus (Central Spain & Portugal) and Segura (SE Spain) interconnected river basins (October 2013). EPI-WATER Project Report (Grant Agreement no. 265213 FP7 Environment). Venice, Fondazione Eni Enrico Mattei.

Grafton, R. Q. (2010). *Economics of water reform in the Murray-Darling Basin*. Crawford School of Economics and Government, Centre for Water Economics Environment and Policy Working Papers (10/04). Canberra: The Australian National University (ANU).

Grafton, R. Q., & Horne, J. (2014). Water markets in the Murray-Darling Basin. *Agricultural Water Management, 145*, 61–71.

Grafton, R. Q., Landry, C., Libecap, G., McGlennon, S., & O'Brien, B. (2010). An integrated assessment of water markets: Australia, Chile, China, South Africa and the USA. NBER Working Paper No. 16203. Cambridge, MA: NBER (National Bureau of Economic Research).

Grafton, R. Q., Libecap, G., McGlennon, S., Landry, C., & O'Brien, B. (2011). An integrated assessment of water markets: A cross-country comparison. *Review of Environmental Economics and Policy, 5*(2), 219–239.

Grafton, R. Q., Pittock, J., Williams, J., Jiang, Q., Possingham, H., & Quiggin, J. (2014). Water planning and hydro-climatic change in the Murray-Darling Basin, Australia. *Ambio, 43*(8), 1082–1092.

Greenhalgh, S., & Selman, M. (2012). Comparing water quality trading programs: What lessons are there to learn. *Journal of Regional Analysis and Policy, 42*(2), 104–125.

Hadjigeorgalis, E. (2009). A place for water markets: Performance and challenges. *Review of Agricultural Economics, 31*(1), 50–67.

Hanemann, M. (2014). Property rights and sustainable irrigation – a developed world perspective. *Agricultural Water Management, 145*, 5–22.

Hansen, K., Kaplan, J., & Kroll, S. (2014). Valuing options in water markets: A laboratory investigation. *Environmental and Resource Economics, 57*(1), 59–80.

Hautakangas, S., & Ollikainen, M. (2011). Making the Baltic Sea Action Plan workable: A nutrient trading scheme. In M. Pihlajamäki, & N. Tynkkynen (toim.), *Julkaisussa "Governing the blue-green Baltic Sea"* (pp. 112–121). Ulkopoliittinen instituutti s. (FIIA Report 31).

Hearne, R., & Donoso, G. (2014). Water markets in Chile: Are they meeting needs? In *Water markets for the 21st century* (pp. 103–126). Dordrecht: Springer.

Hodgson, S. (2006). *Modern water rights: Theory and practice* (FAO Legislative Study 92). Rome: FAO.

Howitt, R. E. (2014). Are lease water markets still emerging in California? In *Water markets for the 21st century* (pp. 83–102). Netherlands: Springer.

Kahil, M. T., Albiac, J., & Dinar, A. (2014). *The debate on water policies: Evidence from drought in Spain*. CITA Working Document 14/03. CITA, Gobierno de Aragón.

Keller, A. A., Chen, X., Fox, J., Fulda, M., Dorsey, R., Seapy, B., Glenday, J., & Bray, E. (2014). Attenuation coefficients for water quality trading. *Environmental Science & Technology, 48*(12), 6788–6794.

Keudel, M. (2007). *Water quality trading systems: An integrated economic analysis of theoretical and practical approaches*. Doctoral dissertation, Universität zu Köln.

Kirby, J. M., Connor, J., Ahmad, M. D., Gao, L., & Mainuddin, M. (2014). Climate change and environmental water reallocation in the Murray–Darling Basin: Impacts on flows, diversions and economic returns to irrigation. *Journal of Hydrology, 518*, 120–129.

Klooster, J. P. G. N., Torenbeek, R., de Vries, C. J., & Wind, M. H. A. (2007). *Water emission trading. An exploration of a new instrument in integrated water management*. CE, ARCADIS, Ecoconstult. The Hague: Ministry of Transport, Public Works and Water Management, DGW.

Lafreniere, K. C., Deshpande, S., Bjornlund, H. (2015). Segmenting and targeting irrigators' preferences regarding proposed water transfers. *Society & Natural Resources* (ahead-of-print), 1–16.

Lankoski, J., Lichtenberg, E., & Ollikainen, M. (2008). Point/nonpoint effluent trading with spatial heterogeneity. *American Journal of Agricultural Economics, 90*, 1044–1058.

Loch, A., Wheeler, S., Boxall, P., Hatton-Macdonald, D., Adamowicz, W. L., & Bjornlund, H. (2014). Irrigator preferences for water recovery budget expenditure in the Murray-Darling Basin, Australia. *Land Use Policy, 36*, 396–404.

McCann, L., & Garrick, D. (2014). Transaction costs and policy design for water markets. In *Water markets for the 21st century* (pp. 11–34). Netherlands: Springer.

Mitchell, G., & McDonald, A. (2015). Developing resilience to England's future droughts: Time for cap and trade? *Journal of Environmental Management, 149*, 97–107.

OFWAT. (2010). *A study on potential benefits of upstream markets in the water sector in England and Wales*. Birmingham: OFWAT.

Oosterhuis, F., & Peeters, M. (2014). Limits to integration in pollution prevention and control. In M. Peeters & R. Uylenburg (Eds.), *EU environmental legislation. Legal perspectives on regulatory strategies* (pp. 91–115). Cheltenham/Northampton: Edward Elgar.

Rosegrant, M. W., Ringler, C., & Zhu, T. 2014. Water markets as an adaptive response to climate change. In *Water markets for the 21st century* (pp. 35–55). Netherlands: Springer.

Shortle, J. (2013). Economics and environmental markets: Lessons from water-quality trading. *Agricultural and Resource Economics Review, 42*(1), 57–74.

Wagnitz, P., Núñez, J., & Ribbe, L. (2014). Cost of environmental flow during water scarcity in the arid Huasco River basin, northern Chile. *Hydrological Sciences Journal, 59*(3–4), 700–712.

Wheeler, S., Loch, A., Zuo, A., & Bjornlund, H. (2014a). Reviewing the adoption and impact of water markets in the Murray–Darling Basin, Australia. *Journal of Hydrology, 518*(Part A), 28–41.

Wheeler, S. A., Zuo, A., & Bjornlund, H. (2014b). Investigating the delayed on-farm consequences of selling water entitlements in the Murray-Darling Basin. *Agricultural Water Management, 145*, 72–82.

Wind, M. H. (2012). Water emissions trading in Europe – A literature overview and discussion of opportunities. *Eco-consult Environmental Economics.* [Version 11-5-2012]. Report presented at the World Congress on Water, Climate and Energy, Dublin, 14 May 2012.

Young, M. D. (2014). Designing water abstraction regimes for an ever-changing and ever-varying future. *Agricultural Water Management, 145*, 32–38.

Zuo, A., Nauges, C., & Wheeler, S. A. (2015). Farmers' exposure to risk and their temporary water trading. *European Review of Agricultural Economics, 42*(1), 1–24.

Chapter 15
Water Quality Trading in Ohio

Mark S. Kieser and Jamie L. McCarthy

Abstract The economic policy instrument (EPI) discussed in this case study involves nutrient credit trading between point source wastewater treatment plants (WWTPs) and non-point (diffuse) sources (agriculture) in the Great Miami River Watershed of Ohio (USA). Commonly referred to as water quality trading (WQT) in the USA, this EPI is a market-based approach to pollution control in which pollutant reductions are treated as commodities. This case study describes how a watershed-based flood control agency in southwest Ohio developed and managed a WQT program to provide a cost-effective alternative for WWTP compliance. WWTPs will soon face more stringent effluent limits as a result of impending numeric nutrient standards being assigned to rivers and streams receiving treated wastewater. Lessons learned from this case study have substantial merit as an EPI because this program provides an economic framework for applying and using WQT in a regulatory setting. This program is one of the largest and most successful WQT programs of its kind in the USA, negotiating nutrient credit trades between five point source buyers and hundreds of non-point source sellers. Of note is the use of a reverse auction for securing lowest-cost credit contracts for point source buyers.

Keywords Trading • Nutrients • Water quality • Wastewater • Cost-effectiveness

15.1 Introduction

The Miami Conservancy District (MCD), a watershed-based flood district with taxing authority, initiated efforts in 2003 to consider development of a point source/non-point source water quality trading (WQT) pilot program for nutrients in the Great Miami River (GMR) in southwest Ohio (USA). This innovative economic

M.S. Kieser (✉)
Kieser & Associates, LLC, Kalamazoo, MI, USA
e-mail: mkieser@kieser-associates.com

J.L. McCarthy
Kalamazoo River Watershed Council, Kalamazoo, MI, USA
e-mail: krwc@kalamazooriver.org

M. Lago et al. (eds.), *Use of Economic Instruments in Water Policy*,
Global Issues in Water Policy 14, DOI 10.1007/978-3-319-18287-2_15

policy instrument (EPI) was developed prior to state-wide rules for WQT and as such, MCD program elements followed the Federal Final Water Quality Trading Policy (USEPA 2003). Starting in 2004, the MCD developed a policy framework for post-compliance WQT and began testing this framework in 2006 by implementing agricultural best management practices (BMPs) through pilot trading.

WQT is a voluntary market-based approach to pollution control compliance in which pollutant removal is traded as a commodity. With WQT, dischargers that reduce their pollutant loading below required levels can sell surplus reductions, called credits, to other dischargers that need to make reductions to meet compliance requirements. In the MCD program, anticipated end-of-pipe wastewater treatment plant (WWTP) load reductions required under future permit limits must be offset by edge-of-field load reductions from agriculture (along with the application of a trading ratio). Trading is considered a compliance option for WWTPs in lieu of expensive technology upgrades. Agricultural participation is voluntary as most farming operations (except large-scale animal operations) are not regulated by permits like WWTPs. Quantitative measures in trading therefore focus on computation of nutrient load reductions and not necessarily in-stream water quality or biology. Trading programs often include monitoring elements, including MCD program. However, in-stream monitoring is inherently limited in its capability to assess in-stream responses for pre- versus post-BMP implementation unless substantial BMP applications across the landscape are made. Such changes would reflect a robust trading program. The MCD program is not considered robust in these regards, even though it has the greatest number of credits being generated by agriculture compared to similar trading programs in the USA.

The intent of the trading program is for agriculture to supply cost-effective nutrient reduction credits in lieu of anticipated point source reductions associated with expensive wastewater treatment plant upgrades. As agriculture is the predominant land use in the watershed, it was originally envisioned that trading opportunities in a water quality market with significant demand will motivate agricultural producers to participate. Robust participation by agriculture in a trading program can overcome common challenges in traditional programs that lack the authority or incentives to engage producers in water quality initiatives. The goal of the program was to establish a unit of credit for nitrogen and phosphorus reductions generated by agricultural BMPs that reduced nutrient loading to local surface water bodies. Cost-effective credits would be sold to downstream wastewater treatment plants looking to offset effluent discharges. The program was developed with pre-compliance incentives to encourage early participation prior to issuance of more stringent wastewater treatment plant (WWTP) effluent limits.

As of this case study report, numeric nutrient standards have not been promulgated at the state or regional level. State government and the regional commission are still in the process of analyzing data and designing approaches to implementing numeric criteria. Despite this setback, the MCD WQT program has been successful in conducting reverse auctions and funding agricultural producers to implement BMPs that improve water quality. As of January 2014, the program has generated more than 1.14 million nutrient credits, amounting to 572 t of reduced nutrient discharges to surface waters for eight participating WWTPs (some with multiple discharges) (WEF 2015) (See Fig. 15.1).

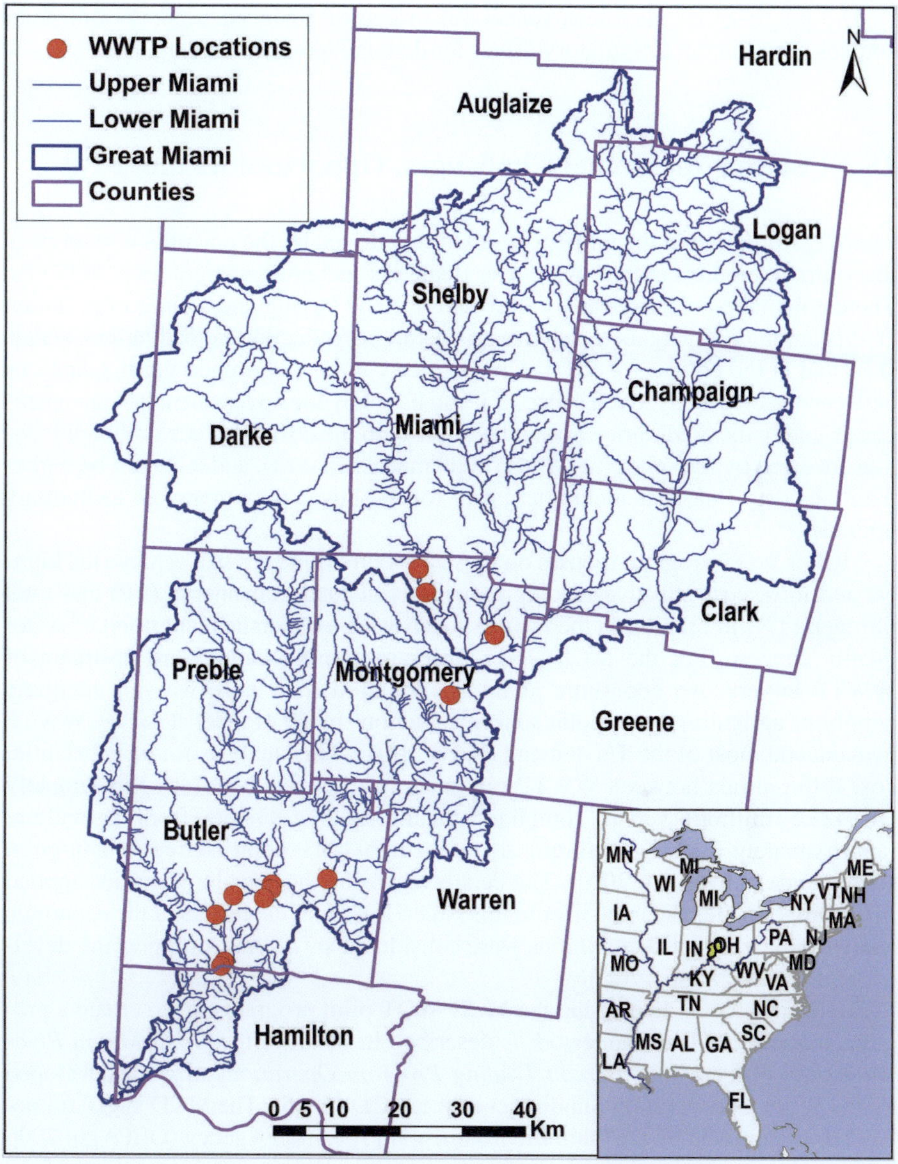

Fig. 15.1 Locations of WWTPs participating in the MCD WQT pilot program

The program has also been successful in getting financial support from point sources even though a regulatory driver for demand has been lagging.

15.2 Setting the Scene: Challenges, Opportunities and EPIs

The WQT program in the GMR watershed was driven by the potential cost-savings that nutrient credits could provide over treatment technology upgrades at WWTPs. The credits take on a monetary value, and it is the buying and selling of pollutant credits among dischargers to achieve an overall net reduction in loading in a watershed that is the essence of WQT. Credit price can be determined via negotiations between the credit buyer and seller, or it can be set by the government or other agencies. Usually the credit price cannot be lower than the cost to reduce pollutant loadings incurred by the seller (i.e., BMP implementation cost), and it cannot be higher than the cost of an abatement alternative for the buyer (i.e., treatment technology upgrade).

MCD's WQT program focuses on those agricultural BMPs that achieve the highest and most cost-effective loading reductions of total phosphorus (TP) and total nitrogen (TN) in relation to the buyer's location in a watershed and point of water quality concern. For the MCD WQT program, these locations are upstream of WWTP buyers. An economic analysis indicated that there was an adequate supply of agricultural non-point source reductions in TP to meet all of the WWTP demand and most of the TN demand (K&A 2004). This analysis found substantial cost differentials between WWTP upgrades for 314 facilities (at approximately US$422.5 million) versus compliance using trading credits from agriculture (approximately US$37.8 million) at an average 1.4:1 point source to non-point source trade ratio (K&A 2004). The details regarding the actual trade ratios applied in the program are discussed later in this paper. Based on the results of the economic analysis, the watershed conditions were considered sufficient to support the development of a trading program.

An initial step in developing the MCD WQT pilot program was to create a program framework. This framework is described in detail in the *Great Miami River Watershed Water Quality Credit Trading Program Operations Manual* developed by the Water Conservation Subdistrict of the MCD (2005). The MCD WQT framework was approved by the Ohio Environmental Protection Agency (OEPA) in 2006 prior to the promulgation of state-wide trading rules in 2007 (see Fig. 15.2). In Fig. 15.2, the elements in the shaded area indicate trading entities such as WWTPs, the MCD and agriculture, which participate through third-party interactions. Compound lines represent credit flow, dashed lines represent the relationship between Ohio Department of Natural Resources (ODNR) and agriculture, and solid lines represent regulatory oversight of OEPA.

MCD now tracks pre-compliance credit use internally to the pilot trading program. In the future, WWTPs purchasing credits through MCD's WQT program will have their National Pollution Discharge Elimination System (NPDES) permits modified to accommodate credit use to meet a future discharge limit. Local county

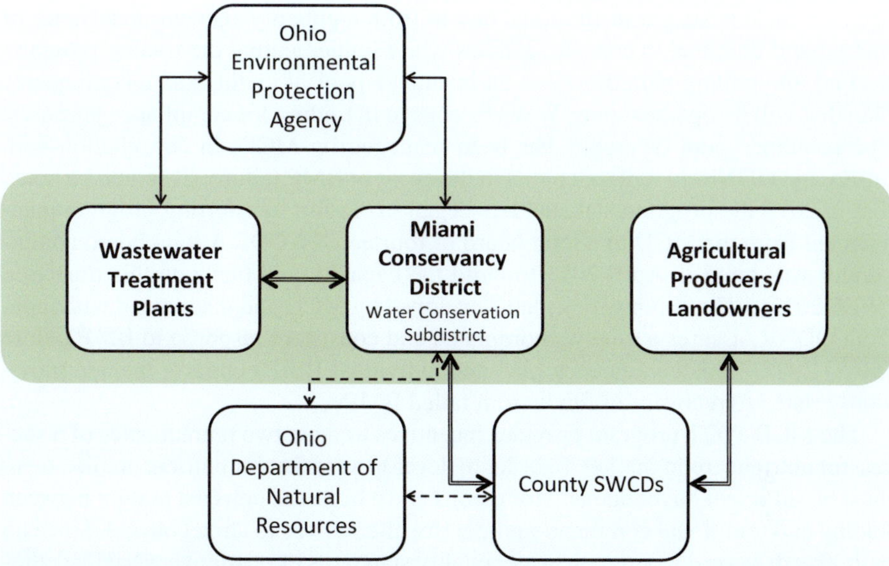

Fig. 15.2 Organizational flowchart of WQT in the Great Miami River Watershed (Adapted from MCD 2005)

Soil and Water Conservation Districts (SWCDs), non-governmental technical assistance agencies for agriculture, play an instrumental role in soliciting and contracting farmers to implement BMPs for nutrient credits. WWTPs are given an additional financial incentive for early pre-compliance participation in the WQT pilot program in the form of a more favourable trading ratio than will be available in the future after the pre-compliance period.

15.3 Water Quality Trading in Action

A number of indicators can be used to measure the success of an EPI. The assessment criteria discussed in this section aims to answer the question of what environmental outcomes this case study has produced and at what cost. Economic efficiency and transaction costs are also explored in more detail.

15.3.1 The EPI Contribution

The WQT program has the potential to increase cost-effectiveness of nutrient reductions for WWTPs, but it also presents an increase in risk when compared to command-and-control alternatives (i.e., plant upgrades). The MCD WQT trading plan was designed to offset this risk using a number of different approaches. First, decision-makers sought institutional buy-in from regulatory agencies in advance of trading and continue to consult agencies while implementing the trading program. Second, the trading plan describes an insurance pool of credits and a contingency plan that will be operated once WWTPs engage in trading for compliance purposes. The insurance pool of credits has been managed by MCD, in consultation with OEPA and ODNR, to replace credits in the event of BMP failure. (It should be noted that in 2013 the program stakeholders began exploring transferring program management from the MCD to a joint board of fourteen SWCDs. A joint board administrator was hired in April 2013 to build the capacity and facilitate this transition (WEF 2015).) The contingency plan, developed by MCD and maintained with input from ODNR, assures a timely, coordinated and consistent response to BMP failure (MCD 2005). This contingency plan and individual BMP contracts include provisions related to recovery of funds from failed BMPs.

The MCD WQT program provides incentives to meet two prerequisites of a successful nutrient credit market. In order to develop a market for nutrient credits, there must be sufficient buyer demand for credits and a large enough cost margin between trading and traditional command-and-control alternatives to attract buyers. Since no definitive drivers (e.g., numeric water quality standards or restrictive wasteload allocations) were in place in the watershed when MCD initially developed the WQT program), MCD provided buyers with a financial incentive to participate in early, pre-compliance trading. The program would operate under this "pre-compliance period" until more definitive drivers were in place in the watershed. MCD negotiated with OEPA that WWTPs that purchased credits prior to a compliance driver would be guaranteed a better trade ratio when purchasing credits in the future. Early participants are now locked into a 1:1 trade ratio for buyers discharging to waters that already meet water quality standards, and a 2:1 trade ratio for buyers discharging to impaired waters (MCD 2005). For WWTPs that purchase credits to address compliance requirements, trade ratios will be 2:1 and 3:1, respectively. The reverse auction method is the second incentive MCD developed to attract buyers. These auctions occur once or twice a year, as funding allows. By purchasing the lowest-cost credits, MCD is able to keep cost margins high enough to make trading more attractive to WWTPs than traditional technology upgrades.

15.3.1.1 Environmental Outcomes

To assess potential water quality benefits, the MCD conducts intensive water quality monitoring at select watershed locations as part of the trading program. Daily composite sampling coupled with real-time flow monitoring provide comprehensive datasets for nutrient loading in the three major tributaries and the mouth of the Great Miami River prior to discharging to the Ohio River. This monitoring provides some of the most robust data on tributary loading in the state of Ohio that would not otherwise have been collected absent the trading program.

Annual nutrient loading from the entire GMR watershed was measured as 2,196,752 lb TP and 17,659,733 lb TN in 2008, a particularly wet year. Contracted load reductions for 2008 in the WQT program were calculated at 16,598 and 44,487 lb of TP and TN, respectively. These nutrient reduction credits constituted only very small proportions (0.76 % TP and 0.25 % TN) of the overall load in the GMR in 2008. Moreover, annually discharged TP and TN tributary loads vary are much as 300 % from year to year in the GMR depending on rainfall amounts in this highly agricultural, non-point source-dominated watershed (MCD 2010). Thus, in-stream monitoring, though informative for watershed management, is not adequate to identify water quality benefits of nutrient trades in the GMR. Establishing the quantitative metric as edge-of-field non-point source load reductions for agricultural BMPs remains the preferred, as well as the only suitable and practical method at present for justifying adequacy of trades. Edge-of-field monitoring has been conducted at a number of crediting locations, and MCD published a water quality report in 2012 (MCD 2012).

In addition to quantitative water quality benefits, the MCD WQT program has promoted (though not measured) several ancillary benefits trading can produce when compared to technology upgrades at WWTPs alone. Many agricultural BMPs reduce sediment loading to local streams and rivers that would otherwise not be produced by or regulated at a WWTP. Streamside BMPs and other upland BMPs result in improved riparian and in-stream habitat. Riparian BMPs can also provide canopy that shades streams and rivers and helps control in-stream temperature. Streambank stabilization and velocity are also improved through select agricultural BMPs that infiltrate or delay agricultural runoff from reaching surface waters. WQT also has the potential of increasing the geographic extent of many water quality benefits when upstream trading in headwater streams is utilized. Instead of implementing an improvement in water quality at the WWTP effluent discharge location, this pilot program requires an equivalent or greater reduction in nutrient loading to occur upstream of the discharge point. In many cases this will result in improved stream conditions and water quality when crediting practices are implemented in sensitive headwater tributaries.

15.3.1.2 Economic Outcomes

The MCD authorized an economic analysis of WQT in the Great Miami River watershed in order to make an informed decision on the economic benefits of WQT before developing a pilot program. The economic analysis reported a substantial cost-savings for WWTPs if they were to purchase nutrient credits rather than upgrade plants to biological nutrient removal (BNR) technology (K&A 2004). Using the best available information at the time of the study and reasonable assumptions, it was determined that treatment plant upgrades to BNR for nearly all of the 314 WWTPs in the Great Miami River watershed would cost approximately US$422.5 million dollars (based on a 20-year investment and 5 % interest rate using 2003 US$). Assuming the WWTPs would have to meet effluent limits of 1 mg/L phosphorus and 10 mg/L nitrogen limit in their discharges, potential demand was computed as kilogram per year using the difference in concentrations above these limits times wastewater flow). The equivalent amount of non-point source nutrient credits on an annual basis (assuming an average trade ratio of about 1.4:1) would cost WWTPs approximately US$37.8 million (based on costs for no-till cropping practices) (K&A 2004). This resulted in a projected cost-savings of approximately US$384.7 million.

One way in which the MCD and program decision-makers worked to make WQT cost-effective was by selecting a reverse auction method for soliciting proposed BMPs projects. This method has been employed in selecting projects in all 11 rounds of Request for Proposals (RFPs). Once all proposals are submitted, MCD selects the lowest "bids" for BMP projects until all of the funds for that particular round are committed. In an ex-post evaluation of the MCD WQT program, Newburn and Woodward (2012) assessed the cost-effectiveness of MCD's reverse auction method from the supply side of the market. According to their cost-savings (CS) metric, the average cost-savings for agricultural BMPs in round one was 32 %. The cost-savings decreased to 19 % when rounds one through six were evaluated. Newburn and Woodward (2012) suggest this was due to SWCDs learning the relatively stable threshold of credit prices MCD would fund after multiple rounds. They point out that over time the WQT program has begun to function more like a fixed-priced program versus a reverse auction (Newburn and Woodward 2012).

While the reverse auction method advances the goal of cost-efficiency, it does not benefit all stakeholders equally. In theory, the WWTP buyers benefit by getting the lowest-cost nutrient credits in the watershed. It is important to note that not all SWCDs in the watershed submitted proposals for BMP projects, so lower cost credits may exist throughout the watershed. In this case, producers located in counties where SWCDs did not participate lost a potential funding source for BMPs. WWTPs are also required to pay into the program proportionally, but because of the trading requirement where WWTPs can only apply upstream credits to their permit, plants located higher in the watershed may not be able to access credits in proportion to what they pay for through MCD.

15.3.1.3 Distributional Effects and Social Equity

WQT provides a flexible and innovative means to realizing pollution reductions in a cost-effective manner. Social equity and equitable distribution of funding are not main tenants of WQT. Stakeholder groups that are at a disadvantage in the MCD program framework include producers already participating in US Department of Agriculture Farm Bill programs (i.e., federal cost-share programs for conservation practices). These producers are ineligible for trading in the MCD program. In addition, BMPs with relatively high costs compared to credit potential would generally not be selected through the reverse auction. Producers that want to be compensated for full opportunity costs are also less competitive in the reverse auction system. Because the reverse auction method favours lower cost credits, BMPs such as vegetative buffers with native plantings are less competitive than those with non-native brome grass, for example. The drawback is that brome grass, which generally comes at a lower cost, has less habitat value than more costly native plantings.

Another important feature of the MCD WQT program is the requirement of upstream trading for credit buyers (i.e., non-point source credits must be generated upstream of a buyers discharge point). Due to this requirement, producers located below WWTPs participating in the MCD WQT program are ineligible to sell credits in this limited pilot market. Another factor that limits producer participation on a spatial basis is the requirement that county SWCDs must complete proposals on behalf of the producer and submit them to MCD for the reverse auction. Of the 15 eligible counties in the GMR, only 10 ever submitted applications. Three SWCDs largely dominated the application process. These tended to have sufficient technical resources and staffing to assist farmers with BMP proposal preparation.

Another factor influencing the sell side of the market (i.e., producers) is whether the producer wants to be compensated for all of their BMP and opportunity costs. The MCD WQT program is voluntary and encourages low cost credits from producers. If a producer wants a particular BMP implemented at their farm because it will increase quality of life or provide an improvement to their operations, the producer can under price their BMP proposal to make it more competitive. Noteworthy is the fact that the MCD program does not allow farmers to participate that have otherwise received federal subsidies from the US Department of Agriculture for conservation practices. This places an emphasis on farmers that are unlikely to have otherwise implemented conservation practices. Despite this program condition, Klang and Kieser (2008) found that farmer cost-share rate payments for federal conservation subsidies were quite similar to payments made to farmers under the WQT program.

15.3.2 The EPI Setting Up

As a watershed-based agency representing the public and various municipalities, MCD recognized the need for less expensive compliance alternatives. Using the WQT feasibility study (K&A 2004) as evidence of need, MCD championed the WQT pilot program development process. MCD obtained stakeholder buy-in through a robust public participation process that introduced the concept of trading to both the WWTP and agricultural sectors. Political conditions in the watershed were also ripe for WQT with the regulatory agency (Ohio Environmental Protection Agency—OEPA) willing to modify WWTP discharge permits as part of the pilot program in the face of backlash from WWTPs arguing against upgrades.

15.3.2.1 Institutions

Several conditions existed or were developed in order for the MCD pilot program to be functional. The first necessary condition was a driver for nutrient reductions. In 2003, the OEPA informed WWTPs that numeric nutrient standards were forthcoming and for the first time these dischargers would be required to reduce TP and TN beyond limits of technology associated with conventional activated sludge wastewater treatment (K&A 2004) (As of this writing, numeric nutrient standards have not yet been promulgated by OEPA.) Thus, WWTPs in the watershed anticipated stricter effluent limits in their discharge permits, and that water quality impairments in the watershed would continue to trigger watershed-based nutrient load reduction requirements through Total Maximum Daily Loads (TMDL) under the federal Clean Water Act (CWA). Second, the feasibility study for WQT in the watershed (K&A 2004) indicated that future WWTP costs to meet new effluent limits via treatment system upgrades were tenfold more expensive than non-point source nutrient credits. The study also reported the capacity for ample agricultural credit supply throughout the watershed, though the duration of these agricultural credits depends on the type of nutrient reduction practice.

In terms of water administration and management, interactions between state regulatory agencies, WWTPs and producers were important to understand and respect when developing the program. In the initial developmental phases, tension existed between state regulatory agencies and agricultural producers. Agricultural producers have traditionally been uneasy about regulators having access to their property as they are largely unregulated under the CWA. Because the WQT program required some level of inspection of BMPs to ensure they are operated and maintained to acceptable standards, MCD had to be sensitive to producer concerns over regulatory agencies playing a role in BMP inspections. To get producer buy-in, MCD developed a system where SWCDs would perform inspections of BMPs. MCD was able to take advantage of the technical expertise as well as the existing and trusted relationship SWCDs had with producers. This avoided direct interaction between OEPA and farmers (see Fig. 15.2).

15.3.2.2 Transaction Costs and Design

Much of the start-up cost and initial program development was heavily subsidized through federal grants (Hall 2011, personal communication). Newburn and Woodward (2012) assessed the transaction costs for both search and bargaining as well as project verification and enforcement in their ex-post evaluation of the MCD WQT program. Overall, they concluded that transaction costs do not create a considerable barrier to market efficiency. Hall (2011, personal communication) has indicated transaction costs were approximately US$1.50/credit, this effectively doubling the lowest average bid pricing of US$1.50/credit. Though largely covered by grant funds, it is expected that these transaction costs would drop both as a percentage of bid price and cost/credit when more participants enter the market. Additional participation will prompt the development of more formal trading infrastructure (crediting and registration) that is systematized and made more broadly available to SWCDs.

Newburn and Woodward (2012) explain that the institutional framework of the MCD WQT program is such that MCD acts as a clearinghouse for nutrient trades. This design lowers the bargaining costs for trading since there is no contract between buyer and seller that needs to be negotiated. In addition, the clearinghouse model eliminates the cost to the buyer and seller incurred when searching for trading partners.

The reverse auction method also reduces transaction costs on the supply side since BMP proposals are accepted or denied based solely on cost and credits generated (Newburn and Woodward 2012).

Another aspect of the trading program that Newburn and Woodward (2012) identify as a way in which MCD gains cost-savings is by using the county SWCD offices to recruit producers. They point out that SWCDs already have similar duties to provide technical services to producers under larger federal conservation programs, therefore SWCD duties are not greatly expanded by participating in trading. The SWCD cost assistance and monitoring represented approximately 3.9 % and 1.0 % (respectively) of the more than US$1.3 million total expenditures of the MCD reported by Newburn and Woodward 2012 after the sixth round of reverse auctions. Verification of BMP project implementation by SWCDs involved site visits and inspections to certify BMPs were functioning and maintained to standards outlined in contracts. Not all SWCDs charged for time spent on verification activities in order to make BMP credit proposals more competitive in reverse auction bid selection.

15.3.2.3 Implementation

Implementation of WQT in the state of Ohio has been relatively successful when compared to similar water quality markets throughout the USA (USEPA 2008). The US EPA, in its 2003 Water Quality Trading Policy, cited estimates that innovative approaches such as WQT could save as much as US$900 million per year in meeting TMDLs. TMDLs established to date, however, have not resulted in trading or savings at these levels. Trading success started with the MCD WQT pilot program based not on the threat of TMDLs which would have varying local applications, but rather on pending state nutrient standards which would affect all dischargers. Thus, the MCD was able to develop a flexible watershed-wide trading program for the entire Great Miami River.

The MCD program provides ubiquitous compliance flexibility to all participating permitted point sources in the watershed. WQT in this manner provides a broader funding source for agricultural conservation practices and watershed management. In addition, the MCD WQT program forced OEPA to promulgate state-wide trading rules. These state-wide trading rules, initially promulgated in 2007, also provide for a flexible and adaptive approach to WQT which allows each individual trading program in the state to be innovative. Additionally, these rules provide assurances to the public that nutrient reductions are real and surplus (beyond what was already required as applicable to each credit seller).

The MCD WQT program prompted atypical cooperation and coordination between OEPA and the Ohio Department of Natural Resources (ODNR—the state's non-regulatory technical agency). These two state agencies now work collaboratively in overseeing different aspects of the WQT program. Buy-in and support from OEPA and ODNR did much to advance the WQT program and helped MCD gain broader buy-in from stakeholders and credit buyers (Hall 2011, personal communication).

In addition to agency support, MCD held more than 100 stakeholder meetings to engage municipal buyers, agricultural sellers, federal agencies, environmental groups and the general public to design a WQT program that would have wide support of stakeholders. This included groups that might typically be opposed to market-based approaches to nutrient reductions.

While MCD has experienced success with its WQT pilot program, there are barriers that have impeded expansion of the program. For instance, there is still no definitive driver in the watershed to create demand for nutrient credits from point source buyers. At the state and regional level, numeric nutrient standards have been lagging. This has inhibited new buyers expressing their interest in the trading program. In addition, a TMDL developed by OEPA for a sub-basin of the Great Miami was determined to be flawed based on early WQT economic modelling (K&A 2004) and later on WQT program monitoring. OEPA eventually rescinded this TMDL. These actions produce uncertainty in the market and work to lower demand for nutrient credits.

15.4 Conclusions

MCD WQT program implementation has been successful in completing nutrient credit trading between point sources and non-point sources. These trades have taken place in a pre-compliance setting through a pilot program. Moving from pre-compliance to post-compliance and increasing trading at scale may present both benefits and challenges to WQT. The program has succeeded in implementing BMPs that have explicit water quality benefits as well as other ancillary benefits. The program has been shown to be cost-efficient due to BMP funding mechanisms and the program framework. The program has been successful in working with existing institutions and adapting to preferences of different sectors in order to build trust between stakeholders (e.g., ensuring producers that site inspections would not be conducted by regulatory agencies, putting them at higher enforcement risk). In addition, Newburn and Woodward (2011) found the program transaction costs to be relatively low. They attributed this to MCD's ability to work with existing agricultural technical service providers in the watershed and MCD's clearinghouse model, which lowered search and bargaining costs.

The specific focus of the trading program was on lowest-cost credits from farmers not previously participating in federal conservation subsidy programs. This feature, along with the reverse auction helped assure competitive credit pricing. Though this excluded explicit consideration of farmer production efficiencies, SWCDs focused on conservation practices that were largely lacking with participating bidders. Inherently, many of the proposed conservation practices for credits provided operational efficiencies in terms of reduced soil losses from fields, improved nutrient management and long-term structural practices that otherwise might not have been affordable to the farmer. These simply were left to the decision of the farmer and the SWCD technician in terms of a decision as to whether or not to voluntarily prepare a bid. A comparison of traditional payments for federal conservation subsidies with WQT credit pricing did, however, show that these were generally comparable (Klang and Kieser 2008).

In terms of economic efficiency, MCD has been successful for a number of reasons. In an ex-post evaluation of rounds one through six of the BMPs bids, Newburn and Woodward (2012) report that MCD funded BMPs at a substantial cost-savings averaging 19 %. This cost-efficiency is due to the reverse auction method used by MCD. A study by K&A (2004) indicated that WWTPs have the potential to realize substantial cost-savings using WQT instead of traditional technology upgrades to meet more stringent effluent limits in permits in the future. While the MCD WQT program has been highly subsidized by federal grants, the transaction costs have been relatively low (Newburn and Woodward 2012). Program staff anticipate costs to continue to decrease in the future once the market grows with future demand (Hall 2011, personal communication).

The pilot program provides useful insight into the general application of the EPI and its transferability to other settings. Lessons learned from the pilot provide guidance for improvements that could be made to the program, especially when transferring the EPI to different watersheds and other local markets. These include:

- The simplified trade ratios in the pre-compliance program may require additional discount factors for nutrient credits if applied at a larger watershed scale where in-stream nutrient attenuation and bioavailability factors may need to be applied for more distant buyer-seller trades.
- Third-party oversight/inspection of completed agricultural BMPs versus SWCD farmer solicitation, BMP design and implementation verification may help with public perceptions over a lack of independent BMP inspection.
- Publicly accessible credit tracking, which currently is internal to MCD for the pre-compliance pilot setting, will be necessary when the program transitions to formal regulatory compliance for WWTPs.

References

Hall, D. (2011). Personal communication via email with Dusty Hall, MCD Program Manager. 7 Apr 2011.

Kieser & Associates, LLC [K&A]. (2004). *Preliminary economic analysis of water quality trading opportunities in the Great Miami River watershed*. Prepared for the Miami Conservancy District. Available online at: http://www.envtn.org/uploads/Great-Miami_Trading_Analysis.pdf

Klang, J. A., & Kieser, M. S. (2008). *An economic comparison of the USDA-NRCS environmental quality incentives program payments and water quality credit trading in the Great Miami River watershed of Ohio*. Prepared for the Miami Conservancy District. Available online at: http://www.envtn.org/uploads/EQIP_WQT_GMR2008.pdf

Miami Conservancy District [MCD]. (2005). *Great Miami River watershed water quality credit trading program operations manual*. Water Conservation Subdistrict of the Miami Conservancy District. Available online at: http://www.watershedconnect.com/documents/files/great_miami_river_watershed_water_quality_credit_trading_program_operations_manual.pdf

Miami Conservancy District [MCD]. (2010). *Water resources report for the Great Miami River watershed*. Report No. 10–23. Available online at: http://newserver.miamiconservancy.org/water/documents/2010FINALWaterResourcesReport.pdf

Miami Conservancy District [MCD]. (2012). *Water quality study – nitrogen and phosphorus concentrations and loads in the Great Miami River watershed, Ohio 2005–2011*. MCD Report 2011-43. Available online at: http://newserver.miamiconservancy.org/water/documents/2012NutrientMonitoringReport_Final.pdf

Newburn, D. A., & Woodward, R. T. (2012). An ex post evaluation of Ohio's Great Miami River water quality trading program. *Journal of the American Water Resources Association*, 156–169. Available online at: http://newserver.miamiconservancy.org/water/documents/NewburnandWoodward.ExpostEvaluation.pdf

USEPA. (2003). Final water quality trading policy. Available online at: http://water.epa.gov/type/watersheds/trading/finalpolicy2003.cfm. Accessed 15 Sept 2014.

USEPA. (2008). *EPA water quality trading evaluation final report, promoting results through evaluation*. Prepared by Industrial Economics, Inc. for EPA Office of Policy, Economics and Innovation. Contract EP-W-04-023. Available online at: http://www.epa.gov/evaluate/pdf/water/epa-water-quality-trading-evaluation.pdf

Water Environment Federation [WEF]. (2015). Chapter 8: Case study of trading in the Great Miami River watershed, Ohio. In, *Advances in water quality trading as a flexible compliance tool*. Water Environment Federation: Alexandria, Virginia, USA.

Chapter 16
Nitrogen Reduction in North Carolina

Andrew J. Yates

Abstract This EPI case study analyzes Nitrogen trading in North Carolina's Neuse River. Under the United States' Clean Water Act, the Neuse River is a section 303(d) impaired water. Typical 303(d) regulation is command-and-control: wastewater treatment plants that emit Nitrogen are required to meet their own emission limits. In the EPI, however, a cap-and-trade program was put in place under which plants are given a permit to emit Nitrogen, and this permit may be sold or temporarily leased to another plant. The EPI met the environmental goal in that emissions were significantly reduced below baseline levels. But the EPI did not meet the economics goal of reducing emissions in the least cost way, because few permits were traded. The design could be improved by restricting trading to occur within zones, rather than having only one single zone. The practice could be improved by encouraging plants to make trades. This case study informs the regulation of water quality in the USA under the Clean Water Act. Moving from the traditional regulation of these point sources to a properly designed EPI with active trading could potentially generate hundreds of millions of dollars in benefits to society.

Keywords Permit trading • Nitrogen • Clean Water Act

16.1 Introduction

The widely acknowledged success of the United States Environmental Protection Agencies cap-and-trade programs for the reduction in emissions of S02 from electric power plants (Stavins 1998), has generated considerable interest in applying cap-and-trade programs to other pollution control problems. In particular, the Environmental Protection Agency (EPA) has encouraged the use of water quality trading to lower the cost of meeting the standards set by the Clean Water Act (USEPA 2003; Stephenson and Shabman 2011). An EPI based on a cap-and-trade

A.J. Yates (✉)
Department of Economics and Curriculum for the Environment, University of North Carolina at Chapel Hill, CB 3305, Chapel Hill, NC 27599, USA
e-mail: ajyates@email.unc.edu

M. Lago et al. (eds.), *Use of Economic Instruments in Water Policy*,
Global Issues in Water Policy 14, DOI 10.1007/978-3-319-18287-2_16

program has many theoretical advantages relative to other types of regulation. Perhaps most importantly, it offers the promise of reaching a given water quality goal in the least cost manner. In this case study, we critically evaluate a cap-and-trade EPI that was applied to Nitrogen emissions into North Carolina's Neuse River Basin.

The impetus for implementing an EPI for nitrogen trading in the Neuse River Basin has its origin in the major fish kills that occurred in 1995 (NCEE 2011). In response, the State of NC government developed a regulatory structure to reduce the flow of Nitrogen into the river. Rulemaking for the reduction of nitrogen was developed by the Environmental Management Commission and administered by the Division of Water Quality (Hamstead 2008). Reduction was targeted from both point and non-point sources, but the rules for point sources contained an interesting provision that generated the EPI. Rather than require that each point source meet an individual emission requirement, the rules allowed polluters to jointly meet an aggregate group emission requirement by forming an association. The members of the association would not be fined by the State of NC as long as the total aggregate emissions of pollution were below the required level (Hamstead 2008). This case study focuses group emissions from 22 point sources known collectively as the Neuse River Compliance Association (NRCA). These are almost entirely wastewater treatment plants (WWTP). This association was formed in 2002 in response to the Nitrogen emission rules described above.

16.2 Setting the Scene: Challenges, Opportunities and EPIs

The Neuse River Basin covers approximately 9 % of the state of North Carolina, or 15,959 km^2. This region has experienced significant growth over the last 10 years with a decrease in forest land and an increase in development. This trend is expected to continue over the next 10 years (UNRBA 2011). All members of the NRCA discharge nitrogen into the Basin. Many of these sources are expected to experience 50–100 % increase in discharges by 2030, due primarily to increases population (all data from NCDWR 2010). In addition to these point sources, emissions from non-point sources also lead to decreases in water quality. The main non-point source emissions are from storm water and agricultural runoff. In particular, there has been a large increase in agricultural runoff from concentrated animal feed operations over the last decade (NCDWQ 2009).

In the USA, emissions of water pollution from point sources are governed by the Clean Water Act. Under the Clean Water Act, an emitter must obtain a National Pollutant Discharge Elimination System (NPDES) permit from the EPA (USEPA 2011). The actual administration of the NPDES permit is usually undertaken by individual states, as is the case in North Carolina. In North Carolina, the Division of Water Quality is the responsible state agency (NCDENR 2011).

Under the Clean Water Act, the Neuse River Estuary in North Carolina has been declared a section 303(d) impaired water. The typical regulation of 303(d) impaired waters can be characterized as command-and-control regulation. The NPDES permit specifies a maximum Nitrogen emission level from that plant. In contrast, for the Neuse, the state of North Carolina, in conjunction with the EPA, has crafted an innovative EPI that gives the WWTP more flexibility. Rather than require that each point source meet an individual emission requirement, the rules allowed polluters to jointly meet an aggregate group emission requirement by forming an association. The members of the association would not be fined by the State of NC as long as the total aggregate emissions of pollution were below the required level (Hamstead 2008). If a WWTP wants to emit more pollution than its individual requirement, it can do so provided that it secures a corresponding decrease in emissions from another plant. In this way the WWTP are effectively trading emissions. Through this process of trade, an individual WWTP may no longer meet its individual requirement, but the aggregate requirement is still met. The intention of the EPI is to lower the costs of compliance with the Clean Water Act by allowing NRCA members to trade emissions.

The trade process is formalized by giving each WWTP a quantity of emissions permits equal to their individual Nitrogen emission requirement. These permits implicitly define a property right, which the plant may permanently sell or temporarily lease. Trading of these rights is approved by a North Carolina statute. The WWTP are often owned by local municipalities, so they are not necessarily profit maximizing firms. Even so, it is not unreasonable to assume that their goal is to trade permits in such a way as to minimize their total costs of abatement activities and permit purchases.

There are two levels of enforcement of the EPI. At the external level, the State of NC imposes fines if the aggregate emissions of pollution of the association exceed the aggregate nitrogen emission requirement. At the internal level, the NRCA has a complicated system for allocating fines to its own members. The internal fine structure reveals that the NRCA has not fully endorsed the emission trading concept. If actual emissions from a WWTP exceed their individual emission requirement, it must pay an internal fine, even if it purchases permits from another WWTP to cover actual emissions. This is at odds with the typical pollution permit trading scheme in which firms are fined only if they do not have enough permits to cover their actual emissions. The goal of these internal fines seems to be to induce the individual members to upgrade their facilities so as to be able to meet their individual requirement in the future, as much of the fine is returned once such improvements are made (Hamstead 2008).

16.3 The Nitrogen Trading Program in Action

The EPI was put in place to lower the aggregate costs of meeting the overall goal in reduction of Nitrogen emissions. The EPI was quite successful in reducing emissions of Nitrogen, but significant cost savings were not realized because the WWTP did not engage in very many trades. In addition, the design of the market could have been improved by breaking the market up into sub-markets and allowing trade to take place between firms in the same sub-market, but not between firms in different sub-markets. A design of this type allows the optimal trade-off between costs to WWTP of reducing emissions and costs to society from damages from these emissions.

16.3.1 The Nitrogen Trading Program Contribution

16.3.1.1 Environmental Outcomes

The fundamental environmental outcome in this case study is the pounds of Nitrogen emitted by the members of the NRCA. This information is given in Table 16.1 (data provided by the NRCA). The EPI was conducted in a setting in which total Nitrogen emitted into the river from all sources is decreasing over time (Lebo et al. 2012). The NRCA was formed in 2002, and using this year as a baseline yields a 35 % reduction in emissions. At face value, this suggests that the EPI has been dramatically successful in reducing emissions. But the aggregate Nitrogen emission requirement assigned to the NRCA is 1,137,171 lb. Thus the members are emitting 52 % less Nitrogen than they are allowed to emit. This indicates significant over

Table 16.1 Yearly emissions of nitrogen by members of NRCA

Year	Total flow (MGD)	Total estimate pounds N to the estuary
1995	83.808	1,784,130
1996	85.675	1,741,492
1997	81.444	1,653,262
1998	93.442	1,387,717
1999	94.659	1,123,169
2000	92.582	1,056,202
2001	86.818	907,381
2002	89.926	797,991
2003	107.463	711,398
2004	101.203	558,553
2005	101.757	566,627
2006	102.970	542,205
2007	92.994	461,322
2008	90.563	489,789
2009	98.570	497,002
2010	101.852	584,192
2011	93.384	513,269
2012	97.248	540,892
2013	102.847	514,847

Source: Data from the Neuse River Compliance Association

compliance, and suggests that there are other reasons for the marked decline in emissions rather than just the EPI.

In particular, it appears that the WWTP are motivated by the dynamic between future population growth and increasingly strict future regulations (Hamstead 2008). Many of the municipalities anticipate significant population growth in the next 20 years. They also anticipate significantly stricter emissions controls over the same time period. In response to this dynamic, they perceive their optimal strategy is to take steps to install abatement capacity now to be ready to meet these future challenges. Hamstead (2008) suggests that this unusually forward-looking behavior on the part of the WWTP is due to a combination of risk aversion, public image incentives, and altruism.

Evidence for this over compliance behavior comes from analyzing capital expenditures by WWTP to reduce Nitrogen emissions over the last two decades. Members of the NRCA spent US$16 million from 1995 to 1998 and they spent US$31 million from 1998 to 2003 (LNBA 2012). More recently, from 2003 to 2006 the City of Raleigh spent US$40 million on upgrades to their WWTP (Yadkin Riverkeeper 2012).

An ex-post assessment of the environmental outcomes of the EPI itself is difficult to perform because, as discussed below, there was very little actual trading of emissions between the WWTP. As Table 16.1 shows, however, there has been a dramatic decline in Nitrogen emissions since the 1995 fish kill. So we can perform a counterfactual of the overall Nitrogen emissions using the 1995 baseline. The baseline level of emissions for members of the NRCA in 1995 was 1.78 million pounds of Nitrogen per year. This Nitrogen was contained in an outflow of 83,000 MGD from the treatment plants. By 2006, the emissions had been reduced to 0.54 million pounds from an outflow of 102,000. Although some of the increase in the flow was due to an increase in membership of the NRCA, we can use this data to approximate the counterfactual level of emissions by simply assuming the pound/gal rate would have remained constant over time. This implies that if "business as usual" had continued from 1995, there would have been 2.19 million pounds of Nitrogen emitted in 2006. This implies there is actually a 75 % reduction in emissions from the counterfactual.

16.3.1.2 Economic Outcomes

The EPI is centered around an aggregate emissions requirement. This specifies the total emissions across all members in the NRCA. As long as the total emissions are below this requirement, the group is considered to be in compliance with the regulation. An important feature, however, is that each member is still given an individual emissions requirement, and, as discussed above, the internal system of fines within the NRCA is based on this individual requirement (Hamstead 2008).

Regulation with an aggregate emissions requirement has the potential to generate significant cost savings for the members of the NRCA relative to the command-and-control alternative. If one WWTP faces high costs of abating pollution, then it can simply buy emission reductions from another WWTP, which presumably has lower costs. This generates abatement costs savings relative to the alternative in which each WWTP has to meet their own individual emission requirement. A simple aggregate emissions requirement, however, is not the least cost EPI available. Yates et al. (2013) describe a system in which the aggregate emissions requirement is further subdivided into zones. WWTP within a zone may trade emissions with one another, but WWTP in different zones may not. The zonal system strikes a balance between reduced abatement costs and increases in "hot spots". (One can think of the actual EPI, with a simple aggregate emission requirement, as a special case of the zone system in which there is only a single zone.) Allowing WWTP to trade within a zone reduces abatement costs in the manner described above: high cost WWTP can trade with low cost WWTP to the benefit of both. Using zones rather than a single aggregate emission requirement allows a greater control over the spatial distribution of emissions. This reduces the likelihood of a large concentration of emissions in a specific part of the river.

In theory, the ability to trade means that some WWTP would not have to undertake costly abatement. In actual practice, there has been very little trading in the EPI. Apparently the WWTP do not view trading as a method for reducing aggregate abatement costs. The only time that permanent trades took place was when a WWTP went out of business. This occurred twice. The WWTP view trading as a short-term measure. If a WWTP is emitting more than their individual emissions requirement, they can use trading as a temporary fix until they can reduce their emissions (Hamstead 2008). There were six of these temporary trades (leases). As a result of the limited trading, the cost savings of the EPI seem to be minimal.

In the absence of abatement cost savings, the primary benefit of the EPI seems to be related to risk reduction for the WWTP, both in the short term and the long term. In the short term, despite the provisions for trading, the WWTP seem to view it as their responsibility to meet their own individual emission requirement. (This is reinforced by the internal fine structure described above.) The few temporary trades that took place appear to have been motivated as "insurance" against the possibility that they might be temporarily out of compliance with their individual requirement. In the long term, due to the increases in population and the stringency of anticipated future regulation, the WWTP like having the option of trading in case they have trouble meeting future emission requirements (Hamstead 2008).

The EPI did not generate any revenues for the local or national government. The two permanent trades and six temporary trades simply transferred money from one WWTP to another (Hamstead 2008). Alternatively, the individual emissions requirements could have been sold to the WWTP at the start of the program to generate revenue for the State of NC.

The EPI seems to have provided the correct incentives in theory, but not in practice. In the case of this EPI, the correct incentives would have led the WWTP to meet the group emission requirement in the least cost way. All of the theoretical requirements for this to happen are found in the EPI. In fact, the EPI seems to be a classic example of a cap and trade permit market. The actual experience, however, shows that there is a subtle requirement needed to insure that the EPI is successful. In particular, the WWTP have to fully accept the group emission concept. It appears that they did not, as they still felt bound to meet their individual requirement. It will be interesting to see if this changes over time. At the current time, in light of the data in Table 16.1, it appears that it is rather easy for the WWTP to meet their individual emission requirements. Thus the WWTP were not really forced to consider how abatement costs could be reduced by moving from individual to group compliance. From Table 16.1, we see that the total allocation of Nitrogen would have to fall well below 500,000 lb before the WWTP will have strong incentives to consider group compliance. This may occur in the future, as the regulations become increasingly stringent. Thus one would expect there to be an increase in trading activity as emission constraints become more binding and WWTP come to realize that trading will enable them to reduce abatement costs.

16.3.1.3 Distributional Effects and Social Equity

This EPI is tightly focused on the WWTP in the NRCA. The distributional effects and social equity are therefore defined with respect to the WWTP. As discussed above, the WWTP made significant capital expenditures to decrease the emissions of Nitrogen. As most of the WWTP are owned by cities or municipalities, these expenditures were typically paid for by a combination of bond issues and tax dollars. As much or all of these expenditures are likely to have taken place without the EPI, we do not provide estimates of the resulting distributional effects.

Based on qualitative interviews with participants in the EPI summarized by Hamstead (2008), we can, however, identify four components of distributional effects and social equity that are directly attributable to the EPI. A more detailed explanation for these assessments is as follows:

1. Public Image. Participants recognized that public image associated with the EPI could be positive or negative, depending on the emissions outcomes. In practice, the emissions have decreased significantly, so the effect is considered to be positive.
2. Information Sharing. The EPI has provided a forum for both formal and informal information sharing between WWTP. The information includes specific abatement practices and technology as well as insight into the regulatory process.
3. Political Representation. The EPI has created a unified group that represents the interests of the WWTP. This group has more political influence than the individual members would have if they acted alone.
4. Social Benefit. Before the EPI, the WWTP had isolated individual relationships with each other. After the implementation of the EPI, the WWTP began to feel united in working toward a common goal. Interestingly, this common goal seems have been viewed as helping each other meet their individual emission requirements.

16.3.2 *The EPI Setting Up*

16.3.2.1 Institutions

In 2003, the EPA officially issued a new water quality policy to encourage trading between point sources in watersheds with an approved aggregate emission requirement (known as the Total Maximal Daily Load, or TMDL) (USEPA 2003). This policy can be viewed within the context of wider use of pollution permits by the EPA after the successful implementation of S02 permit trading in the previous decade (Boyd et al. 2003). In formulating the 2003 policy, the EPA also cited promising results from a trial water quality trading program in Connecticut and a study that suggested that water quality trading could save almost a billion dollars if implemented nationwide (USEPA 2003).

Although concern about water quality in the Neuse started in the 1970s, the real impetus for stricter regulation of Nitrogen emissions was the 1995 fish kill. The TMDL for the Neuse was approved by the EPA in 2002. In that same year, the General Assembly for the state of North Carolina approved a Wastewater Discharge Rule. This rule enabled the formation of the NRCA and allowed it to jointly meet the TMDL rather than comply with the individual NPDES permit (USEPA 2007). Thus the NRCA can be viewed as a new institution that developed from the change in water quality policy. Although these developments pre-date the official EPA policy that supported trading, it is likely that the EPA was already encouraging trading in advance of the official policy statement.

The failure of the EPI to reach the economic goal does not seem to be related to a failure of institutions. Indeed, all the proper institutions to support trading seemed to be in place. This implies that institutions are necessary, but not sufficient for a successful EPI.

The interactions between the EPI and the institutional setting are summarized as follows. The interactions between the EPI and level 2 institutions are positive. The agreement between the EPA and the legislative and executive branches of the NC state government greatly supported the design and implementation of the EPI. As documented above, there were very few trades that took place. But, for the few trades that did take place, prices played their accustomed role in trade. So we rate this a positive interaction for level 4 institutions at the operation phase.

16.3.2.2 Transaction Costs and Design

Unfortunately, little direct information is available about transactions costs of the EPI, so we must rely on indirect evidence. In the absence of the EPI, the DWQ and the NRCA would still have to monitor, report, and enforce emission levels in the Neuse. (A crude estimate of these costs is US$88,000 per year based on expenditures in 1995 (USEPA 1997).) So this analysis focuses on just the incremental transactions costs associated with actual trading of emissions. Miller and Wolverton (2005) qualitatively classify transactions costs (as being either "low", "medium", or "high") in a variety of water quality trading programs. Trading in the Neuse River is classified as having a "low" level of transactions costs. The authors further note that most of the transactions costs are assumed by the State of North Carolina, presumably by the DWQ. Additionally, Breetz et al. (2004) state that the actual transaction costs for point source to point source trading in the Neuse should be small because of the NRCA. Other indirect evidence comes from a study of point/non-point source trading of water quality permits in Minnesota (Fang et al. 2005.) Here the total transactions cost of a single trade across both the permitting and implementing phase is determined to be US$105,000. Of this total, approximately US$19,000 was incurred by the point source and the vast majority of the rest was incurred by the state agency. Given the qualitative estimates above, it is reasonable to interpret the figure from Fang et al. (2005) as a very crude estimate for the upper bound of the costs per trade. As of 2007, there appears to have been only eight total trades in the history of the EPI (Hamstead 2008), giving an upper bound of US$152,000 of total transactions costs incurred by the members of the NRCA. This compares to a price of US$1.7 million for one of the permanent trades.

The EPI design, implementation, and monitoring involved primarily North Carolina's DWQ, although the EPA played an advisory role and supported the development of trading through its policy. The total time for the development of the EPI was 7 years, from the 1995 fish kill to the formation of the NRCA and approval of permit trading by the General Assembly in 2002. The EPI was applied as a particular implementation of the Clean Water Act.

16.3.2.3 Implementation

The EPI is very flexible, and can easily be adopted widely in other river systems. In these other systems, each large point source is typically allocated a fixed level of Nitrogen emissions (a NPDES permit) by the EPA. To implement the EPI, these individual amounts can be aggregated to determine the total cap on Nitrogen among all the point sources. From this a permit trading system can be set up. As discussed above, the EPA has experience with a similar water quality trading program in Connecticut. And there are similar small regional permit markets for other pollution problems, such as the RECLAIM air pollution trading program in California (SCAQMD 2012).

The experience from the Neuse EPI, however, suggests that one must be concerned that the problem of limited actual trading might also appear when the EPI is applied to these other river systems. It may help to move from the internal fine system found in the Neuse to a more typical external fine system. Here each firm must hold enough permits to cover their own emissions after trading or face external fines. Such a system explicitly moves the emphasis from meeting requirements on Nitrogen before trading takes place to meeting requirements on Nitrogen after trading takes place. Perhaps this will lead the WWTP to more fully embrace the group compliance concept.

The EPI can easily be adjusted following a review of its performance or in response to new information about the damages from Nitrogen emissions. For example, if new information reveals that damages are more severe than previously thought, then the size of the aggregate emission requirement can be reduced.

The major stakeholders in the EPI are the WWTP. They were quite successful in influencing the development of the EPI. In particular, the members of the NRCA were instrumental in convincing the EPA and the Division of Water Quality in NC to set up the group permit system rather than using the traditional individual permit system (Hamstead 2008). Their influence seems to stem from the fact that they had cultivated a long relationship with state regulators. Before the NRCA was formed, many of the WWTP belonged to another group called the Lower Neuse Basin Association (LNBA). This group formed in 1994 to collectively monitor emissions of Nitrogen in the Neuse and worked with the state of NC in this capacity (Hamstead 2008). So the step from the LNBA to the NRCA can be seen as the natural extension of group monitoring of emissions to group compliance of emissions.

The EPI would not have been possible without the cooperation of the North Carolina Division of Water Quality (NC DWQ) and the EPA. The EPA provided support for trading through their water quality trading policy statement (USEPA 2003). But the actual administration of the program is conducted by the NC DWQ. Thus these groups had to be in agreement about the usefulness of implementing the group compliance strategy. This strategy allows for more flexibility in meeting the requirements of the Clean Water Act.

16.4 Conclusions

The results of this EPI are decidedly mixed. On one hand, compared with the typical 303(b) regulation, the aggregate emission requirement and attendant trading system is a big improvement. It offers WWTP the opportunity to greatly reduce the total cost of meeting the Clean Water Act regulation. On the other hand, there was not much actual trading. The WWTP never fully endorsed the group compliance concept, and remained focused on meeting their individual emission requirements. Thus there was very little cost savings associated with the EPI.

Moreover, even in theory, the EPI was not the most efficient type of regulation. In the EPI, there is essentially a single market for the entire Neuse River. Any WWTP may trade permits with any other WWTP. A system of trading zones would perform better. In such a system, groups of WWTP are placed into various zones. WWTP within a zone are allowed to trade with each other, but there is no trade across zones. The zones are designed to account for both the abatement costs and the damages from emissions of pollution. Yates et al. 2013 show that a zone system would lead to several million dollars of overall cost savings per year relative to the current design of the EPI, provided of course that the WWTP actually exploited the opportunities for trade.

In theory, pollution permit trading allows a group of emitters to reach an aggregate emission goal in the least cost way. The individual endowment of emissions is merely the starting point. Firms may increase or decrease emissions from this point through trade. In practice, the cost savings from trading will not be realized if the emitters do not actively participate in the market. In this EPI, the WWTP seemed to view the individual endowment of emissions as the desired outcome. Thus the only trades that occurred were temporary transactions when a WWTP found itself out of compliance with their permit endowment.

The support of the EPA for more flexible trading based regulation was a significant enabling factor for the EPI. In addition, the long established relationship between the stakeholders and regulators at the state level was strong positive influence on the EPI design. The stakeholders had already been successfully applying a group monitoring system, so it was not a large step to move to a group compliance system.

References

Boyd, J., et al. (2003). Trading cases: Is trading credits in created markets a better way to reduce pollution and protect natural resources? *Environmental Science and Technology, 37*, 216A–223A.

Breetz, H., Fisher-Vanden, K., Garzon, L., Jacobs, H., Kroetz, K., & Terry, R. (2004). *Water quality trading and offset initiatives in the US: A comprehensive survey.* Working Paper, Dartmouth College.

Fang, F., Easter, W., & Brezonik, P. (2005). Point-nonpoint source water quality trading: A case study in the Minnesota River Basin. *Journal of the American Water Resources Association, 41*, 645–658.

Hamstead, Z. A. (2008). *Water quality trading in the Neuse River Basin.* Masters thesis, UNC Department of City and Regional Planning.

Lebo, M., Paerl, H., & Peierls, B. (2012). Evaluations of progress in achieving TMDL mandated Nitrogen reductions in the Neuse River Basin, North Carolina. *Environmental Management, 49*, 253–266.

LNBA. (2012). http://www.lnba.net/issue.html. Accessed 18 Jan 2012.

Miller, C., & Wolverton, A. (2005). *Water quality trading in the United States.* Working Paper, National Center for Environmental Economics, USEPA.

NCDENR. (2011). NPDES wastewater permitting and compliance program. http://portal.ncdenr.org/web/wq/swp/ps/npdes. Accessed 28 Nov 2011.

NCDWQ. (2009). Nuese River Basinwide water quality plan. http://h2o.enr.state.nc.us/basinwide/Neuse/2008/documents/Chapter17.pdf. Accessed 28 Oct 2011.

NCDWR. (2010). Neuse River Basin water resource plan. http://www.ncwater.org/Reports_and_Publications/Basin_Plans/Neuse_RB_WR_Plan_20100720.pdf. Accessed 28 Oct 2011.

NCEE. (2011). Neuse River Basin. http://www.ee.enr.state.nc.us/ecoaddress/riverbasins/neuse.150dpi.pdf. Accessed 17 Oct 2011.

SCAQMD. (2012). Regional clean air incentives market. http://www.aqmd.gov/reclaim/reclaim.html. Accessed 18 Jan 2012.

Stavins, R. (1998). What can we learn from the grand policy experiment? Lessons from S02 allowance trading. *Journal of Economic Perspectives, 12*, 69–88.

Stephenson, K., & Shabman, L. (2011). Rhetoric and reality of water quality trading and the potential for market-like reform. *Journal of the American Water Resources Association, 47*, 15–28.

UNRBA. (2011). Upper Neuse River Basin. http://www.unraba.org/aunrb.htm. Accessed 13 Oct 2011.

USEPA. (1997). *Monitoring consortiums: A cost-effective means to enhancing watershed data collection and analysis*. http://water.epa.gov/learn/training/wacademy/its03_index.cfm. Accessed 18 Jan 2012.

USEPA. (2003). Water quality trading policy statement. www.epa.gov/owow/watershed/trading/finalpolicy2003.pdf. Accessed 18 Jan 2012.

USEPA. (2007). *Neuse River Watershed, North Carolina*. http://cfpub.epa.gov/npdes/wqbasedpermitting/wspermitting.cfm. Accessed 28 Nov 2011.

USEPA. (2011). Summary of Clean Water Act. http://www.epa.gov/lawsregs/laws/cwa.html. Accessed 28 Nov 2011.

Yadkin Riverkeeper. (2012). http://www.yadkinriverkeeper.org/content/dean-naujoks. Accessed 18 Jan 2012.

Yates, A., Doyle, M., Rigby, J., & Schnier, K. (2013). Market power, private information, and the optimal scale of pollution permit markets with application North Carolina's Neuse River. *Resource and Energy Economics, 35*(3), 256–276.

Chapter 17
Evaluation of Salinity Offset Programs in Australia

Tiho Ancev and M.A. Samad Azad

Abstract This chapter provides an ex-post policy evaluation of three offsetting programs designed to mitigate irrigation induced salinity in Australia. Environmental effects from salinity are substantial in Australia, with the estimated cost of environmental degradation due to salinity of some A$300 million per year. Offsetting, as an economic policy instrument, is cost-effective in comparison to the conventional regulatory approaches (e.g. engineering approaches or mandate based policies) as it allows environmental improvement to be achieved at reduced cost. Salinity offsets are designed to compensate for salinity impacts from a given agricultural activity by providing a commensurate reduction of salinity impact elsewhere. Policy evaluation of salinity offsetting programs was approached by collecting, collating and processing data pertinent to three Australian case studies. A key finding is that salinity offsets in Australia have been reasonably successful since their implementation. While it was not possible to precisely discern the environmental effectiveness of the offsetting programs, there is clear evidence that the salinity problem has subsided in Australia in the time since the introduction of the offsets, and that they can be at least partly credited for this outcome. At the same time, robust findings about the economic effectiveness of salinity offsetting programs emerged from the study.

Keywords Offset programs • Salinity • Economic effectiveness • Murray-Darling Basin Australia

---

T. Ancev (✉)
School of Economics, University of Sydney, 217 Biomedical Building (C81), Sydney, 2006 NSW, Australia
e-mail: tiho.ancev@sydney.edu.au

M.A.S. Azad
University of Sydney and Tasmanian School of Business & Economics, University of Tasmania, 206 Biomedical Building (C81), Sydney, 2006 NSW, Australia
e-mail: Samad.Azad@utas.edu.au

M. Lago et al. (eds.), *Use of Economic Instruments in Water Policy*, Global Issues in Water Policy 14, DOI 10.1007/978-3-319-18287-2_17

17.1 Introduction

Salinity of river water and soil has been a long-standing problem in Australia, in particular in areas with significant irrigation development, such as the lower reaches of the Murray-Darling Basin (MDB). The problem manifested strongly in the 1980s and 1990s, leading to significant research efforts into ways to mitigate it. Around the same time, the use of economic policy instruments (EPI) became prominent in resource management. Salinity offsets have been proposed as an effective EPI for mitigating irrigation induced salinity, and have been subsequently implemented in several areas throughout Australia. This chapter closely examines three salinity offset schemes: the one implemented in the Coleambally Irrigation Area; the Ulan coal mine salinity offset program; and the salinity zoning with offsetting in South Australian portion of the River Murray.

Salinity offsets are designed to compensate for salinity impacts from a given agricultural or other productive activity in a particular area by providing a commensurate reduction of salinity impact elsewhere. The end result is that there is no net increase in the overall salinity impact. The key mechanism of this EPI is to recognise the heterogeneity in abatement cost structures across space and across different enterprises. The main idea is to allow an enterprise with relatively low cost of abatement, or located in an area where the environmental impact is low, to provide an offset for the effects of another, higher cost enterprise located in an area where environmental effects are high. For instance, salinity impact of an irrigated agricultural activity can be offset by establishing new perennial pastures or by revegetation, both of which have an effect of reducing salt loads, and are also low-cost options. In general, salinity offset programs can be used to mitigate salinity at a cost that is an order of magnitude lower than using on-site engineering measures alone to achieve the same reduction (Connor 2004). Salinity offsets can also be an important feature of other policies for irrigation induced salinity mitigation. For example, under an irrigation zoning policy (e.g. the one currently in place in South Australia), salinity offsetting can allow for less costly and more effective reduction of salinity compared to a policy without offsetting (Spencer et al. 2009). This reduces the cost of meeting a given overall salinity load target.

Policymakers in Australia have been active in considering, testing and implementing policy instruments based on economic incentives in relation to water and salinity management. Several policies designed to address increasing water scarcity and salinity problems have been instigated in Australia in general, and in MDB in particular, over the last two decades (Lee et al. 2012; Connell and Grafton 2008). Examples of initiatives within the policy mix to address salinity are: the Joint Works Program (Basin Salinity Management Strategy) and the Natural Heritage Trust, National Action Plan for Salinity and Water Quality, and the current National Water Quality Management Strategy (Lee and Ancev 2009). In addition, many initiatives to explore the possibilities to use various EPIs for salinity mitigation were put in place such as the National MBI (market based instruments) pilot program for natural resource management (BDA Group 2009).

This chapter provides ex-post policy evaluation of three salinity offsetting programs – Coleambally Irrigation Area (CIA), Ulan Coal Mine (UCML), and the South Australian (SA) Irrigation Zoning Policy – with an aim to evaluate their performance since implementation on a range of criteria, and to discern the noted shortcomings of the programs, or the noted features that have been working particularly well. An additional aim is to identify aspects where possible improvements in the existing offsetting programs could be achieved. The literature that reports on evaluation of salinity offset programs (Connor 2008) has been fairly sparse, both in Australia and internationally. This chapter fills that gap by providing a comprehensive evaluation of the considered salinity offset programs.

17.2 Setting the Scene: Challenges, Opportunities and EPIs

In terms of hydrology, Australia is the driest continent in the world on the basis of runoff per unit area. This is due to the high rate of evapotranspiration, the unparalleled temporal and spatial variability of rainfall intensity and frequency, and the generally flat topography across most of the continent (National Water Commission 2005). Nevertheless, significant irrigation activities have been established, mostly throughout the twentieth century: the irrigated area has grown from 350,000 ha in 1941 to more than 2 million hectares in 1997 (ANRA 2008). A large proportion of irrigation – 52 % of total irrigated land – takes place within the Murray-Darling Basin (MDB).

An inadvertent follower of the agricultural and irrigation development, salinity is one of the most significant environmental threats in Australia. It affects the ecological health of rivers, wetlands and streams, and reduces the productivity of crops and pastures. The estimated cost of environmental degradation due to salinity is substantial. Total annual cost of land and water degradation in Australia was estimated at A$1,365 billion, large proportion of which can be directly or indirectly attributed to salinity related degradation (Pigram 2007). Estimated annual costs of salinity include A$130 million in lost agricultural production, A$100 million in infrastructure damage, and at least A$40 million in loss of environmental assets (CSIRO 2008).

In general, offsets can be defined as actions that are undertaken away from the physical location of an activity to compensate for its negative environmental impact. A pollution offset can ensure with some level of confidence that there is no net increase in the load of a particular pollutant entering the environment as a result of a given activity (Tietenberg 2006). Offsetting allows new or expanding pollution sources to commence operations in a given area where there are attainment standards for a particular pollutant, provided they acquire sufficient offsetting credits from existing sources. Offsetting credits can be obtained by certified reduction of environmental impact from existing sources. Salinity offsets were recently used in three separate cases within the Murray-Darling Basin in Australia. These are the Coleambally Irrigation Area (CIA), the Ulan Coal Mine (UCML), and offsets under the South Australian (SA) Irrigation Zoning. The effect that offsetting has had in each of these case studies is briefly presented in the following sections.

17.3 Salinity Offsets in Action

The Coleambally Irrigation Area (CIA) is located in South Western New South Wales (NSW) within the MDB (Fig. 17.1). It was developed for irrigated agriculture between 1958 and 1970. Main crops that are grown are rice and other cereal crops and pastures. CIA has experienced significant problems of waterlogging and salinity (Whitten et al. 2005). Prior to irrigated agriculture, watertables[1] in the CIA were about 20 m below the surface. This was followed by dramatic increases in the period between 1981 and 1991 due to deep drainage of irrigation water below the root zone of the crops, and into the shallow aquifer (Rowe 2005). The extent of area with a

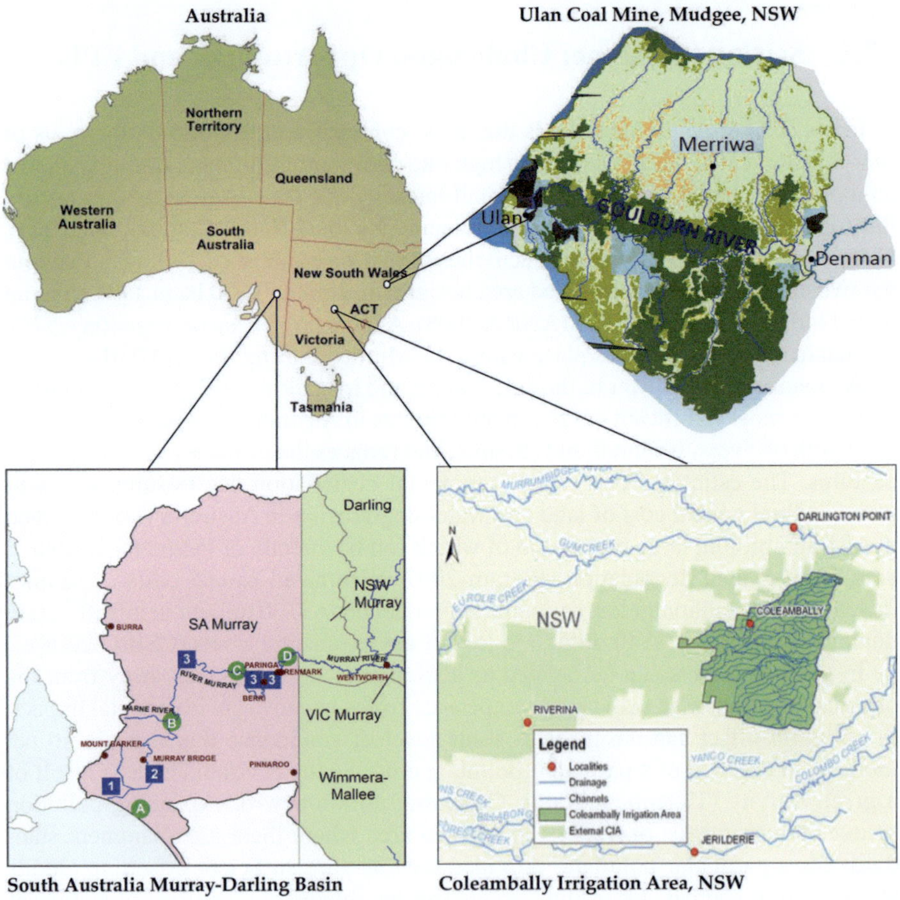

Fig. 17.1 Location of salinity offset case study areas (Source: Own elaboration)

[1] Watertable is the surface where the water pressure head is equal to the atmospheric pressure. Simply it can be visualized as the surface of the subsurface materials that are saturated with groundwater in a given vicinity (Freeze and Cherry 1979). Individual points on the water table are typically measured as the elevation that the water rises to in a well screened in the shallow groundwater.

watertable within 2 m of the surface was about 26,800 ha in 2000/2001. It was predicted that the land area within the CIA under which the watertables are very shallow (less than 2 m from the surface) would rise to 50,000 ha by 2013 and to 60,000 ha by 2023 if no further watertable and salinity management actions were taken (Rowe 2005). To address these problems, a Net Recharge Offsetting Policy has been implemented in the area since 2005 under the auspices of the Coleambally Land and Water Management Plan (LWMP).

Ulan Coal Mine (UCML) is located in the Central West of NSW (Fig. 17.1). It is a 'surplus water' mine: approximately 8.2 ML more water per day is generated through underground mine dewatering than can be re-used through mining activities. This surplus water has historically been released into the Ulan Creek flowing into the Goulburn River, which is a tributary of the Hunter River. As Ulan mine is the only major mine within the Hunter Valley Catchment not involved in the widely known and studied Hunter River Salinity Trading Scheme (Shortle and Horan 2008), it has developed an offsetting program to mitigate salinity impacts resulting from irrigating agricultural crops using the water from the mine. Salinity offsetting is based on the establishment of the Bobadeen Irrigation Scheme (BIS) in 2003. With commissioning of the BIS, surplus mine-water was used to irrigate about 250 ha of land under perennial pasture. As part of the implementation of the BIS, a salinity offset area was established to offset residual salt loads from irrigation activities.

The South Australian Murray-Darling Basin covers 70,000 km^2 (about 7 % of South Australia), and its landscape varies from the low-lying coastal plains of the Coorong to the flat expanse of the Mallee to the steeper slopes of the Eastern Mount Lofty Ranges. Highly saline groundwater naturally flows into the River Murray from the surrounding landscape. Irrigation has accelerated the rate at which the saline groundwater is now entering the River Murray and the floodplain. To address the issue, irrigation zoning policy that restricts the location of new irrigation developments to areas where salinity impact is relatively low has been in place in the irrigation regions along the River Murray in South Australia since 2005 (DWLBC 2005). Salinity offsets are a constituent part of this policy.

17.3.1 The EPI Contribution

17.3.1.1 Environmental Outcomes

The Coleambally Irrigation Area is currently implementing a Net Recharge Policy to mitigate salinity impact of irrigation farms. The offsets under this policy are in the form of planting certain crops that are capable of reducing the level of groundwater recharge, or directly reducing groundwater table. In the period 2002–2008, annual allocations to irrigation water holders have been significantly reduced due to the effects of the prolonged drought (Grafton and Hussey 2007). This period coincides with the time of introducing the Net Recharge Offsets in the CIA in 2005. As a consequence of the dramatic restriction of annual allocations, but also as a result of activities designed to mitigate salinity, including the Net Recharge Offset policy, the area with groundwater levels within 2 m from the surface in the CIA reduced from over 25,000 ha in year 2000 to some 1,700 ha in September 2006. The area of land with watertable within 2 m from the surface reduced further to just 400 ha in 2007 (CICL 2007), and even further to 258 ha in September 2010 (CICL 2010a).

Table 17.1 shows the monthly average salinity over the period 2007–2010, including a benchmark year. It is observed that the salinity level at the two licensed discharge sites and one licensed monitoring site has remained below 200 μS/cm over the period, which indicates a significant improvement in comparison to the benchmark salinity. Lower salinity at the drainage monitoring sites is due to the lowering of groundwater tables within the CIA. The reduction in watertables below the level of the bed (base) of the drainage channels means there is no salt intrusion from watertable into drainage water.

The Bobadeen irrigation scheme and the associated salinity offset program are integrated in the Ulan Coal Mine's environmental management system. The salinity offset program has had positive environmental outcomes. During the period 2009–2010 the average daily discharge of water at Ulan Creek was calculated to be 6.78 ML/day, while the mining activities involved discharging around 11 ML/day before the implementation of the salinity offset program in 2004–2005 (Table 17.2). The pH range for the discharged water was 6.5–8.5 for 2009–2010, with the average pH of 7.41. The average Electrical Conductivity (EC) was 730 μS/cm, with the maximum EC recorded at about 1,000 μS/cm (Table 17.2). The above values are compared to the measurements observed before the offsetting program was implemented, as displayed in Table 17.2.

Table 17.1 Average monthly salinity (μS/cm) at three licensed discharge, and one monitoring point, CIA (CICL 2010a)

Location	Benchmark[a]	2007/2008	2008/2009	2009/2010
Coleambally catchment drain	117	115	161	138
Coleambally drainage channel	510	151	272	232
West Coleambally channel (discharge point)	660	45	167	154
West Coleambally channel (monitoring point)	712	163	108	159

[a]Benchmark includes average data from 1996/1997, 1997/1998 to 1998/1999

Table 17.2 Change in some environmental variables before (2004–2005) and after (2009–2010) the implementation of the salinity offset program, Ulan Coal Mine (UCML 2006, 2010)

Environmental variables	2004–2005	2009–2010
Daily discharge of water (ML/day)	11.0	6.78
pH range	6.7–9.8	6.5–8.5
Electrical conductivity (μS/cm)	1,000–1,200	277–1,013

Natural inflows into the River Murray in South Australia have been at record lows over the last 7–8 years, with an absolute minimum of 360 GL in 2007. Such dismal water availability was paralleled with severe restrictions of water allocations to irrigated agriculture (SADW 2011, Government of South Australia Department for Water. Personal communication, Mr. Christopher Wright). This situation was reflected in significantly reduced interest in establishing new irrigation activities within the SA Murray. The reduced river flows over the last 10 years also had implications on the dynamics of salinity itself. One possible implication is that due to minimal water inflows, which may be insufficient to dilute the natural saline inflows, there could be significant rise in river salinity. On the other hand, as a result of actions taken at the MDB level (e.g. Murray-Darling Basin Salinity and Drainage Strategy implemented 1988–2001 (MDBC 2003), the salinity pressures in the lower parts of the River Murray eased. The trend analysis on the average salinity levels measured at Morgan[2] since 1980 shows that measurements of electro conductivity taken in 2003 were averaging about 525 μS/cm, which was considerably lower than the previous 20-year average (MDBC 2009). Current measurements of electro conductivity at Morgan are around 300 μS/cm (River Murray Data, 2011; http://data.rivermurray.sa.gov.au).

17.3.1.2 Economic Outcomes

The economics of net recharge policy for Coleambally Irrigation Area can be assessed by evaluating the changes in net farm income (gain or loss) that result from changing farming activities due to the net recharge policy. The costs and benefits of the net recharge policy depend on the dynamics of the area of land planted with perennial and annual deep rooted crops, in relation to the area planted with rice. It may be argued that the net recharge salinity offset is more cost-effective than any other available option to reduce groundwater table, in terms of operational and implementation cost. There is evidence that the offset program was considerably less costly than other options for salinity mitigation, including desalination by reverse osmosis, which was seriously considered as an alternative (Whitten et al. 2005).

In case of the Ulan Coal Mine, the salinity offset program required an initial investment by the mine of an estimated A$1.4 million, with annual operating and maintenance costs of about A$94,000 (DEC 2005a). On the other hand, establishing a desalination plant that would have been used to treat the effluent discharge from the mine to the locally acceptable stream ambient concentration levels would have required an initial investment of about A$15 million, with ongoing operational cost

[2]Morgan is a town on the River Murray in South Australia, which is often used as a location for benchmarking water quality, especially salinity, as the salinity readings at Morgan are good indication of the possibility to use river water for drinking water supply to the city of Adelaide. The 'magic' number is 800 EC (electroconductivity) units (or μS/cm), which is the maximum allowed value for the electroconductivity indicator for drinking water.

of about A$6 million per year. This implies that savings of approximately A$91 million in terms of net present value over the next 20 years can be achieved by using the salinity offsets as opposed to installing a desalination plant (DEC 2005a). The cost-effectiveness of the salinity offsetting program for Ulan Coal Mine can be assessed based on the annualised cost of the program, and the estimated residual salt loads that are avoided as a result of the program. Assuming a total productive life period of 20 years for the mine, the annualized cost of the initial investment (A$1.4 million) can be estimated at A$132,150 using an interest rate of 7 %. Adding this to the annual operation costs of A$93,500 gives a figure for the total annualised cost of the salinity offset at A$225,650. Combining this figure with the predicted residual salt load of around 280 tonnes a year avoided as a result of the offsetting program, gives the unit cost for salinity impact reduction through the salinity offsets at A$806 per ton of salt load avoided. This compares very favourably with the costs of any other alternatives.

There is currently no ex-post information available on the value, costs, or prices involved with salinity offsets within the irrigation zoning policy in South Australia. Spencer et al. (2009) compared ex-ante the cost-effectiveness of standalone irrigation zoning policy to an irrigation zoning policy with salinity offsets. Their findings show that offsetting policy provides a better salinity outcome that can be achieved at lower cost than with standalone zoning policy. Average cost of reducing salinity for the salinity offsetting policy is A$148,980/1 EC unit, which is A$48,850/1 EC unit lower than that for standalone zoning policy.

17.3.1.3 Distributional Effects and Equity

The initial salinity problem in the Coleambally Irrigation Area is a clear example of an ownership externality. Each individual irrigator has an incentive to apply irrigation water to their crops, parts of which will drain in the shallow groundwater, raising the water table and aggravating the salinity problem for everyone. Thus, the distributional effects of the offset program are to 'privatise' a 'public bad', which is achieved by requiring each farm to take into account its contribution to the raising water table and, when the circumstances are critical, to offset that contribution. All salinity mitigation programs in CIA, including net recharge offsetting, contribute to long term social equity and sustainability, as they contribute to overcoming the possibility of widespread soil salinisations, which could seriously threaten farming in this region, and consequently threaten the affected rural communities.

In the Ulan Coal Mine the distributional effects of the offsetting scheme are in relation to the transformation of the environmental damage cost to the public (when the salty water was directly discharged in the river system) into abatement cost to the private entity that is the source of the environmental threat (the cost of the offsetting scheme to the UCML). This is a desirable outcome in its own right. The success of this scheme is even more apparent when the magnitude of the abatement costs is considered in relation to other possible alternatives, indicating that improvement of distributional effects from environmental degradation has been achieved in a cost-effective way.

The irrigation zoning policy in South Australia has a clear distributional effect of favouring established irrigation activities over new irrigation activities. Perhaps inadvertently, this policy effectively applies 'grandfathering' to the 'right' to generate salinity impact. The offsetting feature rectifies this bias, by clearly expressing the opportunity cost of irrigation activities in terms of their salinity impact. Standalone irrigation zoning policy provides perverse incentives for old, possibly technologically obsolete irrigation enterprises that may be using irrigation water inefficiently and creating substantial salinity impact to remain in operation, as they will not be able to capitalise on their implied 'right' to create salinity impact, due to the restricted transferability of water rights among salinity impact zones (e.g. without offsetting, an existing enterprise in a high salinity impact zone will not receive any reward should they decide to cease their operation). The offsetting removes this perverse incentive, as an established operation can get a monetary reward by 'selling' their offset, should they decide to cease operation. The institutions of property, or 'use', rights that are implied by the salinity offset in this case have been gaining popularity in water management applications in Australia. These institutions are increasingly better understood and accepted by the public.

17.3.2 The EPI Setting Up

17.3.2.1 Institutions

In Coleambally Irrigation Area, the net recharge offset policy is being implemented under the management of the irrigation cooperative. The use of offsets within the cooperative is an excellent example of institutional innovation, where the community itself (in this case the community of irrigators) recognises the inadequacy of the existing institutions (i.e. open access treatment of the environment), and comes up with a new institution that is designed to deal with an environmental problem. Other institutions partly involved in this program include the Murrumbidgee Catchment Management Authority, NSW Office of Water, Department of Primary Industries (NSW), Coleambally Outfall District Water Users Association, Department of Land & Water Conservation (now DNR), and Department of Environment and Climate Change. The Coleambally Irrigation Cooperative Limited is currently taking part in activities under the "Water Smart Australia" program under the Australian Government's Water for the Future plan to reduce the environmental footprint (including salinity) of irrigated agriculture.

The salinity offset program for the Bobadeen Irrigation Scheme is operated by the Ulan Coal Mine Limited as a part of its environmental protection licence that is issued by the NSW Department of Environment, Climate Change and Water (DECCW). The license stipulates that UCML must develop a program to offset the residual salinity load arising from the irrigation of mine-water generated at the premises so that there will be no net increase in salinity load in the Macquarie and Hunter catchment areas as a result of the irrigation activities. Other institutions such as the Hunter-Central Rivers Catchment Management Authority (CMA), the local municipal council and the community consultation committee were involved to implement the salinity offset program.

Within South Australia, the irrigation zoning policy is administered by the South Australian Department of Water (SADW). Other agencies concerned with management of salinity along the River Murray in SA are the Murray-Darling Basin Ministerial Council, and the South Australia Murray-Darling Basin Natural Resources Management Board.

17.3.2.2 Transaction Costs and Design

Transaction costs are an important factor to consider while assessing the feasibility of EPIs for managing water resources and environmental quality. For example, the initial costs of setting up a cap and trade scheme, including unbundling land and water rights, are thought to be high. In a recent study Ancev (2011) found that the transactions costs of mandating the agricultural sector in a tradeable permit scheme for Green House Gas mitigation would be high. This is in line with previous findings specific to the Coleambally irrigation area (Whitten et al. 2005), which suggested that cap and trade mechanism for salinity mitigation in this case is not feasible, at least partly due to high transactions cost such as early implementation costs, establishing a register of permits, and the costs of trading in salinity permits. There are also ongoing public costs associated with administering salinity permit trades, monitoring water use and maintaining the integrity of the trading system through enforcement. Relatively lower transactions costs under offsets was part of the reason why a salinity offsetting program was preferred to a cap and trade mechanism in the CIA.

Transaction costs of the salinity offsetting program for Ulan Coal Mine Limited are not overly high. These involve mainly the costs of producing reports and other compliance documents; cost of publishing those reports; cost of monitoring of ambient environmental quality; cost of early termination of lease contracts with farmers. Early implementation costs of the salinity offsetting program were estimated at about A$921,000 (Source: DEC 2005b).

The existence of significant transactions costs are possibly a reason for observing limited use of salinity offsets in practice in South Australia. It appears that no activities have been taken by the South Australian government in relation to aiding potential participants in salinity offsetting: there is no register of offsets, trade register, or some sort of clearance house. These usually represent a large proportion of the early implementation costs (Jaraite et al. 2010). However, the absence of registers probably makes transactions costs for potentially interested irrigation developers prohibitively high. Because there is an absence of structured government approach towards salinity offsets within the irrigation zoning policy, the requirements on individual participants willing to buy or sell offsets are very large. This comprises the need to search for a counterparty, the need for adequate contracting, the need to navigate through administrative requirements, and the need to ensure compliance with the policy. The costs of these are likely to be very high, which probably acts as a deterrent for potentially interested parties to engage in offsetting.

17.3.2.3 Implementability

There are number of principles underlying the net recharge policy that serve the purpose of its implementation. The CIA undertakes an annual assessment of farm-based irrigation intensity across all farms within the Coleambally Irrigation Area against two specific criteria (CICL 2010b): (a) If total farm water use (including on-farm bores) exceeds 6.5 ML/ha, the shareholder must demonstrate that net recharge is being controlled by using the Swagman Farm Model or Net Recharge Offsets (Madden and Prathapar 1999), and (b) If the area of the CIA with a watertable within 2 m of the surface is greater than 10,000 ha (based on piezometer data) and if total farm water use (including on-farm bores) exceeds 5.5 ML/ha, then the shareholder must demonstrate that net recharge is being controlled by using the Swagman Farm Model or Net Recharge Offsets. There is a range of prescribed penalties for breaching the above irrigation intensity limit including sanctions against non-compliant rice growers. Within the corporation, rice growers who contravene the environmental policies will be invited to discuss the issue. If a breach is deemed to have occurred, sanctions can be applied, including (i) reductions in rice area and/or refusal to supply water, (ii) mandated soil testing, and (iii) other penalties as determined by the relevant jurisdiction.

In case of Ulan Coal Mine the offsetting program was implemented under the environmental protection licence, which is stemming from the Protection of the Environment Operations Act of NSW. The offsetting was first instigated under a pollution reduction program negotiated between NSW DECCW and Ulan Coal Mine Limited, before becoming the part of the environmental protection licence. The implementability and enforceability of the program is straight forward, as incentive compatibility of the offsetting instrument to the objectives of the mine is evident.

The salinity zoning policy in South Australia has been developed in relation to the salinity management goals of the Water Allocation Plan for the River Murray. This policy ensures that South Australia's salinity management is in line with the salinity management provisions of the Murray-Darling Basin Agreement. Under the Agreement, the states of New South Wales, Victoria and South Australia have committed to keep an up-to-date salinity register, which is used to record all activities that reduce or increase salt loads. Actions that increase salt loads, such as new irrigation developments result in a debit, whereas actions that mitigate salt loads result in a credit (Young et al. 2000). Under the agreement the register needs to be in surplus (credit) at all times. These provisions are directly related to the provisions of the Irrigation Zoning Policy for new developments in the low salinity impact zones.

17.4 Conclusion

The findings that emerged from the collected evidence are mostly consistent across the three considered offsetting programs. In terms of environmental effectiveness, it is not possible to clearly discern the effects of the offsetting programs from the effects pertinent to the climatic and hydrologic conditions over the last 7–8 years. At any rate, the salinity threats in Australia have abated over the period, and various salinity mitigation initiatives, including offsets, can probably claim at least some credit for it. The real environmental effectiveness of the offsets will be tested when the climatic conditions allow for improved irrigation water availability, as is currently the case.

The economic effectiveness of salinity offsetting programs is clear. In all cases, salinity offsets provided a cost-effective way to mitigate salinity when compared to alternative approaches. In addition, salinity offsets have desirable distributional effects, as they transform the costs associated with the environmental damage borne by the public at large, to costs associated with providing the offsets borne by those who cause the environmental damage. The social effects of the offsets are minor, and in principle they can be seen as enhancing social equity in relation to environmental health.

The institutional innovation represented through the implementation of salinity offsets is probably the most exciting and promising feature of these programs. Incentive based approaches to deal with environmental problems, including tradable permits, taxes, and offsets, have become widely accepted in Australia over the last decade. Given that this type of approach effectively corrects for an outdated institution that has governed resource use and environmental management (i.e. the institution of 'open access') in the past, it is satisfying to witness that new institutions that highlight the importance of property rights, are slowly but surely taking the front stage in this domain.

The shortcomings of the reviewed offsetting programs relate to potentially high transactions costs, especially in relation to the environmental outcomes from salinity offsets. While in some cases the transactions costs appear to be acceptable (UCML) due to the small number of affected agents, they are likely to be very high in other cases (Irrigation Zoning in SA). In the latter case, there is clear opportunity for the Government of SA to provide some services (e.g. register of interest for salinity offsets in the high salinity impact zones) that will reduce the transactions costs for the prospective participants in the salinity offsetting. Governments can also be instrumental in improving the performance and uptake of salinity offsets by supporting further research into quantification and management of the uncertainty related to environmental offsets in general, and salinity offsets in particular.

Overall, this chapter finds that salinity offsets in Australia have been reasonably successful since their implementation. Their very existence is a positive development, and an important addition to the policy mix to deal with future environmental and natural resource challenges related to agricultural water use.

References

Ancev, T. (2011). Policy considerations for mandating agriculture in a greenhouse gas emissions trading scheme. *Applied Economic Perspectives and Policy, 33*, 99–115.

ANRA. (2008). *Irrigation overview – natural resource topics*. Australian Natural Resources Atlas (ANRA), Land and Water Australia, Australian Government, Canberra, ACT.

BDA Group. (2009). Final report of the national market based instruments pilot program http://nrmonline.nrm.gov.au/catalog/mql:2417

CICL. (2007). *Annual compliance report 2007*. Coleambally: Coleambally Irrigation Cooperative Limited (CICL).

CICL. (2010a). *Annual compliance report 2010*. Coleambally: Coleambally Irrigation Cooperative Limited (CICL).

CICL. (2010b). *CIA net recharge policy 2010*. Coleambally: Coleambally Irrigation Cooperative Limited (CICL).

Connell, D., & Grafton, R. Q. (2008). Planning for water security in the Murray-Darling Basin. *Public Policy, 3*(1), 67–86.

Connor, J. (2004). Market based policy for River Murray salinity. Contributed paper, 48th conference of Australian Agricultural and Resource Economic Society, Melbourne. Available at: http://www.agric.uwa.edu.au/ARE/AARES/rest2004/Connor%20J.pdf

Connor, J. (2008). The economics of time delayed salinity impact management in the River Murray. *Water Resource Research, 44*, W03401. doi:10.1029/2006WR005745.

CSIRO. (2008). *Salinity fact sheet*. CSIRO Land and Water, Commonwealth Scientific and Industrial Research Organisation. Available from URL: http://www.clw.csiro.au/publications/factsheets/SalinityFactSheet.pdf. Accessed 4 Oct 2010.

DEC. (2005a). *Institutional framework for implementing salinity offsets under the Protection of the Environment Operations Act 1997*. National market-based instrument pilots program. New South Wales: Department of Environment and Conservation (DEC).

DEC. (2005b). *Green offsets for sustainable regional development-ID16*. Final report, national market-based instrument pilots program. New South Wales: Department of Environment and Conservation (DEC).

DWLBC. (2005). *River Murray salinity zoning fact sheet*. Government of South Australia. www.waterforgood.sa.gov.au/wp-content/uploads/2010/07/fs0072_river_murray_salinity_zoning.pdf

Freeze, R. A., & Cherry, J. A. (1979). *Groundwater*. Englewood Cliffs: Prentice-Hall.

Grafton, R. Q., & Hussey, K. (2007). Buying back the living Murray: At what price? *Australasian Journal of Environmental Management, 14*(2), 74–81.

Jaraite, J., Convery, F., & Di Maria, C. (2010). Transaction costs of Firms in the EU ETS: Lessons from Ireland. *Climate Policy, 10*, 190–215.

Lee, L. Y., & Ancev, T. (2009). Two decades of Murray-Darling water management: A river of funding, a trickle of achievement. *Agenda, 16*(1), 5–23.

Lee, L., Ancev, T., & Vervoort, W. (2012). Evaluation of environmental policies targeting irrigated agriculture: The case of the Mooki catchment. *Agricultural Water Management, 109*, 107–116.

Madden, J. C, & Prathapar, S. A. (1999). *Net recharge management – SWAGMAN farm: Concepts and application to the Coleambally irrigation area*. CSIRO Land and Water, Consultancy Report No. 99/11.

MDBC. (2009). *Land and water salinity – Murray-Darling Basin Commission*. Available from URL: http://www.mdbc.gov.au/salinity/land_and_water_salinity

Murray-Darling Basin Commission (MDBC). (2003). Basin salinity management strategy: 2001–2002 annual implementation report. http://www.mdba.gov.au/sites/default/files/archived/mdbc-salinity-reports/2064_BSMS_2001-02.pdf

National Water Commission. (2005). *Australian water resources 2005*. Australian Government, Canberra, ACT 2601.

Pigram, J. J. (2007). *Australia's water resources: From use to management*. CSIRO Publishing, VIC 3168.

Rowe, M. (2005). *Coleambally LWMP five-year review*. IREC Farmers' Newsletter, No. 170, Spring 2005. Hanwood, NSW 2680.

Shortle, J. S., & Horan, R. D. (2008). The economics of water quality trading. *International Review of Environmental and Resource Economics, 2*(2), 101–133.

Spencer, T., Ancev, T., & Connor, J. (2009). Improving cost effectiveness of irrigation zoning for salinity mitigation by introducing offsets. *Water Resources Management, 23*(10), 2085–2100.

Tietenberg, T. (2006). *Environmental natural resource economics* (7th ed.). Boston: Pearson Education.

UCML. (2006). *Annual environmental management report, 2005*. Mudgee: Ulan Coal Mines Limited (UCML).

UCML. (2010). *Annual environmental management report, 2009–10*. Mudgee: Ulan Coal Mines Limited (UCML).

Whitten, S., Khan, S. M., Collins, D., Robinson, D., Ward, J., & Rana, T. (2005). *Tradeable recharge credits in Coleambally irrigation area: Report 7, experiences, lessons and findings*. CSIRO & BDA Group. Melbourne, VIC 3122.

Young, M., Hatton McDonald, D., Stringer, R., & Bjornlund, H. (2000). *Inter-state water trading: A two year review*. Report to the Murray Darling Basin Commission. Adelaide: CSIRO Land and Water.

Chapter 18
Water Trading in the Tagus River Basin (Spain)

Gonzalo Delacámara, C. Dionisio Pérez-Blanco, Estefanía Ibáñez, and Carlos M. Gómez

Abstract Population and economic growth, coupled with rapid and extensive urban development, pushed to the limit the capacity of the upper and middle stretches of the Tagus River Basin to meet an increasing water demand, within the range of available resources and current water regulation infrastructures. In this context, voluntary agreements to transfer water use rights from agriculture to urban uses gained social support and political acceptance as an alternative to cope with the recurrent water supply deficit during dry periods. This was mainly because of their lower cost as compared to the best available alternatives already in place (efficiency improvements, use of strategic reserves, additional water works). Since the early 1990s pioneer voluntary agreements to formally transfer water between water utilities and irrigation districts sprung up for the first time in Spain. This chapter assesses two trades in the Madrid Region (including the capital city, Madrid's metropolitan area and other towns). These trades can be arguably considered as 'embryonic' examples of formal water use right trades in Spain.

Keywords Water trading • Water use rights • Tagus River basin

---

G. Delacámara • E. Ibáñez
Madrid Institute for Advanced Studies in Water (IMDEA-Water Foundation), Punto Net 4, floor 1, Edificio ZYE, Parque Científico-Tecnológico de la Universidad de Alcalá, 28805 Alcalá de Henares (Madrid), Spain
e-mail: gonzalo.delacamara@imdea.org; estefania.ibanez@imdea.org

C.D. Pérez-Blanco (✉)
Centro Euro-Mediterraneo sui Cambiamenti Climatici, Fondazione Eni Enrico Mattei (FEEM), Isola di San Giorgio Maggiore, 30124 Venice, Italy
e-mail: dionisio.perez@feem.it

C.M. Gómez
Departamento de Economía, Universidad de Alcalá and IMDEA-Water Foundation, c/ Plaza de la Victoria s/n, 28803 Alcalá de Henares (Madrid), Spain
e-mail: mario.gomez@uah.es

M. Lago et al. (eds.), *Use of Economic Instruments in Water Policy*, Global Issues in Water Policy 14, DOI 10.1007/978-3-319-18287-2_18

18.1 Introduction

Shaped by a poor and uncertain natural supply the need to manage water resources on a collective basis has been a persistent political and social concern in the history of Spain. Responses to the increasing water gap have too often come from the supply side. Nevertheless, informal trades of water, mostly to meet short-run needs have also been pervasive (Hernández-Mora and De Stefano 2013). Formal water markets in Spain, however, are still incipient (Gómez et al. 2013).

Population and economic activity in the international Tagus River are concentrated in Madrid and Lisbon at both ends of the river basin district. Scarcity problems, though, are in principle more significant in the former.

Although the quality of surface water is relatively good in the upper basin, only one tenth of rainfall and runoff is available to cope with more than two thirds of the demand for urban uses in the entire watershed. In clear contrast to that, the intermediate reach of the Tagus River basin is dominated by the use of water for extensive agriculture (TRBA 2014), disputed by the more productive[1] irrigated agriculture in the Segura River basin (SE Spain), close to the Mediterranean coastline and which is connected to the Tagus by a major diversion project (Tagus-Segura Water Transfer).

Within the last three decades water management in Madrid has been a clear example of a gradual adaptation towards a more efficient use of infrastructures, together with pricing schemes and other incentives designed to adjust water demand (TRBA 2014). However, in the two decades before the current downturn, the combination of intense demographic change, economic expansion, and rapid and extensive urban development pushed to the limit the capacity to manage an increasing water demand within the range of available resources and current water infrastructures.

Since the early 1990s, voluntary agreements to transfer water use rights from agriculture to urban uses emerged as an alternative to cope with the recurrent water supply deficit during a number of dry periods since then (especially the two drought events of 1990–1991 to 1994–1995, and 2004–2005 to 2007). This option gained social and political momentum since the costs of the best available alternatives already in place (efficiency improvements, use of strategic reserves, additional water works) grew in the margin (Estevan and Lacalle 2007). In fact, efficiency in water treatment and distribution in Madrid is already above 80 %; a high percentage of wastewater is currently being re-used for watering public gardens, for high-pressure street washing and to maintain environmental flows. In addition, some strategic groundwater reserves have only been used (when not strictly preserved) for extreme events, but their water stock is limited by definition. Finally, the construction of new infrastructures in the Tagus has been ruled out because of its economic and political cost (TRBA 2014).

This is the context where pioneer voluntary agreements, albeit actively supported by the water authority, to transfer water between irrigation districts and water utilities

[1] Average direct water productivity in the Tagus River Basin is estimated at EUR 0.18/ha, and EUR 0.72/ha in the Segura River Basin (Gómez et al. 2013).

and sprung up for the first time in Spain. Needless to say that, as above, in a drought-prone and water-scarce country as Spain, informal water transfers have always spontaneously occurred and even some meaningful trades have taken place amongst farmers. The difference in the latter derives from the volume of exchanges, the parts involved in the bargaining process, the purpose of exchanges, and their importance to foster the adaptation of the institutional framework in order to allow for a wider use of water use trades.

This chapter aims at illustrating the performance of a diversity of water transfer arrangements in which economic incentives were used to tackle challenges posed by drought events. Two different transfers are to be assessed: the public water utility (*Canal de Isabel II*, *CYII*) taking water (for which it holds rights) from the Alberche River to supply Madrid city's domestic uses (also including the water transfer from Las Parras stream in the Middle Tagus to the Alberche Canal to compensate irrigators); and the water right transfer from the Henares Canal irrigators to the *Mancomunidad de Aguas del Sorbe* (hereafter MAS) (Sorbe River Water Community) to supply water and sanitation to different towns in the Henares' Corridor. Each of them is full of nuances and different features, which will shed light on formal and informal practices of water exchanges.

18.2 Setting the Scene: Challenges, Opportunities and EPIs

The study site is located in the upper section of the Tagus watershed and includes parts of the Alberche, Jarama and Guadarrama catchments (three of the main tributaries of the Tagus), an area of 8,022 km^2 (see Fig. 18.1). It includes the densely populated Madrid metropolitan area (4,567,190 inhabitants) (Eurostat 2014) and its sprawl along the Henares Urban Corridor, and the two irrigated areas involved in the assessed water trades, respectively located in the Alberche catchment and on the banks of the Henares River.

In the pre-crisis period 1996–2008 Madrid increased its real GDP by more than 50 %, its number of employed people by more than 1.25 million and its real GDP per capita from EUR 19,755 to EUR 23,636, thus generating a pull effect which increased population even during the first year of the current economic crisis (INE 2014). The municipalities of the MAS, with most of its served population and industry located in the Region of Madrid, showed a very similar trend, with high economic and population growth in the towns and villages of the Henares Urban Corridor (mainly in Alcalá de Henares and Guadalajara). These trends propelled water demand and gave rise to the two water trades assessed.

Regarding the **first trade**, the Alberche Canal irrigation area spreads out in 10,000 ha, with an approximate irrigation demand of 75 million cubic meters per year (hm^3/yr). Irrigation water supply depends on the Picadas Dam, connected to the Valmayor Dam, operated by CYII. Due to the joint effect of scarce water inputs and consumption rises in the preceding years, water supply for Madrid city could have actually become uncertain.

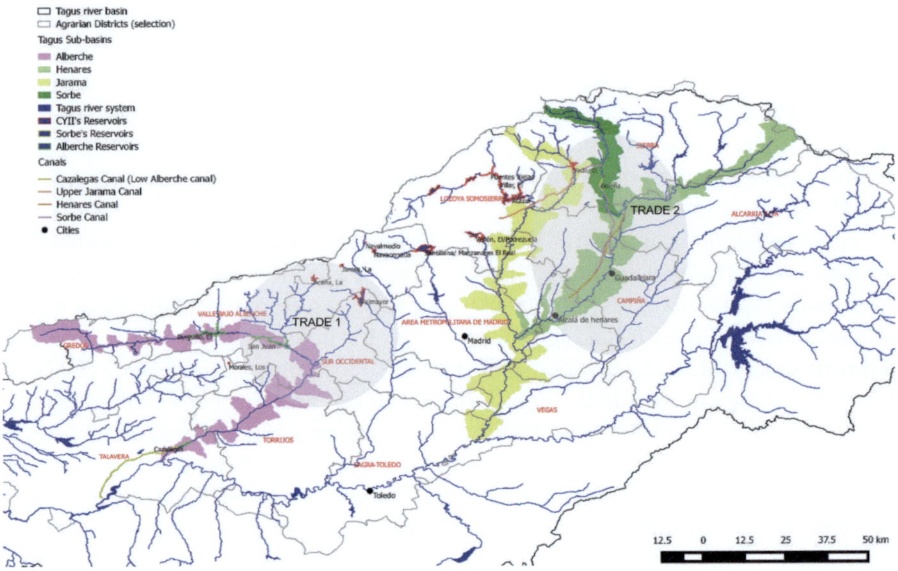

Fig. 18.1 The study site (Source: Own elaboration)

In March of the hydrological year 1993–1994 water storage in the reservoirs of the Alberche System was less than 128 hm^3, almost equal to the volume of the Alberche River flow allocated for Madrid's supply[2] (119.8 hm^3). Given the hierarchy of use in the Spanish Water Act (i.e. the priority of domestic supply) and the legal water allocation of the CYII in the Alberche, this implied that irrigated crops depending on the water of the Alberche Canal would not have received enough water for the cropping season (TRBA 2008). CYII used their entitlement to supply Madrid city; to compensate local farmers, a water transfer was conceived from the middle stretch of the Tagus River (Las Parras Stream) to the Alberche, exclusively for irrigation purposes.

As to the **second trade**, the MAS was created to supply water to the towns of Alcalá de Henares, Guadalajara and other municipalities of the Henares River valley, downstream Beleña's reservoir. Total population supplied was 68,000 inhabitants (CYII 2011), and the volume of water used was 6.8 hm^3 (MAS 2014). The project for the aggregation of water and sanitation services in the area included growth projections that have been largely exceeded. These included a maximum population for Alcalá de Henares and Guadalajara of 100,000 inhabitants each and 25,000 for the other municipalities at stake for the purposes of this assessment. Nowadays, total population supplied by MAS is 363 126 inhabitants (INE 2014) (which includes the seven original municipalities and other six that joined later) plus 20,000 inhabitants from municipalities which are not MAS members. In terms

[2] The public water utility holds most use rights on water resources of the Alberche River; 119.8 hm^3 as the initial entitlement, plus 100 hm^3 added in 2006 (TRBA 2013).

of supply for water and sanitation services, that can be translated into an increase of the total water use, which in 2005 was already at 46 hm^3 (MAS 2014).

MAS held water rights for 1,300 l/s and water supply mainly stemmed from the Beleña Dam, with only 50.3 hm^3 of effective capacity (Estevan and Lacalle 2007). By the end of the 1990s it was already supplying water to a much larger population than originally projected (TRBA 2014). The river had an average contribution of 168.68 hm^3/year (TRBA 2008). The Sorbe River system demands an average of 75 hm^3 per annum from the Beleña Dam, out of which 51 hm^3 are for domestic supply, leakages amount to 12 hm^3 on primary and secondary mains, 9 hm^3 help maintain the environmental flow and the remaining water went to filtration and evapotranspiration (CYII 2011).

Before 2001–2002, MAS managed to provide services to all municipalities every year. However during that hydrological year, Sorbe's contributions were 13 % of average levels during the twentieth century. At the beginning of February 2002 the reservoir level was under 9 hm^3, equivalent to 2 months of consumption. Given the risks for spring and summer seasons, and at the request of the TRBA, MAS contacted the irrigators of the Henares Canal to negotiate the purchase of a certain amount of water, within the framework of article 67 of the Water Law (on lease contracts).

The Henares Canal has a total irrigation area of 7,500 ha placed in 15 municipalities at southeastern Guadalajara. It holds water rights up to 5,600 l/s from the Henares River and an upper bound of 66.18 hm^3/year. The dams of Palmaces (with a total maximum storage capacity of 31 hm^3), Alcorlo (180 hm^3), and El Atance (35 hm^3) regulate the canal. An additional advantage was that the canal is parallel to MAS pipes and is just 2 km away from the Mohernando wastewater treatment plant (WWTP).

18.3 Water Markets in Action

Nowadays both transfers are deemed successful examples of drought adaptation. They are part of the institutional developments that have moderately boosted the use of voluntary transfers of water use rights as a water security mechanism avoiding other costly or politically challenging alternatives such as new water infrastructures or new expensive water sources.

However, sharp increases in water demand revealed the need for a more flexible approach (the Spanish model is based upon a concessional regime, not a water market at such), so that users could meet their demand but not necessarily at the expense of a higher use of the resource. This was even more compelling considering that water administrative mechanisms to re-allocate water (such as administrative procedures for water rights expiry, water concession audits, etc.) proved to be ineffective. For instance, the river basin authority would take a year and a half to process water right applications, even when water was available. Lease contracts, on the other hand, allow water users to get exclusive water rights in just 2 months (Vázquez 2010).

18.3.1 Main Outcomes of the EPI Implementation

18.3.1.1 Environmental and Economic Outcomes

High demographic and economic growth have led to increased water demand and the upturn of physical capital in the upper and middle stretches of the rivers of the region: as a result of that, Madrid is now amongst the Spanish regions with more heavily modified rivers (Alcolea and García 2006) and, in turn, in the world (Gómez 2009). No significant surface water supply increases can be obtained through additional human-made capital, and in fact there have not been new investments on relevant dams since the 1970s (CYII 2011). While demand increased and surface resources grew unable to meet water demand during droughts, groundwater has been used as a buffer stock. Although some strategic reserves have been kept to couple with drought events, overall groundwater withdrawals during these years have scaled up, exhausting aquifers and leading to a poor ecological status (TRBA 2014). Under these circumstances, the solution to scarcity problems came through significant rises in water efficiency that made it feasible to bridge the gap between water use and withdrawals.

Within this scheme, economic policy instruments such as water rights transfers were called to play a key role provided they did not contribute to exacerbate increasing demand trends. In the first water trade analysed, water from the Middle Tagus compensates irrigators in the Alberche Canal, which in turn provide water to guarantee Madrid reserving water flows, crucial for the provision of water and sanitation services to the households in the Madrid Region. However, the lack of a formal previous agreement with the irrigators of the Alberche Canal to replace water from the Alberche River with water from the Middle Tagus sub-basin hampered the right operation of Las Parras-Alberche Canal connection, as it happened in the 2004–2005 drought, thus leading to overexploitation of Alberche River's resources (TRBA 2008). Moreover, while the Alberche-CYII water transfer supplied high quality freshwater to Madrid, irrigators from the Alberche complained about the low quality of the water diverted from Las Parras stream.

Unlike in Madrid city, in the MAS water supply problems were related to a lack of regulation capacity as compared to increasing water demand. There had been flawed attempts to upgrade the regulation capacity of the system. Several proposals had actually been analysed to transfer water to the Sorbe catchment from the Alcorlo Dam (180 hm^3) in the Bornova catchment (TRBA 2008). However, the procedure was slow and the last investment alternative (a transfer of 80.9 hm^3/year from the Beleña Dam to the Alcorlo Dam) is still today at a standstill due to its high environmental impact. Consequently, the solution had to come from a water right transfer from the Henares Canal (TRBA 2014), which provides water of a lower quality than that contained for example in the Alcorlo Dam. To some extent, this is an idle and derelict infrastructure, initially planned to transfer water to the Beleña Dam in several undone projects.

In turn, the Henares Canal takes its resources from the Henares River. Economic and population growth in the surrounding areas had increased water demand in the Henares catchment, thus leading to a reduction of water supply guarantee during scarcity junctures. As a result, groundwater abstraction in the area had increased, and although the quantitative status of the Guadalajara aquifer was (and is) still fair, its qualitative status was poor (TRBA 2014). Indeed, non-point pollution resulting from agricultural activities is still a problem of major concern, also affecting the Alcorlo Dam; its resources are used for urban supply as well as the provision of amenities (TRBA 2014).

In both cases, the water transfer directly responds to the need to guarantee water supply to the population. Indirectly, it also contributes to guarantee water supply to other urban uses, such as the service sector, the most relevant economic sector in the study area (INE 2014). Water trading in both areas has been only made possible through the development of water infrastructures to transfer water from agricultural districts to urban areas.

In the first water trade assessed, the Alberche – CYII water transfer has been working since 1967, although the expansion that is relevant to our analysis was installed in 1993. The expansion allowed the CYII to use 119.8 hm^3/year at a cost of 10 billion pesetas (equivalent to 67 million ECU_{1993}[3]). The Region of Madrid was responsible for the payment, which had to be effective within a period of 25 years. The compensatory water transfer from Las Parras stream in the Middle Tagus sub-basin to the Alberche River was built in 1991 with an initial capacity of 5 m^3/s. It was used for the first time in 1993 when, following a drought, 35 hm^3 were transferred from the Middle Tagus to the Alberche Canal. In 2006, the TRBA allocated an additional amount of 100 hm^3 of the Alberche's resources to the CYII water utility.[4] This increase was followed by an expansion in the capacity of Las Parras – Canal de Alberche water transfer, up to 7 m^3/s at a cost of EUR 2 million. In 2008, the Las Parras-Alberche infrastructure had to be used again to solve a water shortage in the Alberche River. The cost of the intervention was EUR 1.48 million, and the water transferred had a lower quality than that of the Alberche River. An additional projected measure consists in the modernization of the irrigation systems in the Alberche Canal, with an estimated cost of EUR 50 million, which is expected to save 25 hm^3/year through more efficient irrigation.

In the second water trade (MAS), the limited capacity of the Beleña Dam fostered an agreement between the irrigators of the Henares Canal and the MAS to transfer water rights for urban water supply. The agreement entered into force in February 2002 and transferred the use of 20 hm^3/year. from the irrigators of the Henares Canal to the MAS. This agreement was extendable for 2-year periods up to a maximum of 10 years. The infrastructural cost amounted to EUR 3.5 million and mainly consisted in the construction of the Maluque-Mohernando connection, with a length

[3]The ECU was the unit of account of the European Community and was replaced by the euro, at parity, in 1999.

[4]This increase should have only been effective after additional regulation in the river (which did not actually happen) and subject to compensation to the parties.

of 2 km and a capacity of 1.3 m^3/s. The maximum flow rate was variable throughout the year: between September and April the transfer could work at its maximum capacity, but from May to August the maximum flow was 300 m^3/s. The fixed costs of the water transfer for the MAS was EUR 38,000/year, plus EUR 0.01 for the first 4 hm^3 and EUR 0.02 from that amount onwards. During the summer months (June–August) each additional hm^3 was paid at EUR 0.03/hm^3. Besides, MAS had to pay the pumping costs to the TRBA, which during 2005–2006 amounted EUR 388,000 (CYII 2011).

The guarantee of water provision through the use of EPIs such as water right transfers has contributed to consolidate economic growth in the most dynamic urban area of Spain: Madrid and its sprawl along the Henares Urban Corridor. Coupled with economic growth, urban water productivity in the Madrid Region has experienced a remarkable growth during the period 1997–2006. The service sector (80.5 % of region's GDP) more than doubled its water productivity in this period. Overall, water productivity has increased in the secondary and tertiary sectors as GVA and GDP grew (INE 2014), showing a consistent pattern that can be described as Verdoorn's Law[5] for water (Pérez-Blanco and Thaler 2014). In those sectors, the apparent productivity of water is well over EUR 1,000/m^3, while in the building sector it shoots up to over EUR 13,000/m^3. On the other hand, irrigation productivity is under EUR 1/m^3 for many crops in the Talavera and La Campiña agricultural districts (roughly corresponding to the areas supplied by the Alberche and Henares canals, respectively), which show an average water productivity of EUR 3.57/m^3 in La Campiña and EUR 3.79/m^3 in the Talavera agricultural district (INE 2014; MAGRAMA 2009).

Although transfers from agricultural to urban sectors unambiguously result into higher water productivity values, these results need to be taken with caution. Agriculture may be a minor economic sector for the aggregate of the study area but in certain rural areas it is actually the main activity, and reducing water availability could lead to substantial economic losses and depopulation in these areas. Noteworthy, rainfed agriculture shows much lower income than irrigated agriculture in the two agricultural districts at stake (MAGRAMA 2009).

18.3.1.2 Distributional Effects and Social Equity

Farmers growing irrigated crops in La Campiña Agricultural District get an average income of EUR 1,123.06/ha, with important variations across crops. Corn, which covers the widest area, produces an average income of EUR 2,000/ha while barley obtains less than 900 and peas less than 200. These three crops altogether represent 95 % of the study site area; the remainder of the area is covered by more profitable and water demanding vegetables. As an indicator of the value of water, it can be said

[5] According to the Verdoorn's Law, faster growth in output increases factor (e.g. water) productivity due to increasing returns in certain blocks of the economy prone to technological improvements and efficiency gains (such as the manufacturing industry) (Verdoorn 1949).

that the average income amounts to EUR 0.19/m^3, but 23 % of the irrigated area using 30 % of the water might be generating an income lower than EUR 0.04/m^3 (MAGRAMA 2009).

The importance of water for irrigation can be observed if comparing the previous numbers with those of rainfed agriculture, covering around 120,000 ha and earning EUR 490/ha on average. For example, as per the most common crop, both under irrigated and rainfed agriculture, irrigation facilities and water represent a shift from EUR 460 to 2,000/ha of income and an increase from 2,300 to 11,000 kg of average yield (MAGRAMA 2009). From a social viewpoint, water does make a difference in the study site indeed.

The average income obtained in the Talavera agricultural district (Alberche catchment) is about twice that of the Henares (EUR 2,180/ha) and although cereals still account for three quarters of the irrigated area, crops are more diversified than in the Henares. Average income is EUR 0.32/m^3 with lower variations than in the Henares. The dominant crop is also corn, which covers 40 % of the irrigated area, uses 45 % of water, and yields EUR 0.22/m^3 (MAGRAMA 2009).

The assessed water use right trades have not had significant impacts on material living standards, since water was guaranteed for both water supplying irrigation districts holding stakes. From stakeholder consultation, it can be inferred that the irrigators of the Henares Canal did not suffer noteworthy losses (Gómez et al. 2011). The fact that farmers in the Henares Valley accepted to give their water up at a price lower than EUR 1/m^3 is but an indication that probably (at least part of) those water resources were not being used at all for irrigation.

As above, irrigators from the Alberche Canal complained about the low quality of the Middle Tagus water received via Las Parras stream. A study carried out by the public utility (Estevan and Lacalle 2007) states that the conductivity of the Tagus River up to Talavera can reach 2,000 μS/cm, which basically means that it is semi-brackish water. This does not seem to have led, though, to critical production losses or major protests.

This idea that no major equity impacts were found is reinforced by the fact that compensation payments were implemented in both water transfers (in the Henares – Sorbe transfer, as part of contractual explicit clauses). In 1993, the Region of Madrid paid for the energy costs of pumping (50 million pesetas; that is, 0.335 million ECU$_{1993}$) (Estevan and Lacalle 2007) in which the irrigators would incur to divert water (35 hm^3) from the Tagus River to the Lower Alberche Canal trough Las Parras stream. In the agreement the irrigators did contract the energy supply, which was thus partially paid by the CYII. Despite a number of interviews with stakeholders (representatives from the River Basin Authority, on one side, and CYII water public utility, on the other), no significant evidence has been obtained as to why this compensation was implemented in spite of the utility holding water rights in the Alberche. It seems part of a compromise between the company and the basin authority.

Regarding the transfer from the Alberche to supply Madrid city, households did not face an increase in water tariffs due to additional expenses for the public utility (power for pumping).

In November 1996, the energy company Unión FENOSA claimed the payment of 1 billion pesetas (6.7 million ECU_{1996}) from CYII arguing that some production losses would occur after the construction of the Alberche – CYII transfer. Since the transfer started working, Unión FENOSA considered that its water concessional rights were being affected. The transfer reduced the volume of water that the company could turbine, and hence, its capacity to generate electricity. These damages were estimated at roughly 1.2 billion pesetas (7.3 million ECU_{1996}). CYII managers argued that they were acting under the safeguard of the water rights they held, granted by the TRBA, in which there was neither specific constraints to the transfer nor compensations to other third parties.

In terms of positive equity impacts, local communities served by MAS and benefitting from the water transfers from the Henares River, managed to elude water restrictions. There are records of some complaints regarding the quality of water, but this was always in accordance with regulations. MAS, on the other hand, does not only supply households but also industries; for some of them water is an essential input.

18.3.2 The EPI Setting Up

18.3.2.1 Institutions

A significant number of institutions were involved in the implementation of the two analysed water transfers. The first transfer (that from the Henares Canal to the MAS) describes a situation in which the transfer itself was formally feasible. The other (CYII using their entitlements to supply Madrid with water from Alberche River) is an example of a situation where the water transfer is viable but only under more specific circumstances. The former is a case in which a water right transfer is performed in strict sense. The latter, on the contrary, is an example of a water right holder (the public water utility) using their rights and the affected irrigators being compensated through a decision by the TRBA.

In the first transfer, the irrigators of the Alberche Canal[6] do not hold rights although they are beneficiaries of the allocation of public water flows for irrigation. These farmers have not been granted a formal entitlement and thus their rights are not registered, which implies that they cannot be part of a water transfer contract. Nevertheless, as an exception, the RDL 15/2005 allowed water users adjoin to public irrigation land to sign lease contracts provided some conditions were met (BOE 2005). Resources allocated for this irrigated area (hydrological plan, 1999) were 7,500 m^3/ha and year.

[6]The Irrigation District of the Lower Alberche Canal started working in 1953, with 3,000 commoners and 4 municipalities (San Román de los Montes, Pepino, Talavera de la Reina, and Calera y Chozas).

In the transfer from the Henares to the Sorbe, it is interesting to point out that although in those two events in which water right transfers were needed they were implemented following the conditions required by the MAS, the contract did not include a single binding clause for the Henares Canal irrigation community to set aside water resources for water transfers; neither did it contain any provision that could shed light on the relative priority of water leasebacks or trades over the Canal users' risks (Estevan and Lacalle 2007).

18.3.2.2 Transaction Costs and Design

It should be clear that water transfers generate benefits and costs relative to what the situation would have been in the absence of these transfers (in other words, as compared against a baseline scenario). If stakeholders had not been aware of those impacts and there had been no compensation for those bearing negative impacts, there would have been a clear economic incentive to politically contest these water transfers. Should that have been the case, the effort that stakeholders would have devoted to lobby decision-making should have been added to the explicit costs of the EPI implementation; these costs are generally assumed away. Yet, there was compensation in both cases, as it has been explained in previous sections. Compensations followed a number of meetings which entail non-negligible transaction costs.

For the **first trade**, in March 11th, 1993, the Bo of the Irrigation District of Alberche River held a meeting to analyse alternative solutions to provide water for irrigators of the Lower Alberche Canal. Because of the urgency in building the Alberche – CYII connection, the transfer started working in November of that very year.

As per the **second trade** (2002), the agreement was signed in February 8th, 2002 and in July of the same year it started working for 4 months. The connection between the Henares Canal and Mohernando's treatment plant did not work again until June 2005 for a period of 8 months (up to January 2006). The fact that it took less time to implement the second transfer is an indicator of lower transaction costs, due to the prior effort, which provided some institutional assets for ulterior water trades.

There is no evidence of the time devoted to these meetings, since, as above, these two water transfers were designed as ad-hoc urgent measures to tackle drought consequences. However, despite this emergency feature, the process was longer (and consequently transaction costs higher) than one could infer. Once irrigation was established in the Alberche, different drought periods threatened water availability for farmers. This motivated, in 1991, the construction of an emergency infrastructure, thanks to an intake from the Tagus. This means that although no significant transaction costs may be linked to the decision to transfer water in 1993, some ex-ante costs may need to be taken into account (no available information has been made available to the authors), regarding the construction of the infrastructure for water transfers.

18.3.2.3 Implementation

After centuries of enforcement of the appurtenance principle[7] a number of factors led to the transfer of water rights among users. Economic development and urban growth, as well as droughts themselves, could no longer be managed within the limits imposed by a water law based on the needs of a formerly agricultural society and the limits of traditional administrative procedures. As a result of these parallel developments Spanish water legislation was upgraded (1999) to allow for the transfer of water rights.

The Spanish scheme accepts the transfer of water rights, but this faces somewhat significant restrictions. One of them is that trades may only take place between and among uses of similar or superior ranking in the hierarchy of uses.[8] The ranking of preferences in the law is based on custom (BOE 2001), thus hampering efficiency gains that may be attained through exchange, since it does not take account of water productivity in different alternative uses.

The drawbacks of the system are best illustrated by the fact that legislation issued to cope with droughts, and related emergency measures include exceptions to the above-mentioned order of preferences. If allowances are needed for the system to perform there is an indication that structural, permanent rules, may need some amendment if water markets are to be developed.

Whilst the coordination between parties (the public water utility and the irrigators from the Alberche Canal on one side, and the MAS and the irrigators from the Henares on the other), was realistic, especially for the facilitating role of the Tagus River Basin Authority, some flaws can be observed regarding the implementability of these transfers in other contexts. Under current water legislation and institutional set-up, request for a transfer may be approved by default, if the administration does not approve or refuse it within 1 or 2 months, depending on the reach of the transfer (within or outside the same community of users). That legal provision provides a powerful incentive for administrators to reject transfers outright should they be complex and time consuming.

Water trading, as assessed in these two water trades, may face additional challenges. Allowing transfers from agricultural to urban uses (as in the two cases that have been assessed) may bring to the negotiation process water resources that are not being effectively used, unless strict monitoring provisions are implemented. Not surprisingly, given the low quality of soil in the region of Madrid, agriculture is a waning activity and in some areas water allowances are higher than the effective demand for irrigation water. Once subsidies from the Common Agricultural Policy

[7] Permanent and rigid bundling of water to a piece of land and a single use, preventing exchanges and reallocation.

[8] The ranking of preferences set in the water law prioritises urban use, in which low water consuming industries located near urban areas and connected to municipalities are included. Next use is water for irrigation and other agricultural uses. After that, industrial use for power generation, other industrial uses not included in previous categories, aquaculture, recreational uses, navigation and other transport and other uses, respectively (TRBA 2014).

(CAP) have been phased out and agricultural markets have been liberalized, the irrigation sector in some areas may be in excess water supply.

Both trades were the result of measures to cope with droughts. Because of this, all procedures and public works were executed on an emergency basis (time laps between decisions and effects were 8 months for the 1993 transfer, and 3 months for the 2002 one), which hampered whatever participation process. This lack of participation, though, does not seem to have damaged the acceptance of water transfers, nor has it conditioned its design. Yet, these instances have created a favourable context (built on protocols and formal procedures), which opens up space for further transfers.

Despite the fact that no participation took place regarding the assessed transfers, there have been ulterior consultation processes within the context of the water planning drafting stages. As part of those, a number of potential conflicts were identified (regarding the ecological status of river Tagus, as pacing the town of Talavera; the availability of water resources in the Alberche – mainly water for irrigation; the use of Sorbe's resources; etc.).

18.4 Conclusions

The emergence of water markets, as many other once innovative EPIs, is a gradual adaptive and learning-by-doing process that must be judged by its ability to push water institutional development rather than by the failure or success of the experience itself. The water transfer from the Alberche River although useful to manage the supply deficit in the 1990s would not be an alternative nowadays anymore and many doubts exist as to the real prospect of repeating the 2002 water trade from the Henares Canal to the MAS in the same formal terms. The actual value of these examples is in the lessons that can be drawn and its importance to furthering agreements on reallocating water use rights as an instrument for water security.

Both examples also illustrate the critical importance of managing water use conflicts. It is well known in economic analysis that water management is essentially conflict management. In fact, according to the Spanish law, households have a priority over irrigators in water use, and there is no need for a voluntary agreement to take water away from farms in order to guarantee a sufficient supply of drinking water in dry periods. The real buffer for drinking water in Spain is the irrigated agriculture whose use rights are defined every year depending on the rainy season. Moreover, instead of just taking water or forcing farmers to let water flow, the agreement is easier to reach if alternative resources are available, the harvest is protected and third-party effects are avoided.

This is the real meaning of the 2002 transfer. The existence of these alternative resources is precisely what makes the replication of this trade almost impossible in 2011 (as there is evidence of overallocation or water rights in the middle Tagus river). Nevertheless, lessons learnt can be important to understand how, instead of paying for water, agreements are easier to reach when alternative sources are provided to guarantee existing uses, particularly in irrigated agriculture. Nowadays, alternative resources can either come from re-used or desalinated water.

Water trading also faces some important challenges. Allowing transfers from agricultural to urban uses may bring to the negotiating table water resources that are not being effectively used. In fact, given the low quality of soil in the Madrid area, agriculture is a receding activity and in some areas water allowances are higher than the effective demand for irrigation water. Paradoxically, once EU subsidies for agriculture have been phased out and agricultural markets have been liberalized, the irrigation sector in some areas may be in excess of water supply. The fact that farmers in the Henares valley accepted to give their water up at a price lower than one euro-cent per cubic meter is but an indication that probably those water resources were not being used for crops. Hence, water trading might not be a means to reduce water scarcity but rather to increase it and would not be instrumental to re-allocate water but to effectively increase its use. This would be a real risk should water saved after the publicly supported shift towards more efficient irrigation systems, becomes part of the water trading system rather than being left in already degraded aquifers.

Both transfers were designed, as emergency measures, for severe drought situations that threatened water supply of important cities and towns including Madrid. It is therefore evident that a more systemic consideration of non-structural alternatives to water management in Spain, rather than a drought-based-emergency resort to market-like solutions, may be requested on economic efficiency grounds.

References

Alcolea, M., & García, J. M. (2006). El agua en la Comunidad de Madrid. *Observatorio Medioambiental, 9*, 63–96.

BOE. (2001). Real Decreto Legislativo 1/2001, de 20 de julio, por el que se aprueba el texto refundido de la Ley de Aguas.

BOE. (2005). Real Decreto-Ley 15/2005, de 16 de diciembre, de medidas urgentes para la regulación de las transacciones de derechos al aprovechamiento de agua (Vigente hasta el 30 de Noviembre de 2009), Royal Decree.

CYII. (2011). Infraestructuras de Captación (Report). Canal de Ysabel II.

Estevan, A., & Lacalle, A. (2007). Transferencias de Derechos de Agua entre Demandas Urbanas y Agrarias. El caso de la Comunidad de Madrid (Cuaderno No. 1), Cuadernos de I+D+i. Canal de Ysabel II, Madrid.

Eurostat. (2014). Eurostat database [WWW Document]. Eurostat database. http://epp.eurostat.ec.europa.eu/portal/page/portal/statistics/themes. Accessed 10.5.13.

Gómez, C. M. (2009). La eficiencia en la asignación del agua: Principios básicos y hechos estilizados en España. *Información Comercial Española Revista de Economia, 847*, 23–39.

Gómez, C. M., Delacámara, G., Pérez-Blanco, C. D., & Ibáñez, E. (2011). Water transfers in the Tagus River basin (Spain) (Deliverable No. 3.1), work package 3 – ex-post case studies. 7th framework contract project EPI-water project (GA 265213).

Gómez, C. M., Delacámara, G., Pérez-Blanco, C. D., Ibáñez, E., & Rodríguez, M. (2013). Droughts and water scarcity- Tagus (Central Spain & Portugal) and Segura (SE Spain) interconnected river basins (Deliverable No. 4.3), work package 4 – ex-ante case studies. 7th framework contract project EPI-water project (GA 265213).

Hernández-Mora, N., & De Stefano, L. (2013). Los mercados informales de aguas en España: una primera aproximación, in: Usos Del Agua: Concesiones, Autorizaciones y Mercados Del Agua. Cizur Menor: Thomson Reuters Aranzadi.

INE. (2014). Instituto Nacional de Estadística – INEBase [WWW Document]. http://www.ine.es/inebmenu/indice.htm. Accessed 15 Sept 14.

MAGRAMA. (2009). *Resultados Técnicos-económicos de Sistemas de Explotación Agraria (Report)*. Madrid: Ministerio de Agricultura, Alimentación y Medio Ambiente.

MAS. (2014). Mancomunidad de Aguas del Sorbe [WWW Document]. Aguas Sorbe. http://www.aguasdelsorbe.es/web/

Pérez-Blanco, C. D., & Thaler, T. (2014). An input-output assessment of water productivity in the Castile and León Region (Spain). *Water, 6*, 929–944. doi:10.3390/w6040929.

TRBA. (2008). Esquema de Temas importantes de la demarcación hidrográfica del Tajo (Report). Tagus River basin authority.

TRBA. (2013) Proyecto de Plan Hidrológico de cuenca de la Parte Española de la Demarcación Hidrográfica del Tajo (noviembre-2013) (Report). Tagus River basin authority.

TRBA. (2014) Plan Hidrológico de la parte española de la Demarcación Hidrográfica del Tajo (Report). Tagus River basin authority.

Vázquez, D. (2010). El contrato de cesión de derechos al uso privativo del agua: otra forma de adquirir derechos sobre las aguas públicas más allá de las concesiones administrativas. D. Ley 3–.

Verdoorn, P. J. (1949). Fattori che regolano lo sviluppo della produtività del lavoro. *L'industria, 1*, 3–10.

Chapter 19
Chilean Water Rights Markets as a Water Allocation Mechanism

Guillermo Donoso

Abstract Chile is illustrative of a transition from command and control to market based water management policy, where economic policy incentives (EPI) play a significant role in water rights allocations. The enabling factor that allowed for the implementation of water rights markets in Chile was Chile's tradition and culture, dating back to colonial times, of managing water resources with water rights. The Chilean Water Code of 1981 established that water rights are transferable in order to facilitate markets as an allocation mechanism. The framers of the 1981 Water Code sought to achieve efficient water allocations with this EPI. The existence of water markets has been documented. A key conclusion of these studies is that water markets are more prevalent in areas of water scarcity. They are driven by demand from relatively high-valued water uses and facilitated by low transactions costs in those valleys where Water User Associations and infrastructure present assist the transfer of water. In the absence of these conditions trading has been rare and water markets have not become institutionalized. A major challenge of water rights markets in Chile is how to ensure optimal water use without compromising the sustainability of rivers and aquifers. The implementation of this EPI did not establish new institutions; however, it significantly modified their existing powers. Nevertheless, in order for it to deliver its full potential as an efficient allocation mechanism, Chile requires an institutional reform in order to respond to the country's actual water challenges.

Keywords Water rights markets • Water allocation mechanism • Chile • Market based water management policy

G. Donoso (✉)
Department of Agricultural Economics, Pontificia Universidad Católica de Chile,
Vicuña Mackenna 4860, Macul, Santiago, Chile
e-mail: gdonosoh@uc.cl

M. Lago et al. (eds.), *Use of Economic Instruments in Water Policy*,
Global Issues in Water Policy 14, DOI 10.1007/978-3-319-18287-2_19

19.1 Introduction

Within the global context, Chile as a whole may be considered privileged in terms of water resources. The average total runoff is on average equivalent to 53,000 m^3/person/year (World Bank 2011), a value considerably higher than the world average (6,600 m^3/person/year). However, there exist significant regional differences: from Santiago to the north, arid conditions prevail with average water availability below 800 m^3/person/year, while south of Santiago the water availability is significantly higher reaching over 10,000 m^3/person/year.

Water withdrawals in Chile average approximately 4,000 m^3/s/year (World Bank 2011). Of this almost 85 % is used in non-consumptive hydroelectric generation. Consumptive water use in Chile is dominated by irrigation with 73 % of consumptive water use. Industrial use of water is 12 % of consumptive withdrawals, mining and potable water supply account for 9 % and 6 % of total water consumptive water use, respectively. It is interesting to note that all consumptive water uses have increased since 1990; total consumptive water use has increased 13 % between 1990 and 2006. Industry is the sector with the highest consumptive water use increase (79 %), followed by potable water and mining (48 % and 46 %, respectively).

Growing water scarcity puts more pressure on policy makers to improve water allocation, make irrigation systems financially sound, and provide incentives for adoption of water-saving technologies. The different water policies in existence prior to 1980 were limited in their ability to reach an economically efficient water allocation. These limitations were primarily related to the definition of water rights, the information available to users, and transaction costs. The objective of the governmental action in this field was to create solid water use rights in order to facilitate the proper operation of the market as an allocation mechanism. The Water Code of 1981 (WC 1981) established transferable water rights and facilitated water rights markets as a water allocation mechanism. Hence, Chile's 1981 water law is illustrative of a transition from water management based on command and control to one based on a mix of command and control and economic policy instruments (EPIs), where economic incentives play a significant role in water allocations.

Water Rights (WR) markets in Chile, have helped to (i) facilitate the reallocation of water use from lower to higher value users (e.g., from traditional agriculture to export-oriented agriculture and other sectors such as water supply and mining), (ii) mitigate the impact of droughts by allowing for temporal transfers from lower value annual crops to higher valued perennial fruit and other tree crops, and (iii) provide lower cost access to water resources than alternative sources such as desalination (Donoso et al. 2010, 2014; Grafton et al. 2011; Jouravlev 2010; Hadjigeorgalis 2009; Hadjigeorgalis and Riquelme 2002; Rosegrant and Gazmuri 1994).

The problems that water use rights market have not been able to resolve are water use inefficiency in all sectors, not only in the agricultural sector, environmental problems, and the maintenance of ecological water flows. Additionally, WR trades from agricultural users to water and sanitation and mining companies, leads to greater water use. This occurs since agriculture does not demand water all year round as these industries. Since WR specify a total water flow and not effective water consumption, water use increases when WR are transferred from agriculture to more water-intensive economic sectors. The elements that have hindered WR market effectiveness are the lack (i) of WR and WR market information; (ii) of securitization of customary WR; and (iii) of a rapid, efficient controversy resolution system.

19.2 Setting the Scene: Challenges, Opportunities and WR Markets in Chile

The average total runoff is on average equivalent to 53,000 m^3/person/year (World Bank 2011), a value considerably higher than the world average (6,600 m^3/person/year). However, there exist significant regional differences: in the Northern Dry Pacific area, arid conditions prevail with average water availability below 800 m^3/person/year, in Central Chile, water availability is on average 2,500 m^3/person/year, while in the Southern Humid Pacific area, water availability is significantly higher reaching over 10,000 m^3/person/year (see Fig. 19.1).

Average annual recharge of groundwater resources in Chile also varies geographically. In the Dry Pacific area, aquifer's recharge is approximately 55 m^3/s, while it is three times that level in the Southern Humid Pacific (160 m^3/s). However, estimated groundwater extractions in the Dry Pacific reaches an average of 88 m^3/s (Salazar 2003); therefore, groundwater use in this area is unsustainable.

Water withdrawals in Chile average approximately 4,000 m^3/s/year (World Bank 2011). Of this total, 85 % is used for non-consumptive hydroelectric generation. Consumptive water use in Chile is dominated by agriculture with 73 % of consumptive water use. Industrial use of water is 12 % of consumptive withdrawals, mining and potable water supply account for 9 % and 6 % of total water consumptive water use, respectively. Thus, agricultural production is the greatest consumptive water user in Chile, which is the case in most undeveloped nations (Molden et al. 2007).

In the last 30 years Chile's real GDP has grown at an annual growth of 6.2 % (Banco Central de Chile 2013). During the same period, total consumptive water use has increased 13 %; industry is the sector with the highest consumptive water use increase (79 %), followed by water and sanitation services and mining (48 % and 46 %, respectively). These increased water demands due to increased economic growth,[1] together with population growth, urbanization, water contamination and pollution, are putting considerable pressure on available water resources. Decoupling of economic growth from water demands in Chile has thus, not been an automatic by-product of growth in national incomes and requires dedicated policies to improve water allocation between competing uses so as to not limit future economic growth.

[1] The Chilean economy is highly dependent on exports from water-intensive commodities such as copper and molybdenum, vegetables, fruits, wine, salmon, and pulp and paper, among others. Thus, economic growth is coupled to water use.

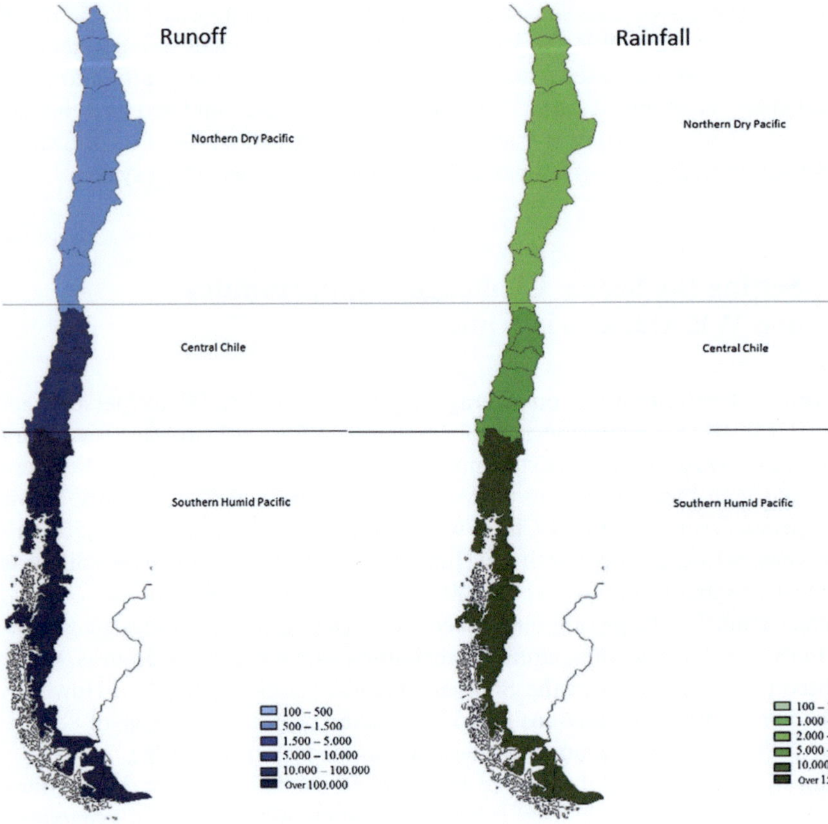

Fig. 19.1 Surface water runoff and rainfall (mm/year) (Based on DGA 2014; Peña et al. 2011)

The different water policies in existence prior to 1980 were limited in their ability to reach an economically efficient water allocation. These limitations were primarily related to the definition of water rights, the information available to users, and transaction costs. Additionally, these policies were not consistent with the many neo-liberal reforms introduced by the military government. During this period, the establishment and defence of property rights and the restriction of state interference in markets drove reforms in the Chilean water sector. The first step towards reforming the National Water Code occurs in 1979 with the Executive Decree 2.603, which recognized customary and historical water rights. This decree strengthened the security of private ownership of water rights, separating water rights from land ownership. Article 19 number 24 of the Chilean Constitution of 1980, which distinguishes between constituted and recognized water rights, reinforces this.

The 1981 National Water Code established transferable water rights and facilitated water markets as a water allocation mechanism. Hence, Chile's 1981 water law is illustrative of a transition from water management based on command and control to one based on economic policy instruments (EPIs), where economic incentives play a significant role in water allocations.

19.3 The Chilean Water Rights Markets in Action (1800–2800 Words)

The Water Code of 1981 (WC 1981) maintained water as "national good for public use," but granted permanent, transferable water rights (WR) to individuals so as to reach an economically efficient water allocation[2] through market transactions of WR; these WR were granted free of charge and without requiring a specification on intended use. The WC 1981 allowed for freedom in the use of water to which an agent has WR; thus, WR are not sector specific. Similarly, the WC 1981 abolishes the water use hierarchy of use lists, present in the previous Water Codes of 1951 and 1967. Additionally, WR do not expire and do not consider a "use it or lose it" clause.[3]

The WC 1981 established that WR are transferable in order to facilitate WR markets as an allocation mechanism. The framers of the WC 1981 sought to achieve the efficiencies of market reallocation of water, the objective of the governmental action in this field was to create solid water use rights in order to facilitate the proper operation of the market as an allocation mechanism (Buchi 1993). Thus the WC 1981 was designed to protect traditional and customary WR and to foster economically beneficial reallocation through market transfers (Bauer 2004; Buchi 1993; Hearne and Donoso 2005).

The WC 1981 specifies consumptive WR for both surface and groundwater, and non-consumptive WR for surface waters. Non-consumptive WR allow the owner to divert water from a river with the obligation to return the same water unaltered to its original water source.[4] Consumptive use rights do not require that water be returned once it has been used. Consumptive and non-consumptive WR are, by law specified as a volume per unit of time. However, given that river flows are highly variable in most basins, these WR are recognized in times of scarcity as shares of water flows. This characteristic of WR[5] has proven to be appropriate, given that the use of a system of WR defined as pure shares precludes any excess water use for other uses such as environmental objectives since it would lead to full use of water by the current holders of WR (World Bank 2011). However, total granted water flows are

[2] An economically efficient water allocation is reached when marginal net benefits are equated across individuals, satisfying Jevon's Equimarginality Principle.

[3] At present, these two characteristics are highly questioned, and Congress is debating a water policy reform to establish expiration dates on new granted WR and to introduce a use it or lose it clause (Ministerio Secretaria General de la Presidencia 2014).

[4] Water use in thermal electric generation plants require consumptive WR.

[5] Which combines volumetric maximum amounts per unit time in times of plenty, with shares in times of scarcity.

greater than average water supply in most basins in the north of Chile and, thus, there was no water provision for environmental objectives.

Additionally, consumptive and non-consumptive WR can be exercised in a permanent or contingent manner and in a continuous, discontinuous or alternating mode. Permanent WR are specified as a volume per unit of time, unless there is water scarcity in which these WR are recognized as shares of water flows. Total permanent WR are determined by the water flow that is satisfied at least 85 % of the time. Contingent rights are specified as a volume per unit of time and only authorize users to extract water once permanent rights have extracted their rights. These rights are determined by average flows of the basin that exceed those assigned to permanent rights. Continuous rights are those use rights that allow users to extract water continually over time. On the other hand, discontinuous rights are those that only permit water to be used at given time periods. Finally, alternating rights are those in which the use of water is distributed among two or more persons who use the water successively.

New WR are granted free of charge, and the petition procedure for a new WR starts with an application that had to meet the following requirements:

(a) Identification of the water source from which the water is to be extracted, specifying whether the source is surface water or ground water;
(b) Definition of the quantity of water to be extracted, expressed in litres per second;
(c) Yield and depth must be specified in the case of groundwater;
(d) Specification of the water extraction points and the method of extraction; and
(e) Definition of whether the right is consumptive or non-consumptive, permanent or contingent, continuous, discontinuous or alternating.

The administrative procedure requires that this application be published in the *Diario Oficial*, in a daily Santiago newspaper, and in a regional newspaper, where applicable. Previous to the WC 1981 reform of 2005, the DGA could not refuse to grant new water rights without infringing a constitutional guarantee, provided there was technical evidence of the availability of water resources and that the new use would not harm existent rights holders.[6] If there is competition for solicited water rights, they are to be allocated through an auction with an award to the highest bidder. This allocation rule between competing WR petitioners was designed so as to allow water to be allocated to its highest use value. The allocated WR is registered in the DGA's Public Water Registry (PWR).

Peña et al. (2004) and Bitrán and Sáez (1994) point out that the absence of an obligation to use WR led to a proliferation of WR requests for speculation and hoarding[7] purposes, that led to non-real water shortages and created obstacles to the development of new investment projects due to the impossibility of acquiring new WR. This was particularly evident in the case of non-consumptive WR where entry

[6] But, the DGA can declare certain aquifers to be fully exploited and refuse to grant new groundwater rights.

[7] This is a strategic entrepreneurial action, rather than a matter of speculation, per se.

barriers were created for new hydroelectric plants, discouraging competition in hydroelectric power generation. In fact, Riestra (2008) points out that of 15,000 m^3/s granted in non-consumptive WR, only 2,800 m^3/s were being effectively used. There is little concern about unused consumptive rights for water, given that, under a system of proportional use, all water is eventually distributed to users (Hearne and Donoso 2005). Dourojeanni and Jouravlev (1999) estimate the percentage of consumptive use rights that are unused to be less than one percent of the total allocated consumptive WR.

The State, concerned about monopolistic behaviour and supported by the antimonopoly commission, refused to grant new non-consumptive WR. In fact, the Constitutional Court established that the State could impose additional conditions on petitions for new WR by reformulating the WC 1981. This led to an amendment of the dispositions of the WC 1981 in 2005. The Law N°. 20,017 of 2005 amended the procedure to grant new WR of the WC 1981 and introduced a non-use tariff (*patente de no-uso*). The WR petitioner must now justify the water flow that is petitioned and clearly indicate the use that will be given to the water. Additionally, WR are only granted in accordance with the requirements of the use the WR is solicited for.

Due to the difficulties of monitoring the effective use of all WR, the non-use tariff is applied to all consumptive WR that do not count with water intake infrastructure and to all non-consumptive WR that do not have water intake and return infrastructure (Law N°. 20,017 of 2005, art. 129 *bis* 4–6). Non-use tariff (τ) for consumptive and non-consumptive WR is calculated as $\tau = \gamma Q f$ and $\tau = \gamma Q H f$, respectively, where γ is a constant that varies geographically,[8] Q represents the average water flow that is not used, f is a temporal factor,[9] and H is the difference between the water intake level and the level where the water is returned.[10]

19.3.1 *Assessment of Chile's WR Markets*

19.3.1.1 Environmental Outcomes

Water quality was not an objective of the WC 1981, its focus was and still is on water quantity and allocation. For example, the non-use tariff is biased towards productive uses since non-use is defined as the lack of water extraction infrastructure. Thus, all in-stream uses are subject to the non-use tariff.

---

[8] γ takes different the value of 0.1 for all regions between Magallanes and Los Lagos, 0.2 for regions between O'Higgins and Araucanía, and 1.6 for all regions north of the Metropolitana. γ is greater in those regions located further north, in order to reflect increased water scarcity.

[9] f=1 for years 1–5, f=2 for years 6–10, and f=4 for over 11 years without effective use. Thus f is a temporal factor that increases the non-use tariff if the water use right remains without use.

[10] H is only applied to non-consumptive WR and starts at a minimum value of 10 m.

Law No. 19,300 of 1994 introduced the main instruments available for water quality management; these instruments are: (a) environmental water quality standards, (b) decontamination plans and strategies, (c) emission standards, (d) environmental impact assessments for new investments, and (e) minimum ecological flows. Therefore, changes in water quality in the past 30 years cannot be attributed to WC 1981.

Before the WC 1981 reform of 2005, most river basins located in the Dry Pacific and Central Chile Regions were fully allocated and, thus, it has not been possible to implement minimum ecological flows due to the lack of water. River basins that have protected minimum ecological flows are mainly located in the Southern Humid Pacific Region where water is more abundant and presents lower use values.

The main regulatory measure established in the WC 1981 to control for potential negative effects on third parties and/or the environment due to the transfer of WR between water users is when the transfer implies a change of water intake location, the transfer must be authorized by the DGA. The analysis of potential third party or environmental effects associated with WR transfers between water users is conducted by the DGA. Transfer requests, as well as new WR petitions, are broadcast three times and published in a newspaper at the national and provincial levels. Additionally, the SEIA introduced in 1994 by the Law 19,300 requires water users to mitigate or compensate environmental damages that may result from the transfer of WR. It is important to note that transfers of WR that do not require a change in water intake location are not regulated.

A major challenge of WR markets in Chile is how to ensure optimal water use without compromising the sustainability of rivers and aquifers. The sustainability of northern rivers and aquifers is compromised due to the over-provision of WR related to the practice of allocating WR based on foreseeable use. The foreseeable use considers the probable effective water extraction of different sectors. For example, an agricultural WR does not extract water in winter months, whereas a mining WR extracts water all year round. In this case, the authority would consider a lower pressure on water resources of an agricultural WR with respect to the pressure of a mining WR. This practice commits the mistake of not considering the transferable nature of WR. Thus, when water scarcity increases and inter-sectoral WR transactions increase, water resources will be overexploited and unsustainable.

The WC 1981 did not pay much attention to the sustainable management of groundwater because at that time, groundwater extraction was marginal during the early 80s. Recognizing the need to improve groundwater management regulation due to increased groundwater pumping, the 2005 amendment of the WC 1981 introduced procedures to reach a sustainable management of underground water resources. However, World Bank (2011) concludes that these groundwater regulations have not been fully implemented over time and thus, there exist various problems associated with groundwater management.

Table 19.1 Consumptive WR transactions and prices for the period 2005–2008 (World Bank 2011)

Region	Total transactions (Number WR)	WR transactions independent of land	WR Transaction values (Only WR transactions independent of land) (10^6 US$)	Average WR transaction price (US$/WR)
Dry Pacific	12,221	11,223	3,623	512,243
Central Chile	8,835	8,522	1,160	228,737
Southern Humid Pacific	793	784	31	50,863
Total	**21,849**	**20,529**	**4,814**	**215,623**

19.3.1.2 Economic Outcomes

Although market reallocation of water has not been common throughout most of Chile, the existence of water markets has been documented. Studies have shown active trading for WR in the Limarí Valley, where water is scarce with a high economic value, especially for the agricultural sector (Hearne and Easter 1997; Donoso, et al. 2002; Hadjigeorgalis 2004). Inter-sectoral trading has transferred water to growing urban areas in the Elqui Valley (Hearne and Easter 1997), the upper Mapocho watershed and the first section of the Maipo Basin, where water companies and real estate developers are continuously buying water and account for 76 % of the rights traded during the 1993–2003 period (Donoso et al. 2002, 2014). Other studies have shown limited trading in the Bío Bío, Aconcagua, and Cachapoal Valleys (Bauer 1998; Hadjigeorgalis and Riquelme 2002).

A key conclusion of these studies is that water markets are driven by relative scarcity of water resources, demand from relatively high-valued water uses and facilitated by low transactions costs in those valleys where WUAs and infrastructure present assist the transfer of water. In the absence of these conditions trading has been rare and water markets have not become institutionalized in most valleys (Hearne and Donoso 2005).

Table 19.1 presents consumptive WR transaction data based on the PWR of the DGA, for the period 2005–2008.[11] The results for this 4-year period show 21,849 WR transactions, of which 94 % were independent of other property transactions, such as land. As expected, WR markets are more active in areas where the resource is scarce; WR transactions decrease from the Dry Pacific Region towards the Southern Humid Pacific. In fact, the Dry Pacific region accounts for 56 % of total transactions in this period.

The value of WR transactions independent of other property transactions is US$ 4.8 billion, which on average is US$ 1.2 billion per year. As water scarcity increases so does the value of each WR. The Dry Pacific Region, which has an average water

[11] The PWR of the DGA has data only for the period 2005–2009. The data for the year 2009 is incomplete.

availability below 800 m^3/person/year, presents an average WR price of US$ 512,243, which decreases to US$ 50,863 in the Southern Humid Pacific, whose water availability is significantly higher.

WR prices present a large dispersion, with a coefficient of variation of 470 %. This large price dispersion is due, in great part, to the lack of a WR prices revelation mechanism, and reflects, in fact, that welfare gains from trade have not yet been exhausted. Each WR transaction is, thus, the result of a bilateral negotiation between an interested buyer and seller of WR where each agent's information, market experience and negotiating capacity is important in determining the final result (Donoso et al. 2014).

Jouravlev (2010) notes that as a result of the WC 1981 reform of 2005 (together with other measures), consumptive WR that still are not used are, in general, no longer a major obstacle to the development of the water basin. Additionally, it is likely that non-use of WR will continue to reduce in the future due to the projected increase in the non-use tariff. Along the same lines, Valenzuela (2009) notes that the non-use tariff has operated as a small incentive for the return of non-consumptive WR; an equivalent of 65 m^3/s has been returned, which represents 1 % of both the total WR affected by the non-use tariff.

The elements that have hindered WR market effectiveness are the:

(a) Lack of WR and WR market information;
(b) Lack of regularization of customary WR;
(c) Existence of transaction costs;
(d) Lack of a rapid, efficient controversy resolution system.

19.3.1.3 Distributional Effects and Social Equity

Research in Chile on the impact of WR markets on small farmers, has been limited and no reliable conclusions have been reached to date. Hadjigeorgalis (2008) is a notable exception. She conducted research in the northern Limarí water basin in order to study the impact of water markets on small farmers. Results indicate that WR markets have been equitable with respect to offer prices; resource-constrained farmers receive the same offer prices for their water and water rights as wealthier farmers. Additionally, these markets represent a safety net for small farmers.

Future research on the equity impacts of WR markets is required to clarify the distributional effects of WR markets.

19.3.2 The Setting Up of Chile's WR Markets

19.3.2.1 Institutions (or Institutional Set-up)

The WC 1981 did not establish new institutions; however, it significantly modified their existing powers established in the WC 1967. Under the WC 1981, the State reduced its intervention in water resources management to a minimum and increased the management powers of water use rights holders that are organized in WUAs.

The *Dirección General de Aguas (DGA)*, part of the *Ministerio de Obras Públicas* (MOP), is the main public institution and is responsible for monitoring and enforcing the WC 1981. With its 15 regional offices, it collects and maintains hydrological data and PWR. As the leading government agency in water resources management, it develops and enforces national water policy. In this role, it has led efforts to amend the 1981 Water Code and developed a National Water Policy. In general, the DGA has maintained a limited role in accordance with the paradigm of limited state interference on which the WC 1981 is inspired.

However, multiple central authorities (ministries, departments, public agencies) are involved in water policy making and regulation at central government level. In Chile the number of actors involved in water policy making are 43, the highest of Latin American and OECD countries (Akhmouch 2012; OECD 2011; World Bank 2013). The overall performance evaluation of Chile's water institutionality is low, due to a high level of fragmentation, insufficient budget and qualified personnel, and problems in horizontal and vertical coordination (Akhmouch 2012; OECD 2011; World Bank 2013).

The WC 1981 establishes that WR owners are responsible for local water management. User management has existed in Chile since the colonial era, and currently there are more than 4,000 Water User Associations (WUAs) (Dourojeanni and Jouravlev 1999). Three types of WUAs exist in Chile and are recognized by the WC 1981: *comunidades de aguas* (water communities), *associaciones de canalistas* (canal user associations), and *juntas de vigilancia* (river user committees).[12]

Many of these WUAs have professional management (Hearne and Donoso 2005). The effectiveness of some of these institutions in managing irrigation systems and reducing transactions costs for water market transactions has been noted (R. Hearne and Easter 1997). However, according to the DGA and the *Dirección de Obra Públicas* (DOH), a large percentage of these institutions have not updated their capacity to meet new challenges. Additionally, (Bauer 1998) points out that vigilance committees have not been effective in resolving inter-sectoral conflicts. To address some of these concerns, the *Comisión Nacional de Riego* (CNR) and DGA have implemented programs to train WUA managers and directors (Peña et al. 2011).

Thus, in order for Chile's WR markets to deliver its full potential as an efficient allocation mechanism, Chile requires a significant institutional reform.

19.3.2.2 Transaction Costs and Design

Transaction costs associated with the transaction of WR includes legal costs to study WR and elaborate transaction contracts, broker costs, notary costs and registration of WR in Real Estate Registry (*Conservador de Bienes Raíces, CBR*), and costs

[12] Water communities are any formal group of users that share a common source of water. Canal user associations are formal associations with legal status that can enter into contracts. River user committees are comprised of all the users and canal associations on any river, river section, or stream; they are responsible for administering water and allocating water to different canals.

regarding the DGA's authorization of water extraction points changes. The last of these are the most significant; recent studies estimate that transaction costs associated to DGA's authorization of extraction point changes represent, on average, between 20 % and 50 % of the WR's value, depending on the geographic location of the WR.

The costs associated with the use of water acquired are those that are required to modify water distribution infrastructure.[13] These transaction costs due to infrastructure modification have been estimated to be approximately 10 % of the WR's value in the Maipo River (Donoso et al. 2002).[14] On the other hand, transactions carried out in river basins with flexible-pipe distribution systems occur with much greater frequency.

19.4 Conclusions

Compared to the situation in most countries in Latin America and the Caribbean, Chile's water policies are unusually conducive to efficient resource use and development (Southgate and Figueroa 2006). Secure and transferable property rights are the salient feature of the Chilean regime. In Chile, water use rights markets guide the use of water, including its reallocation when and where appropriate.

This review of Chile's WR markets and WC 1981 regulations leads to the identification of lessons that must be considered in order establish an effective water allocation mechanism based on a WR market. The main lessons are the following:

(a) A cultural context of the society consistent with the economic paradigm of solving inefficiencies of free access goods based on the establishment of property rights (WR);
(b) The existence of water scarcity; when water is not scarce, there is no need to reallocate WR;
(c) Clearly specified WR, secure ownership, and formally registered WR;
(d) Explicit and transparent conditions for WR trade and transfers;
(e) Clear legislation respect to unused WR;
(f) Environmental and in-stream needs addressed prior to the introduction of trade;
(g) Adequate regulations that address externalities and potential damage to third parties due to WR transactions;
(h) A complete registry of WR holders;
(i) An efficient information system that considers an efficient flow of market information such as data on transactions and a price revealing mechanism;
(j) Detailed information and models of both surface and groundwater resource availabilities;

[13] These transaction costs will be involved in any type of water reallocation and are independent of water management institutions. They are present under administrative or market based allocation mechanisms.

[14] This percentage diminishes as the total volume of water transferred increases.

(k) Flexible water distribution infrastructure that allows for the transfer of WR at low costs;
(l) Strengthening and capacity building of WUAs.

The elements that have hindered WR market effectiveness in Chile are the;

(a) Lack of WR and WR market information;
(b) Lack of regularization of customary WR;
(c) Existence of transaction costs;
(d) Lack of a rapid, efficient conflict resolution system.

References

Akhmouch, A. (2012). *Water governance in Latin America and the Caribbean: A multi-level approach*. Paris: OECD Publishing.

Banco Central de Chile. (2013). *Síntesis de estadística de Chile 2008–2012*. Santiago: Banco Central de Chile.

Bauer, C. (1998). *Against the current: Privatization, water markets, and the State in Chile*. Boston: Kluwer.

Bauer, C. J. (2004). Results of Chilean water markets: Empirical research since 1990. *Water Resources Research, 40*(9), W09S06. doi: 10.1029/2003WR002838.

Bitrán, E., & Sáez, R. (1994). Privatization and regulation in Chile. In B. P. Bosworth, R. Dornbusch, & R. Labán (Eds.), *The Chilean economy: Policy lessons and challenges*. Washington, DC: The Brookings Institution.

Buchi, H. (1993). *La transformación económica de Chile. Del Estatismo a la libertad económica*. Bogota: Norma.

DGA. (2014). *Boletines Hidrológicos*. Retrieved from: http://www.dga.cl/productosyservicios/informacionhidrologica/Paginas/default.aspx

Donoso, G., Montero, J. P., & Vicuña, S. (2002). Análisis de los Mercados de Derechos de Aprovechamiento de Agua en las Cuencas del Maipú y el Sistema Paloma en Chile: Efectos de la Variabilidad de la Oferta Hídrica y de los Costos de Transacción. In A. Embid (Ed.), *El Derecho de Aguas en Ibero América y España: Cambio y Modernización en el Inicio del Tercer Milenio* (pp. 109–151). Spain: Civitas.

Donoso, G., Cancino, J., Melo, O., Rodríguez, C., & Contreras, H. (2010). *Análisis del Mercado del Agua de Riego en Chile: Una Revisión Crítica a través del caso de la Región de Valparaíso*. Santiago: ODEPA, Ministerio de Agricultura de Chile Retrieved from http://www.odepa.cl/wp-content/files_mf/1369772320Analisis_agua_riego_Valparaiso.pdf

Donoso, G., Melo, O., & Jordán, C. (2014). Estimating water rights demand and supply: Are non-market factors important? *Water Resources Management, 28*(12), 4201–4218. doi:10.1007/s11269-014-0739-3.

Dourojeanni, A., & Jouravlev, A. (1999). El Código De Aguas De Chile: Entre La Ideología Y La Realidad *Serie Recursos Naturales e Infraestructura N°. 3*. Santiago: CEPAL.

Grafton, R. Q., Libecap, G., McGlennon, S., Landry, C., & O'Brien, B. (2011). An integrated assessment of water markets: A cross-country comparison. *Review of Environmental Economics and Policy, 5*(2), 219–239. doi:10.1093/reep/rer002.

Hadjigeorgalis, E. (2004). Comerciando con Incertidumbre: Los Mercados de Agua en la Agricultura Chilena. *Cuadernos de Economia, 41*(122), 3–34.

Hadjigeorgalis, E. (2009). A place for water markets: Performance and challenges. *Review of Agricultural Economics, 31*(1), 50–67. doi:10.2307/30224846.

Hadjigeorgalis, E., & Riquelme, C. (2002). Análisis de los Precios de los Derechos de Aprovechamiento de Aguas en el Río Cachapoal. *Ciencia E Investigacion Agraria, 29*(2), 91–99.
Hearne, R., & Donoso, G. (2005). Water institutional reforms in Chile. *Water Policy, 7*, 53–69.
Hearne, R., & Easter, W. (1997). The economic and financial gains from water markets in Chile. *Agricultural Economics, 15*, 187–199.
Jouravlev, A. (2010). *Código de Aguas de Chile: Desafíos Pendientes.* Presentation to Dirección General de Aguas, Ministerio de Obras Públicas de Chile.
Ministerio Secretaria General de la Presidencia. (2014). *Formula Indicación Sustitutiva Al Proyecto De Ley Que Reforma El Código De Aguas (Boletín N° 7.543-12).* Santiago: Retrieved from http://www.camara.cl/sala/verComunicacion.aspx?comuid=12733
Molden, D., Oweis, T. Y., Pasquale, S., Kijne, J. W., Hanjra, M. A., Bindraban, P. S. et al. (2007). Pathways for increasing agricultural water productivity. In: Water for food, water for life. A comprehensive assessment of water management in agriculture. In D. Molden (Ed.), *Water for food, water for life: A comprehensive assessment of water management in agriculture* (pp. 279–310). London: Earthscan.
OECD. (2011). *Water governance in OECD countries: A multi-level approach*: OECD Studies on Water. Paris: OECD Publishing.
Peña, H., Luraschi, M., & Valenzuela, S. (2004). *Water, development, and public policies – strategies for the inclusion of water in sustainable development.* Santiago: South American Technical Advisory Committee, Global Water Partnership.
Peña, H., Brown, E., Ahumada, G., Berroeta, C., Carvallo, J., Contreras, M. et al. (2011). *Temas Prioritarios para una Política Nacional de Recursos Hídricos*. Santiago: Instituto de Ingenieros de Chile.
Riestra, F. (2008). *Pago de patente por no uso de derechos de aprovechamiento.* Santiago: Dirección General de Aguas, Ministerio de Obras Públicas de Chile.
Rosegrant, M., & Gazmuri, R. (1994). *Reforming water allocation policy through markets in tradable water rights*. Washington, DC: International Food Policy Research Institute (IFPRI).
Salazar, C. (2003). *Situación de los recursos hídricos en Chile*. México: Centro del Tercer Mundo para el Manejo del Agua.
Southgate, D., & Figueroa, E. (2006). Reforming water policies in Latin America: Some lessons from Chile and Ecuador. In K. Okonski (Ed.), *The water revolution*. London: International Policy Press.
Valenzuela, S. (2009). *La patente por la no-utilización de las aguas en Chile: origen, diseño y primeras experiencias en su implementación*. Santiago: Universidad de Chile.
World Bank. (2011). *Diagnóstico de la gestión de los recursos hídricos*. Santiago: Dirección General de Aguas, Ministerio de Obras Públicas, Chile.
World Bank. (2013). *Estudio para el mejoramiento del marco institucional para la gestión del agua*: Dirección General de Aguas, Ministerio de Obras Públicas de Chile.

Chapter 20
Unbundling Water Rights as a Means to Improve Water Markets in Australia's Southern Connected Murray-Darling Basin

Michael D. Young

Abstract Australia has defined its water entitlement and allocation arrangements in a manner that has made it possible to establish one of the world's most sophisticated water marketing systems. Entitlements are defined in perpetuity as an entitlement to a proportion of any allocations assigned to a water resource pool. Entitlements and allocations are tradable and in the Southern Connected River Murray system a vibrant water market has emerged. The functioning of this market is reviewed in this chapter. Overall the assessment from an individual water use perspective is that the introduction of this EPI has succeeded. From a national perspective, most experts also describe it as a success. As a Nation however, Australia would have been better off if it had solved the water accounting and over-allocation problems before it introduced water trading. An important conclusion is that unbundling has made it easier to resolve issues step by step. It also makes it much easier for individuals to adjust and innovate. New business and new technology must be expected to emerge with each reform that is made. The chapter concludes by highlighting relevant policy lessons for the practical application of water markets.

Keywords Water markets • Unbundling water rights • Australia • Water entitlement and allocation arrangements

20.1 Introduction

Australia has defined its water entitlement and allocation arrangements in a manner that has made it possible to establish one of the world's most sophisticated water marketing systems. This system is best developed in the Southern Connected Murray-Darling System which sits within Australia's Murray-Darling Basin (Fig. 20.1).

M.D. Young (✉)
Water Economics and Management, School of Earth and Environmental Sciences, The University of Adelaide, Adelaide, SA, Australia
e-mail: Mike.Young@adelaide.edu.au

M. Lago et al. (eds.), *Use of Economic Instruments in Water Policy*, Global Issues in Water Policy 14, DOI 10.1007/978-3-319-18287-2_20

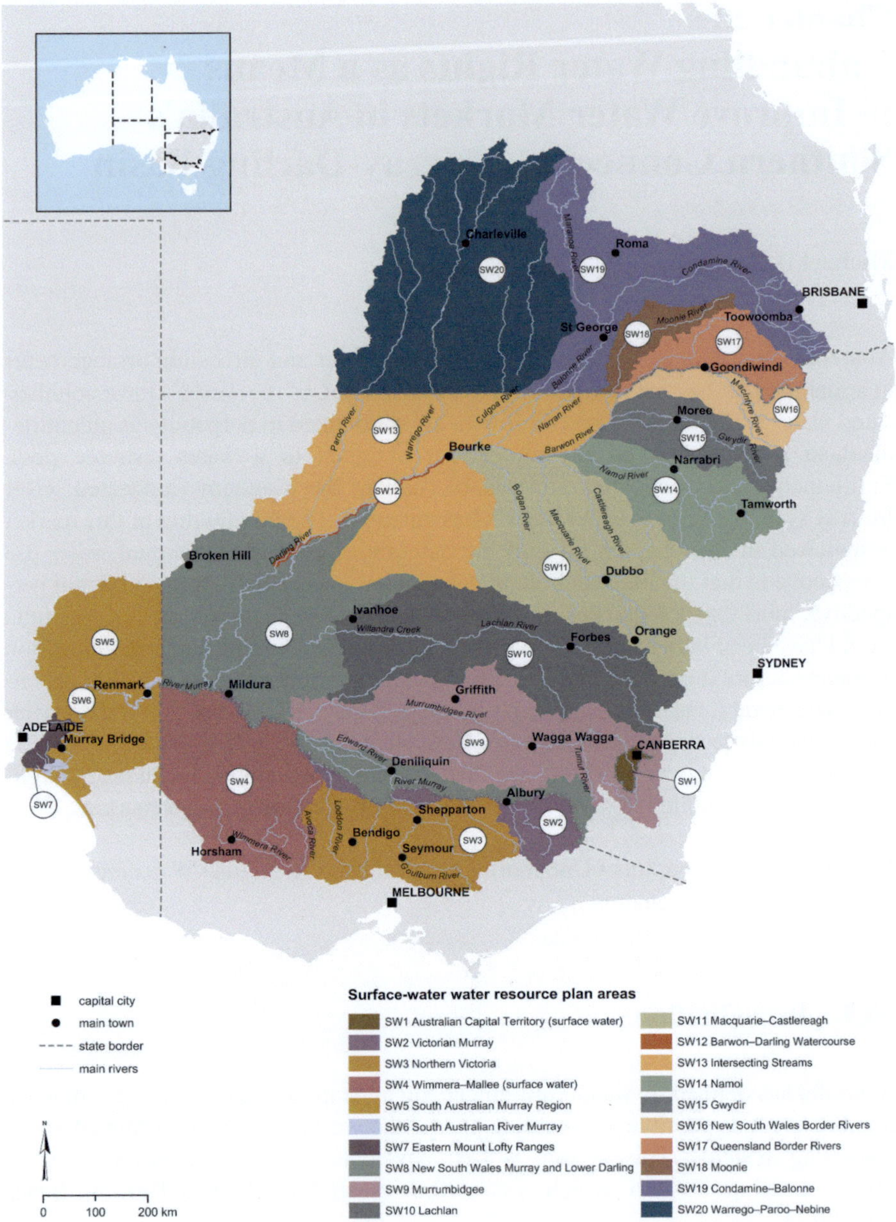

Fig. 20.1 The Murray-Darling Basin. The southern connected portion of this Basin, where water markets are most developed, includes the Murray, Lower Darling, Murray Murrumbidgee, Ovens, Goulburn Broken, Campaspe and Loddon Catchments (MDBA Guide to the proposed Basin Plan, 2011)

Initially, irrigators were issued licences to irrigate a maximum area of land. These licences were converted into licences to take up to a defined maximum volume of water each year. As irrigation expanded, a way to ration water use became necessary. At the entitlement level, initially, two types of licence were introduced

- High security licences – which nearly always received their full allocation; and
- General security licences – which only received a full allocation in wet years.

When it is not possible to give users their full allocation, water is allocated first to High Security Licences and then to general security licence holders on a proportional basis. Eventually, it was realised that no more licences should be issued and a cap was placed on water use in an attempt to prevent over-use and maintain environmental assets. At the same time that this limit – known as the cap – was introduced a suite of water reforms were put in place to enable water trading. The initial objective of trading was to make water use more efficient and enable it to move to its highest and best use at any point in time.

Initially trading was administratively complex and slow. Gradually, however, experience increased and the benefits of trading became more and more apparent. There have been dramatic increases in water use efficiency and considerable innovation.

One of the key innovations that made it possible to trade large volumes of water efficiently is the introduction of what is now known as "unbundling". Unbundling involves the conversion of one property right into a bundle of separate instruments each designed to pursue a different objective and, often, operate at different scales. Today, two markets exist – one for water shares and one for allocations. All water use is metered.

20.2 Rationale for Choosing the Case Study

While it would be possible to present the "Australian" water entitlement, allocation, use control, distribution management and trading system as a case study, for the purposes of this chapter it is judged more useful to focus on one of the key features of this system. The feature chosen is the "unbundling" of the licensing system.

Unbundling is chosen because it demonstrates one of the necessary conditions for the development of market-based approaches to the management of natural resources that can be expected to remain efficient through time and deal equitably and fairly with large numbers of water users.

The underpinning goal of water trading was to increase economic growth by allowing water to be moved to places where it could make the greatest contribution to economic development. The initial argument was that water should be put to its "highest and best use."

In retrospect, however, Australia has learned that water trading enables efficient and rapid adjustment to extreme water scarcity. The "unbundling" innovation identified in this case study has been critical to the development of this capacity to adjust quickly to water scarcity problems.

Australia began with a water allocation system that issued a single property right (a licence) to a water use. Each licence consisted of a "bundle" of entitlements to use water, conditions about how it may be used, etc. Unbundling involves the separation of this bundle of rights into a number of separate parts.

Prior to the introduction of unbundling, the amount of water used by irrigators was administered using a licensing system that made it difficult to transfer water allocations from one location to another. Transaction costs were high and, typically, it took months to complete a trade. The approach taken was to temporarily transfer the licence from one water user to another, then take the water off the licence and then, after the water had been taken from the licence, the licence was transferred back to the original owner. The process was slow and administratively complex. To this day, this type of transfer is known as a temporary trade because the trade used to involve the temporary transfer of a licence from one person to another.

To simplify this process, a decision was taken in 1994 to allow people to hold water licences without owning any land. In order to facilitate this and increase investment security formal water entitlement registers were established and procedures put in place to enable landholders to obtain permission to irrigate an area of land without knowing where the water would come from. As reforms progressed further, it was decided to define water licences as shares and issue them in perpetuity.

Separate bank-like water accounts were then set up and structured so that water could be allocated to each shareholders account in proportion to the number of shares they held. In parallel with these arrangements, any landholder who wished to use some water in an account needed to have a use approval that authorised the government to deduct water from an account as it was used. Whilst complex, the result was the emergence of extremely efficient water trading arrangements.

In parallel with these reforms, efforts were made to improve system-wide planning processes so that irrigators could make investments with greater confidence.

20.3 Legislative Setting and Economic Background

In Australia, the degree of protection from competition in the production of agricultural products is low.

Significantly, in 1994 Australia established a National Competition Policy that sought to use markets as the prime mechanism to make water use and many other services provided by government more efficient. This commitment, nearly 20 years, has forced many changes. Productivity and water use efficiency are now much greater (Young 2008).

With regard to the legislative setting used to enable water management:

- Each component of the unbundled set of arrangements is defined in legislation and in a suite of plans approved by parliament.
- A key feature of the resultant suite of institutional arrangements is a process that uses the approved plans to manage third party impacts.

- If a third party is aggrieved by a water trade and the trade is in accordance with the rules set out in the plan, the only course of action available for a third part to prevent the trade from occurring is to arrange for the rules in the plan to be changed. There is no opportunity for a third party to prevent a transaction that is consistent with rules set out in the plan.
- An independent regulator is used to minimise opportunities for regions to find ways to impede trades from occurring. A complex set of rules, for example, are used to define the maximum fee that a person may be charged for trading water from one district and into another.
- As each stage in the development of the current unbundled system of property rights was introduced, a pragmatic decision was taken to begin by defining formally defining each dimension of the emerging system in a manner that mimics the status quo. (This is known as grandfathering.)

Figure 20.1 shows the location of the Murray-Darling Basin in Australia and its prime water resource management regions. Water trading arrangements are most developed in what is known as the Southern Connected River Murray System. This southern system contains a suite of large dams at the top of the system coupled with a series of locks and weirs that makes a high degree of flow regulation possible.

20.4 EPI Background

The Australian approach to the development of an unbundled water entitlement and allocation system has evolved over many years. Many mistakes have been made and many lessons learned. In a paper prepared for the OECD, Young (2010) identifies 17 lessons of particular importance to the development of systems like this.

In retrospect, a number of the key features of the Australian approach were developed without any expectation that an EPI would ultimately be established.

An historical decision to define all licences within a region in a similar way has made the development of low cost water trading arrangements possible. In effect, each water region is treated as a pool of water available for use. Within any defined pool, all licence holders are treated equally and, unlike the USA, no licence holder is more senior than any other licence holder. This also made it possible ultimately to define water entitlements as shares and make allocations in proportion to the number of shares held.

A decision in 1994 to commit Australia, through a National Competition Policy, to the development of more competitive approaches to the development of the economy by bringing market disciplines to the delivery of many services provided by state governments and "fine" states who did not implement the required policy reforms within an agreed timeframe. In water this required, among other things,

1. The separation of water licences from land titles so that it would be possible for people to hold a water licence even if they did not own any land.

2. The corporatisation of all water supply arrangements so that those responsible for water allocation and policy decisions would not be responsible for delivery of water. In practice, this meant that each state had to transfer ownership of its water supply and delivery infrastructure to a company and appoint a board to make all decisions associated with the operation of this infrastructure.
3. A requirement that each water supply business recover at least the full marginal costs of operating that business and move progressively towards full cost recovery including the cost of environmental externalities.
4. That it become possible to trade water from one location to another. In the same system and that it be possible any one to own a water entitlement – even if the y don't own land.

A parallel decision in the 1994/1995 to place a limit of the total amount of water that could be diverted from all surface water resources in the Murray-Darling Basin – known as the "cap" was also taken.

Federal and State agreement to implement a National Water Initiative in 2004 that added a lot of detail to the 1994 competition arrangements and, in particular, required:

(i) clear and nationally-compatible characteristics for secure *water access entitlements*;
(ii) transparent, statutory-based water planning;
(iii) statutory provision for *environmental and other public benefit outcomes*, and improved environmental management practices;
(iv) complete the return of all currently over-allocated or over-used systems to *environmentally-sustainable levels of extraction*;
(v) progressive removal of barriers to trade in water and meeting other requirements to facilitate the broadening and deepening of the water market, with an open trading market to be in place;
(vi) clarity around the assignment of risk arising from future changes in the availability of water for the *consumptive pool*;
(vii) water accounting which is able to meet the information needs of different water systems in respect to planning, monitoring, trading, environmental management and on-farm management;
(viii) policy settings which facilitate water use efficiency and innovation in urban and rural areas;
(ix) addressing future adjustment issues that may impact on water users and communities; and
(x) recognition of the connectivity between surface and groundwater resources and connected systems managed as a single resource.

A series of attempts to resolve over-allocation and water accounting problems in the Murray-Darling Basin first by a decision to secure 500 GL of water for the environment under a Living Murray Initiative and second by the transfer of Basin wide water planning responsibilities to an independent Murray-Darling Basin Authority and the commitment of A$3.1 billion for the purchase of water entitlements from

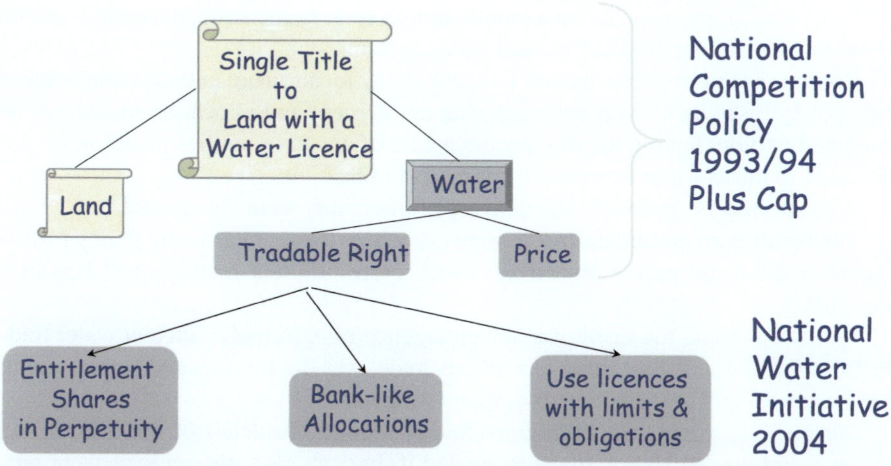

Fig. 20.2 An overview of the way that water licence arrangements in the Murray-Darling Basin have been unbundled (Own elaboration)

irrigators and the transfer of these entitlements to a Commonwealth Environmental Water Holder coupled with the commitment of A$5.8 billion for investment in so-called water savings projects to improve water use efficiency in a manner that enables half of the savings made to be transferred to the Commonwealth Environmental Water Holder.

Figure 20.2 provides an overview of the unbundling process. Prior to the introduction of unbundling, the amount of water used by irrigators was administered using licences that made it difficult to move water allocations for one location. The approach taken was to temporarily transfer the licence from one water user to another, then take the water off the licence and then, after this had been done, the licence was traded back again. The process was slow and administratively complex.

The formal proposition that it made sense to unbundle water licences was first made by Young and McColl (2002) and followed from Young's involvement in drafting amendments made to administrative arrangements in New South Wales in 2000. In particular, the legislation required licences to be defined as shares of water allocated in proportion to the number of shares held.

As shares had no water use conditions attached to them, they could be defined as rights in perpetuity. The definition of shares in perpetuity proved to be particularly important. It meant they could never be taken away. Under this new arrangement, the only way an aspiring water user could gain access to water was to convince an existing water user to sell water or sell a water access entitlement to them.

In the process of unbundling it became necessary to establish formal registers that define each licence holder's share of any water allocated to a region.

Separate bank-like water accounts were then set up to record the amount of water allocated to each shareholder and track use and sales of that water. Typically, allocation announcements are made twice a month and, as soon as the announcement is made, these allocations are credited to each water account.

Conditions that regulate the use of water at any location are defined using a separate policy instrument with the result that entitlement and allocation trades can be executed without having to consider the nature of any externalities resulting from a decision to move water from one location to another.

Separate works approvals and delivery entitlements were also issued.

The result is an administrative framework where there are as many policy instruments as there are policy objectives. Much more efficient management becomes possible.

Whilst complex, the result was the emergence of extremely efficient water trading arrangements. Today water allocations trade over the Internet and water trading has become a business that involves many brokers.

Surprisingly, there was little consultation around the detail of the unbundling reforms and the legislation that surrounded it. In each case, the reforms were presented as a win-win opportunity for licence holders. From the outside, the reforms looked like an attempt to simplify administrative procedures and define licensing arrangements with rigour.

20.5 Environmental Outcomes

These apparent benefits of the unbundled approach to water allocation used in the Southern Connected River Murray System hide an important oversight. Unbundling drove structural adjustment, investment and innovation but unless the system-wide water allocation system is designed to adjust for these changes, the system must be expected to trade into trouble (Young 2014a).

In Australia's Murray-Darling Basin, this is exactly what happened. A massive over-allocation problem has emerged because system managers and the agreements they had negotiated did not anticipate the extent of change that the EPI would induce.

In retrospect it can be seen that it is critical to establish robust water accounting arrangements and allocation arrangements that are consistent with hydrological realities. When these arrangements are not in place the introduction of an EPI can make the nation as a whole, many communities and many individual irrigators worse off.

The unbundling of water entitlements in Australia made the low cost and rapid trading of water allocations possible. Today, most water allocation trades are executed in less than 2 days. Trade is possible across state jurisdictions and during the irrigation season occurs on a daily basis.

The sequence of reforms is important to understand (see Box 20.1). In the Murray-Darling Basin, these reforms began, in the late 1980s, with a series of negotiations that introduced a cap on diversions in 1994. This "cap", as it was called, was acknowledged as an interim cap and was expected to prevent an increase in

Box 20.1: An Overview of the Sequence of Water Reforms in the Murray-Darling Basin

1994: Introduction of an interim cap on diversions

1994: National Competition Policy requires states to introduce policies that require full cost pricing, the introduction of water trading in rural areas and arrangements that allow water entitlements to be held by legal entities that do not hold an interest in land

1996: Within-state trading allowed

1998: A 2 year pilot interstate water trading trial commenced between NSW, Vic and SA but limited to areas close to the South Australian border

2000: Review of interstate water trading results in a decision to expand trading to cover most surface water use in the connected Southern Connected River Murray System

2002: Various proposals for the reduction of water use in the Basin by reducing allocations by as much as 1,500 GL which eventually resulted in a decision to take a first step towards solving the "problem" by returning 500 GL to the environment over the next 5 years

2004: National Water Initiative introduced

2007/2008: Commonwealth Government passes a Water Act that attempts to transfer responsibility for development of a water use plan for the Murray-Darling Basin and the resolution of over-allocation problems in this system to the Commonwealth. Subsequent negotiations between the Commonwealth and State Governments eventually resulted in a decision to establish an independent, expertise based Murray-Darling Basin Authority coupled with arrangements that gave State Ministers and officials a larger say in the development of the Basin Plan

2010: A guide to the Basin Plan released

2011: A proposed Basin Plan released

2012–2013: Basin Plan finalised and approved by Commonwealth Parliament

Source: Own elaboration

water use. If the cap had been introduced without the introduction of trading the volume of water used in the basin would have remained the same.

In 1994, however, as part of a National Competition Policy, state governments were required to allow water entitlements to be held separately from land titles and traded. The result was a dramatic increase in the volume of trading (see Fig. 20.3).

Trading stimulated widespread investment in technologies designed to improve water use efficiency. These investments, however, significantly reduced return flows and, also, in the use of ground water that previously flowed unused into the river (Young and McColl 2003; Young 2010). There was also a significant increase in the capture of overland flows that previously flowed to the river. In short, the introduction of water trading worsened the extent of the Basin's over-allocation problem that

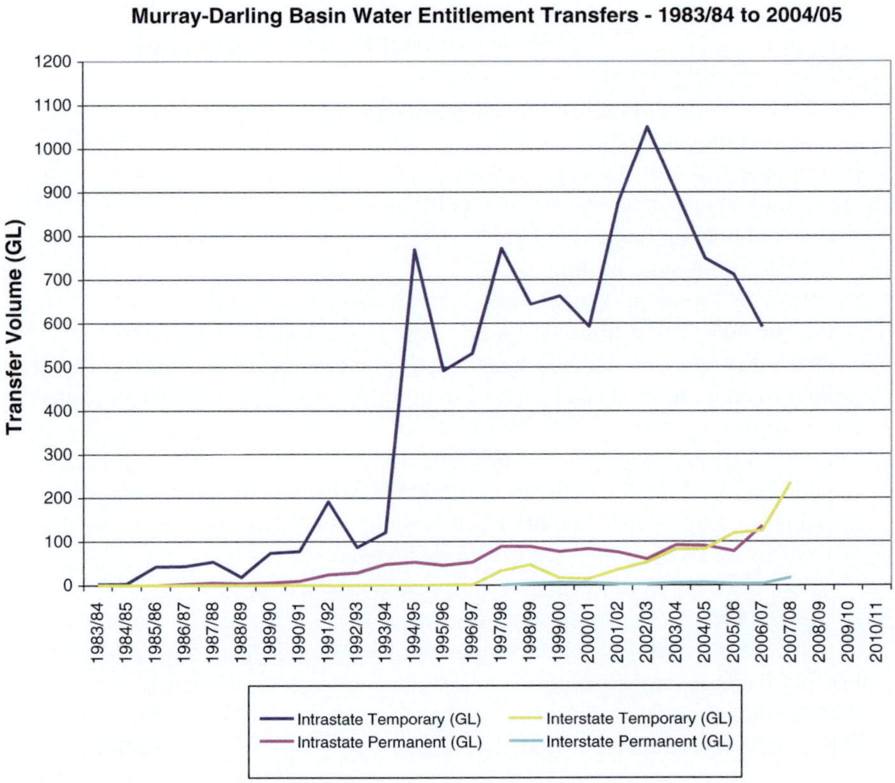

Fig. 20.3 Development of Murray-Darling Basin water market. Allocation trades are known as temporary trades. Entitlement trades are known as permanent trades (Young 2010)

was identified when the cap was introduced. In retrospect, the cap should have been a cap on *nett use* rather than a cap on diversions which allowed those who improved irrigation efficiency to expand water use (Young 2010).

In 5 years immediately after the introduction of water trading, use of water increased by 29 %. The area irrigated increased by 22 % (Bryan and Marvanek 2004) and nearly all of this new area involved the establishment of new vineyards and orchards. None of the water allocation plans, however, made any allowance for this increase in water use. Allocations continued as if no increase in water use had occurred. As a result, late in 2002 the River Murray stopped flowing and in November 2003 dredges had to be put into the mouth of the River to keep it open.

Officials were aware of these problems but were unable to find a politically acceptable way to manage the adverse effects of these processes on the health of the river. By 2002, it had been estimated that, at least, 1,500 GL of cap equivalent would be needed to restore health to the Basin and estimates of the economic and social impacts of securing this and other amounts of water for the environment where being made (See for example Young et al. 2002). Whilst the increasing environmental costs of not fixing the Basin's over-allocation problems were appreciated, governments found difficulty in agreeing about what to do. Ultimately, it was decided that a Living Murray program would be implemented as a first step towards solving the over-allocation problem. Under this program, it was decided that 500 GL of water would be secured for the environment over the 4 years between 2004 and 2009. This amount was, however, insufficient to cover the losses being caused by the expansion of irrigation and investment in new technology (Young and McColl 2003).

Nett progress in the resolution of the over-allocation problem was negative and, in 2007, the Commonwealth Government decided to step in and introduced a new Commonwealth Water Act coupled with a commitment to purchase A$3.1 billion of water entitlements and invest a further A$5.8 billion in improving the efficiency of irrigation on the condition that half to the savings were returned to the river. Progress still proved difficult and in 2010 the Murray-Darling Basin Authority in a guide to the development of a new plan for the basin estimated that entitlements in the entire Basin had to be reduced by over 3,000 GL (MDBA 2010). Whilst the benefits of trading were apparent it was becoming increasingly clear that the costs of not fixing the Basin's over-allocation problems before introducing water trading were rising. A problem that could have been fixed in 1994 – at little cost to taxpayers – had evolved into a problem that would cost over A$8.9 billion of tax revenue to fix. In retrospect, Australia got the sequence of the reforms it implemented wrong (Young 2014b).

20.6 Economic Efficiency

As shown in Fig. 20.3, the decision, taken in 1995, to enable water entitlements to be held by people who did not own an irrigation property was critical in reducing the transaction costs associated with water trading. Once implemented, investors could decide when and how to buy water and many innovations followed. The old command and control approach where permission to change where water was used was difficult to obtain was abandoned.

To the surprise of many, but as expected by the architects of this reform program, the result was a large degree of innovation and new investment in water use. Water use efficiency has increased dramatically. As shown in Fig. 20.4, the return on investment in water entitlements has averaged well over 12 % per annum. During the long dry period in the MDB from 2002/2003 to 2008/2009 all assessments of Basin productivity have shown that trading was critically in minimising the economic impact of this period on the irrigation community (NWC 2010). The National Water Commission has estimated that the introduction of water trading has increased Australia's Gross Domestic Product in the 2008/2009 irrigation year by A$220 million.

Adoption rates for water trading are high. In the 3 years to 2010/2011, ABARES estimates that 43 % of irrigation farms in the Southern Connected River Murray

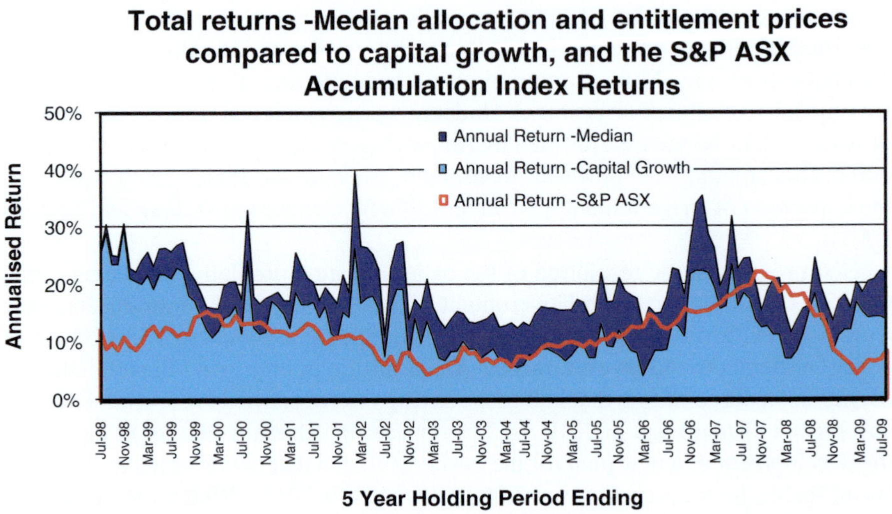

Fig. 20.4 Annual returns from selling allocations (*dark blue*) and capital growth (*light blue*) in the value of a water entitlement compared with an index of the value of shares in the Australian Stock Exchange (S&P ASX), Goulburn Murray System, Murray-Darling Basin (Bjornlund and Rossini 2007)

System traded water. The majority of irrigators indicated that they found the process of trading temporary water allocations to be easy (89 %), reliable (84 %) and affordable (72 %) (Fargher and Olszak 2011).

When water trading was introduced, however, the new policy signal given to irrigators was that if you could not profitably use any water allocated to you, you should sell it someone who could. Irrigators responded accordingly and water that would have previously been left unused in the systems main dams was sold to someone who could use it. As a result, too much water was used and dam storages were run down too quickly. So much so that Brennan (2007) estimates that the apparent annual benefits of water trading were less than the cost of the increased drought-like impact of trading on the amount of water available for use in subsequent years. As soon as officials appreciated the importance of allowing the carry forward of water from one season to another allocation policies where changed (Young 2010).

In retrospect, the golden rule, now realised by all Australian governments, is that if water trading is introduced, it must be possible for irrigators to decide that the optimal strategy is to carry forward water from 1 year to the next – especially when water supplies are low.

20.7 Cost Effectiveness

There has never been a formal assessment of the administrative costs of unbundling the water licence systems maintained in each Australian State. A case study has, however, been completed for the Gwydir Valley (Young and Esau 2013). The first step in this process, involved building water entitlement registers and running the processes necessary to register them. Prior to this step, licences were attached to land titles and often lacked clarity as to who really "owned" the water licence. Whilst the department may have issued the licence to a farmer, the land title on which the irrigation occurred may be held jointly in the names of three people. To make matters even more complicated, one of the people on the title may have deceased or be in the process of going through a divorce. On a case by case basis, each licence had to be examined and, once all issues resolved, placed on a register.

The interests of banks also had to be considered. Prior to the separation of water entitlements from land titles, the value of land included the value of all the water licences associated with it and banks used these titles as security. As water entitlements were separated from land titles, registers had to be built in a manner that enabled third parties to formally register an interest in a water entitlement. Once this had been done, each register had to be validated in terms of ownership and banks given the chance to renegotiate an appropriate level of security. In each state, this process took several years.

At the same time, bank-like water allocation accounts had to be established and arrangement put in place to ensure that these accounts had integrity. Today, every entitlement is linked to a water account and the holders of these accounts can transfer water from their account to another account. In the most sophisticated systems, these transfers can be executed over the Internet in a manner that is similar to the processes used to transfer money from one account to another (Young and McColl 2002).

In all cases, the government picked up the costs of establishing registers, building water accounting systems, etc. at the State level. Within some irrigation areas, however, in a parallel set of reforms ownership of the water distribution systems where transferred at no charge from the government to water supply companies owned entitlement holders. Whilst this enabled irrigators to take control of "their" water supply system, it meant that they, not government would be responsible for the full marginal costs of water supply. The result, once again, was a dramatic increase in the efficiency of water delivery. In the case of the Murrumbidgee Irrigation System, for example, the transfer of responsibility for management of this supply system to irrigators in 1999 resulted in a real reduction in management costs for each of the next 6 years. The NSW government, however, found it necessary to almost continuously increase bulk water charges over this period (see Fig. 20.5).

Throughout the Murray-Darling Basin, water now trades on a daily basis and a complex array of water supply and information systems have been developed by government and by industry. A water broking industry has been established. Figure 20.6 provides an overview of the relationship between water trading and the volume of water available for use. As theory predicts, in times when allocations are low, trading is high and vice versa.

The extent to which water trading has also improved water use can be seen from Fig. 20.7. As a result of the long dry in the first decade of this century, the amount of water diverted for irrigation in the Murray-Darling Basin declined from nearly 12,000 to 2,000 GL but the Gross Value of Irrigated Production only declined from

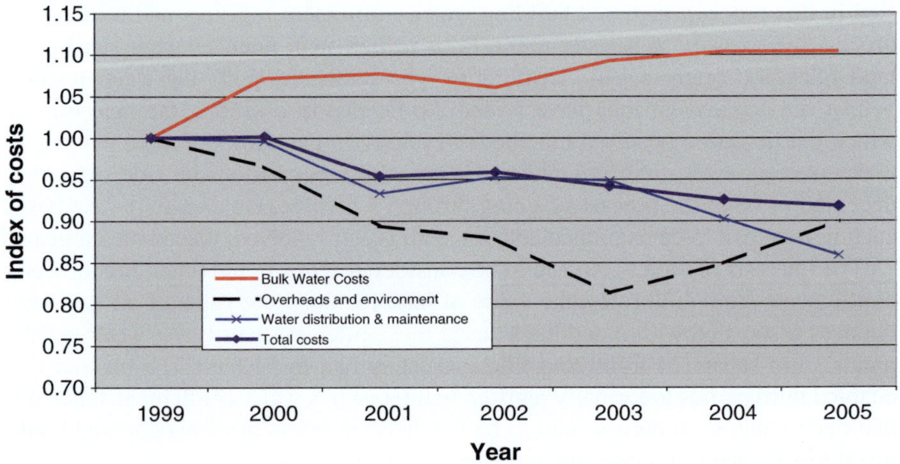

Fig. 20.5 Example of the efficiency obtained by transferring ownership and responsibility for operating the Murrumbidgee Irrigation system to irrigators (Young et al. 2006)

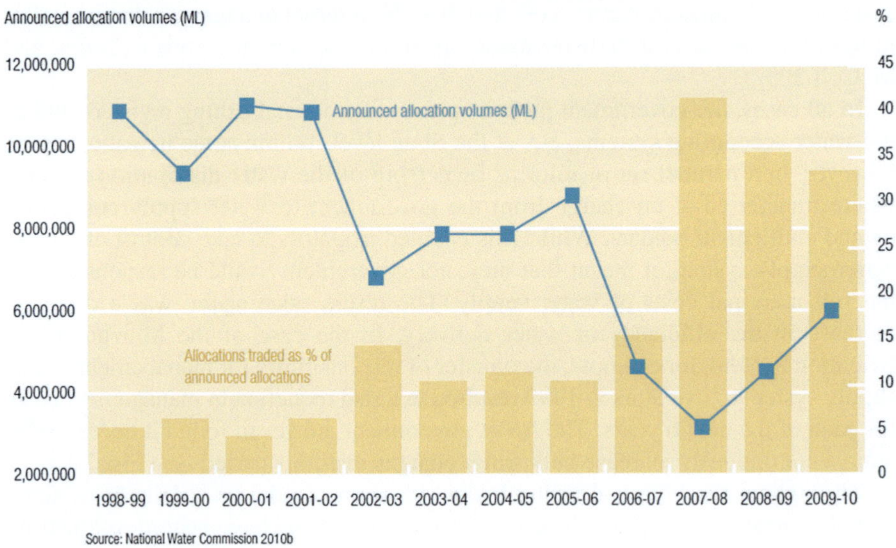

Fig. 20.6 Relationship between announced water allocations and the volume of water traded (Fargher and Olszak 2011)

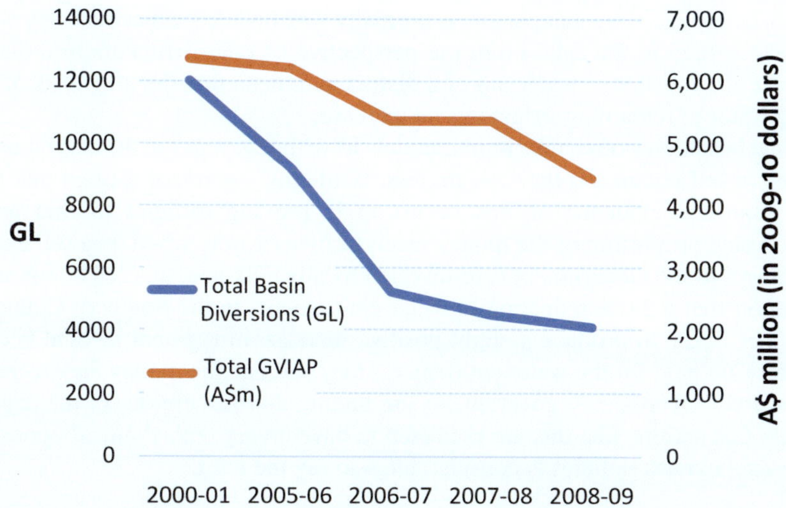

Fig. 20.7 Change in the total value of irrigated agricultural production and the amount of water used in the Murray-Darling Basin, 2000–2001 to 2008–2009 (Gooday pers. com., 2011; presentation to ACCC Conference Brisbane)

A$7.5 billion to A$4.5 billion. That is, an 80 % drop in water availability only caused a 40 % decline in the gross value of production.

20.8 Distributional Effects and Social Equity

Until recently, all the distributional effects of the introduction and development of water trading were found to be positive. Few people were made worse off and many were made better off as the value of water entitlements increased and people voluntarily choose to sell water because they could make more money from selling this water than using it. Many also chose to sell water to finance investment in new more efficient irrigation technology.

Towns and local communities also appeared to benefit from these changes even though local shop keepers complained that the trade of water out of their region was not in their best interests. Case study and focus group work, however, has found it very difficult to find concrete examples of situations where this was the case (Young et al. 2006).

As the extent of the Murray-Darling Basin's over-allocation problems have become more apparent, however, a new suite of distributional effects have become apparent. As already mentioned in an attempt to resolve the over-allocation problem, the Australian Government has been buying water entitlements for the environment from irrigators willing to sell some or all of their water to them. From the perspective of a person selling a water entitlement the transaction, given the circumstances faced by that person, is normally worthwhile – otherwise they would not have agreed to the sale. From the perspective of other irrigators in a district, however, the transfer of water out of a district can mean that the unit costs of supplying water to remaining irrigators can increase.

Local businesses often take a similar view as with less water in the district opportunities to sell goods and services are less. While this argument is often put, however, empirical evidence of this occurring is proving difficult to find as the considerable proportion of the money received by irrigators when they sell a water entitlement to the Government is re-invested locally. Dixon et al. (2011), for example, report that a 23 % reduction in water entitlements in the Southern Connected System is likely to produce a slight positive increase in regional income because irrigators are paid for the water entitlements they sell and the money they receive is re-invested. Nevertheless, governments are finding that perceptions of the negative impacts that actions like this are predicted to have a very real. Political opposition to current buyback policies is considerable – to say the least.

20.9 Institutional Context

A range of different institutional arrangements underpin Australia's approach to water reform. A recent political imperative was the emergence of an 8-year long dry period in the last decade right throughout Australia. Water – at least water shortage – rose to the top of the political agenda. Every mainland capital city, except Darwin, was placed on major water restrictions. Irrigation allocations to many irrigation entitlement holders was zero. In such an environment, the public is looking for and expects water policies to change. In the middle of this dry period, the Australian government was able to produce a National Water Initiative that set the context for many of the reforms that followed. It also made it possible for Australia's Federal Government to propose to "take over" management of the Murray-Darling Basin and establish a new Murray-Darling Basin Authority.

20.9.1 Unbundling

Unbundling commenced in 2000 in the State of New South Wales and is now required under the National Water Initiative. It has now been implemented in all States in the Murray-Darling Basin.

A quite complex suite of administrative arrangements had to be put in place to enable the unbundling of the water licensing system originally used to allocate water in Australia. In most cases, a new water act was drafted and then approved by Parliament. Under the new regime, water entitlements are a special form of a property right. The term "property right", however, is rarely used by Australian administrators as they have found it easier to talk about the nature of each person's entitlement and avoid getting tangled up in debates about the nature of people's rights. The right issued is officially described as a "Water access entitlement."

Access entitlements take the form of a share and are usually issued in perpetuity. Once the system is set up the only way to secure an entitlement to a share of water in a system is to purchase a share from an existing share holder.

Ownership of entitlements is vested in individuals and arrangement put in place to enable water to be traded from one irrigation district to another.

Water supply companies are allowed to charge people who permanently transfer water entitlements from one irrigation district to another an exit fee.

To prevent unfair behaviour the maximum fee that may be charged is regulated by a national market regulator (the Australian Competition and Consumer Commission).

Allocation trades are implemented by debiting one person's water account and crediting another person's water account.

Entitlement trades are implemented by amending names on a water entitlement register. Entitlements can be mortgaged.

Brokers are used to bring buyers and sellers together and settle each trade.

Administrative arrangements have also been unbundled. Under a new National Water Act (2007), a Murray-Darling Basin Authority has been established and this Authority given responsibility for developing a new Basin Plan which amongst other things will define the maximum amount of water that can be diverted from each water resource in the Basin. State Governments are then responsible for developing and implementing a water sharing plan for each region. Use approvals are managed locally.

20.10 Transaction Costs

A large amount of attention has been given the development of arrangements that reduce transaction costs associated with trading. In particular, a series of rules have been developed in an attempt to prevent irrigation districts for discouraging trade out of their district and also to prevent States from preventing the transfer of water out of their state. Tables 20.1 and 20.2 below summarise the water allocation and trade service standards that government now try to comply with.

Table 20.1 Water allocation trade service standards (implemented from 1 July 2009)

State and territory	Intrastate trade approval	Interstate trade approval
New South Wales, Victoria Queensland, Australian Capital Territory	90 % of allocation trades within 5 business days[a]	90 % of allocation trades within 10 business days[a]
South Australia	90 % of allocation trades within 10 business days	90 % of allocation trades within 20 business days

[a]All interstate trades except for trades with South Australia, which would be consistent with standards set out above for South Australia

Source: Own elaboration

Table 20.2 Water entitlement trade service standards (implemented from 1 July 2009)

State and territory	Intra and interstate trade approval	Intra and interstate trade registration
Trade approval/rejection time	Total amount of time taken by the relevant water authorities to approve or reject a trade application received from the buyer or seller	
	The time excludes the duration when the application is back with the buyer or seller due to incorrect/incomplete information and include the approval/rejection times for all water authorities involved in processing the trade	
Trade registration time	Total amount of time taken by the relevant water authorities to register a water entitlement trade in the water register after receiving the relevant transfer documents/registration application from the buyer or seller	
	The time excludes the duration when the application is back with the buyer or seller due to incorrect/incomplete information and include the times for all water authorities involved in adjusting the water accounts and registering the trade	

Source: Own elaboration

20.11 Policy Implementability

The choice of the EPI in this case derives from an initial decision to develop water markets as part of National Competition Policy. At the time, the Australian Government decided that it was critically important the Australia became more competitive. Water was included as part of this agenda. If this commitment had not been made then it is likely that much less progress would have been made. Significantly, any state that failed to comply with the Australian government's competition policy agenda was fined many millions of dollars. Implementation of water reform, in political practice, was mandatory.

One of the driving factors underpinning this policy reform was a significant and early increase in the value of water entitlements. Although many problems emerged, and had to be dealt with, all understood that abandonment of this new policy would result in a significant decline in the personal and newly found wealth that the increase in the value of water entitlements generated. Soon after the reform was implemented, it became clear that Australia would probably always have water markets – at least in the Southern Connected River Murray system. Any government that stopped water trading would be accused (rightly) of causing a massive decline in the wealth of a significant group of people.

20.12 Conclusions

The main conclusion and arguably most significant observation that can be made from the development of water trading in Australia is that it takes time. The development to this EPI has taken over 20 years and, at least, another 10 years of reform is expected as progress is made in the resolution of over-allocation issues and improving water markets.

A second conclusion is that unbundling has made it easier to resolve issues one by one. It also makes it much easier for individuals to adjust and innovate. New business and new technology must be expected to emerge with each reform that is made.

20.12.1 Lessons Learned

Over all the assessment from an individual water use perspective is that the introduction of this EPI has succeeded. From a national perspective, most experts also describe it as a success. When one looks carefully, however, it is clear that Australia got the reform sequence wrong. As a Nation, Australia would have been better off if it had solved the water accounting and over-allocation problems before it introduced water trading.

In a report to the OECD (Young 2010) draws attention to the following lessons:

- *Lesson 1*: *Unless carefully managed, the legacy of prior licensing decisions can result in markets causing over-allocation problems to emerge in a manner that erodes the health of rivers, aquifer and the water dependent ecosystems associated with them.*
- *Lesson 2*: *Transaction and administrative costs are lower when entitlements are defined using a unit share structure and not as an entitlement to a volume of water.*
- *Lesson 3*: *Market efficiency is improved by using separate structures to define entitlements, manage allocations and control the use of water.*
- *Lesson 4*: *Early attention to the development of accurate licence registers is critical and a necessary precondition to the development of low-cost entitlement trading systems.*
- *Lesson 5*: *Unless water market and allocation procedures allow unused water to be carried forward from year to year, trading may increase the severity of droughts.*
- *Lesson 6*: *Early installation of meters and conversion from area based licences to a volumetric management system is a necessary precursor to the development of low cost allocation trading systems.*
- *Lesson 7*: *It is difficult for communities to plan for an adverse climate shift and develop water sharing plans that deal adequately with a climatic shift to a drier regime. More robust planning and water entitlement systems are needed.*
- *Lesson 8*: *The allocation regime for the provision of water necessary to maintain minimum flows, provide for conveyance and cover evaporative losses need to be more secure than that used to allocate water for environmental and other purposes.*
- *Lesson 9*: *Unless all forms of water use are accounted for entitlement reliability will be eroded by expansion of un-metered uses like plantation forestry and farm*

dam development, increases in irrigation efficiency, etc. *and place the integrity of the allocation system at risk.*

- *Lesson 10*: *Unless connected ground and surface water systems are managed as a single integrated resource, groundwater development will reduce the amount of water available that can be allocated to surface water users.*
- *Lesson 11*: *Water use and investment will be more efficient if all users are exposed to at least the full lower bound cost and preferably the upper bound cost of supplying water to them. One way of achieving this outcome is to transferring ownership of the supply system to these users.*
- *Lesson 12*: *Manage environmental externalities using separate instruments so that the costs of avoiding them are reflected in the costs of production and use in a manner that encourages water users to avoid creating them.*
- *Lesson 13*: *Removal of administrate impediments to inter-regional trade and inter-state trade is difficult but necessary for the development of efficient water markets.*
- *Lesson 14*: *Markets will be more efficient and the volume of trade greater if entitlements are allocated to individual users rather than to irrigator controlled water supply companies and cooperatives.*
- *Lesson 15*: *Equity and fairness principles require careful attention to and discipline in the way that allocation decisions and policy changes are announced.*
- *Lesson 16*: *Water markets are more effective when information about the prices being paid and offered is made available to all participants in a timely manner.*
- *Lesson 17*: *Develop broking industry and avoid government involvement in the provision of water brokering services.*

20.12.2 Enabling/Disabling Factors

At the highest level, these lessons and the framework that emerges from them are readily transferable to other countries. In many cases, however, the first step is likely to require significant property right reform. Australia was lucky. It started, accidently, with an approach to the development of its water entitlement and allocation system that made it relatively easy to introduce a market. The starting point was a property right system that was fungible or at least through unbundling made in to a fungible asset. If Australia had started with a seniority allocation system, such as that used in much of the USA, this would not have been possible.

References

Bjornlund, H., & Rossini, P. (2007). An analysis of the returns from an investment in water entitlements in Australia. *Pacific Rim Property Research Journal, 13*(3), 344–360.

Brennan, D. (2007). Missing markets for storage and their implications for spatial water markets. Paper presented to AARES conference, Queenstown, 14 Feb 2007.

Bryan, B., & Marvanek, S. (2004). Quantifying and valuing land use change for integrated catchment management evaluation in the Murray-Darling basin 1996/97–2000/01. Stage 2 report to the Murray-Darling Basin Commission. CSIRO Land and Water.

Dixon, P. B., Rimmer, M. T., & Wittwer, G. (2011). Saving the southern Murray–Darling Basin: the economic effects of a buyback of irrigation water. *The Economic Record, 87*(276), 153–168.

Fargher, W., & Olsvak, C. (2011). Water markets: A national perspective. In: J. Langford, & J. Briscoe (Eds.) with assistance from Taylor, N. *The Australian water project: Volume 1*. Melbourne: Committee for Economic Development of Australia.

Murray–Darling Basin Authority (MDBA). (2010). *Guide to the proposed basin plan*. Canberra: Murray–Darling Basin Authority.

National Water Commission. (2010). *Australian water markets report: 2009–10*. Canberra: National Water Commission.

Water Act. (2007). Commonwealth of Australia.

Young, M. D. (2008). The effects of water markets, water institutions and prices on the adoption of irrigation technology. In J. Albiac & A. Dinar (Eds.), *The management of water quality and irrigation technologies* (pp. 227–248). London: Earthscan.

Young, M. D. (2010). Environmental effectiveness and economic efficiency of water use in agriculture: The experience of and lessons from the Australian water policy reform programme. Consultant report prepared for the OECD, Paris.

Young, M. (2014a). Trading into trouble? Lessons from Australia's mistakes in water policy reform sequencing. In W. Easter & Q. Huang (Eds.), *Water markets for the 21st century: What we have learned?* Boulder: Westview Press. Chapter 11.

Young, M. (2014b). Designing water abstraction regimes for an ever-changing and ever-varying future. *Agricultural Water Management, 145*, 32–38.

Young, M. D., & Esau, C. (Eds.). (2013). *Investing in water for a green economy. Services, infrastructure, policies and management*. New York: Earthscan from Routledge. 296p. ISBN 9780415501262.

Young, M. D., & McColl, J. C. (2002). *Robust separation: A search for a generic framework tosimplify registration and trading of interests in natural resources*. Adelaide: Policy and Economic Research Unit, CSIRO Land and Water.

Young, M. D., & McColl, J. C. (2003). Robust reform. The case for a new water entitlement system for Australia. *Australian Economic Review, 36*(2), 225–234.

Young, M. D., Young, D., Hamilton, A., & Bright, M. (2002). *A preliminary assessment of the economic and social implications of environmental flow scenarios for the River Murray System. A report prepared for the Murray Darling Basin Commission, Policy and Economic Research Unit*. Adelaide: CSIRO Land and Water.

Young, M. D., Shi, T., McIntyre, W. (2006). Informing reform: Scoping the affects, effects and effectiveness of high level water policy reforms on irrigation investment and practice in four irrigation areas. CRC for irrigation futures technical report no. 02/06.

Chapter 21
The Development of an Efficient Water Market in Northern Colorado, USA

Charles W. (Chuck) Howe

Abstract An efficient water market has been established in a large water district in northern Colorado, USA. This is the most active water market in the USA in terms of number of transactions per year. The typical trade is from agriculture where marginal net values lie in the range of US$20– US$50 per acre-foot to towns where willingness-to-pay is closer to US$500 per acre-foot. The water being traded is imported from another basin, a feature that, under western US water law, allows the importer to consume the water completely without concern for downstream impacts. The ownership instruments are homogeneous shares that allow the owner to share proportionally in water available to the District. Transfers of the shares must be within the District and require approval only by the District Board (as opposed to typical State level administration of transfers). These two features result in low transaction costs that stimulate frequent small trades. Since irrigated agriculture consumes 85 % of Colorado's total supply, typical transactions involve permanent share transfers from agricultural uses to industrial and urban uses but temporary leases for 1 year are frequent, especially among agricultural users. Environmental groups and some towns have increasingly contributed or loaned their shares to instream flow and riparian ecosystem maintenance. Prices of these shares have risen rapidly with high population and commercial growth of the region.

Keywords Water law • Water markets • Inter-basin transfers • Transaction costs • Indirect impacts • Colorado

21.1 Introduction: The Region and the Legal Framework

The EPI of this chapter is the market for water shares that has been established in Northern Colorado, USA. This water market allows owners of shares in the Northern Colorado Water Conservation District (NCWCD) to trade shares with other water users within the boundaries of the District that covers roughly 1,000 square miles.

C.W. (Chuck) Howe (✉)
Environment & Society Program, Institute of Behavioral Science (IBS), University of Colorado-Boulder, 483 UCB, Room 281, 1440 15th Street, Boulder, CO 80309, USA
e-mail: charles.howe@Colorado.EDU

M. Lago et al. (eds.), *Use of Economic Instruments in Water Policy*, Global Issues in Water Policy 14, DOI 10.1007/978-3-319-18287-2_21

The market is unusual in that transfers of shares can take place quickly with low transactions costs, thus facilitating the transfer of water to its highest-valued uses and resulting in frequent small trades. Typical trades are from agriculture to urban uses, motivated by low marginal values in irrigated agriculture and high "willingness-to-pay" in urban areas.

Colorado established its "prior appropriation" water law in 1876 that defined water rights as personal property subject to purchase and sale (Getches 1997). Informal water trading has existed since that date. At the State level, the administration of water rights and transfers is through a system of water courts that supervise transfers to guarantee that the water will be put to "beneficial use" (avoiding speculation) and "no injury" to other water users as a result of the trade. This process is frequently time consuming and costly because of required engineering and agricultural analyses.

The water supplies being traded in the NCWCD market are those produced by the Colorado-Big Thompson Project (C-BT), a federally funded project initiated in the 1930s as relief from an extended drought that affected all of the western US. The C-BT Project transfers water from the headwaters of the Colorado River on the western side of the Rocky Mountain Range to the much drier South Platter River Basin on the eastern side of the mountains where the most productive agriculture and the majority of the State's population exist. The diverted water is distributed through an extensive network of canals, pipelines and natural rivers to the owners of shares in the NCWCD. The shares (known as allotments) being traded represent proportional shares in the water available to the Project each year, each share representing 1/310,000th of the water available.

The water is thus "inter-basin water" or "foreign water", i.e. new to the South Platte Basin. As noted in the abstract, under state water laws in the western US, imported "foreign water" "can be fully consumed by the importing agency, implying that return flows from any use are owned by the importing agency and thus cannot be legally claimed as water rights by downstream users. The importer and subsequent users thus are not responsible for protecting return flows when transfers take place, i.e. not subject to the "no injury rule" that is enforced by the Courts in typical transfers elsewhere in the State. However, all transfer applications are analysed carefully by the District Board to avoid significant negative third party and environmental effects.

No water market was contemplated in the original C-BT Project plans. The market for shares in the NCWCD evolved through trial and error to provide flexibility in the allocation of water over time. The C-BT Project was completed in 1957 following wartime interruptions and the Northern Colorado Water Conservancy District (NCWCD) was established under Colorado law to distribute the Project water and to take responsibility for the repayment of a portion of construction costs to the Federal Government. The funds for this repayment were to come from fees imposed on NCWCD share owners plus property taxes on all agricultural and urban lands within the District. As financial arrangements of the District evolved through discussions with water users, user charges were kept at a low level while property taxes have provided the majority of revenues, implying that the District's charges to water users do not begin reflect the scarcity value of the water. *That important function is provided by the water market that confronts each user with the opportunity cost of the water being used.*

The US Bureau of Reclamation that built the Project had insisted on the property taxes in addition to user charges to assure sufficient revenues for construction cost repayment. In addition, the Bureau pressed the District to attempt to sell or lease the return flows from the initial users (Howe et al. 1982), again to assure sufficient revenue. The District wisely declined to do this because of the complexity of identifying and quantifying the return flows. Transfers are allowed to take place subject only to the approval of the District Board and thus do not have to be overseen by the State Water Courts.

The effect of these arrangements has been to allow the evolution of a continuous smoothly working market in the District shares. Typical transfers involve small numbers of shares moving from agriculture to other uses since transaction costs are low and buyers historically have known that shares would be available on the market when needed, guaranteed by the willingness of marginal agricultural users to sell some of their shares. This easy availability may be changing as the volume of C-BT water owned by agriculture decreases, currently 33 % and falling each year).[1] The market continues to permit small farm operations, businesses and towns to acquire water in needed quantities and assure towns of available supplies for growth and during drought (Howe et al. 1990).

The existence of the NCWCD market means that all users of Project water know that they can buy and sell shares easily and quickly. They are continually confronted with the *opportunity cost* of the water they are using, which is many times the minimal user charge made by the District. This is particularly important in agriculture since which accounts for 80 % of consumptive water use in the District and throughout the western US.

The largest volume of transfers of NCWCD shares has been from agriculture to municipal and industrial uses. Because of low transaction costs and the speed of market transactions in this market, the typical size of share transfers is small in comparison with the size of transfers in traditional water rights markets. This reduces the negative impacts on the agricultural economy and minimizes needed adjustments in agriculturally- linked business and social sectors. The region served by the NCWCD market is quite diversified and prosperous, so that *agriculture-to-urban transfers reinforce regional economic growth.*

In the western United States, towns typically protect against drought by buying water rights in excess of average annual use so that supplies, while curtailed during drought, will be adequate to serve priority needs. The existence of an active efficient water market means that urban utilities can usually acquire added water even during drought, reducing the need for excess water rights as drought protection. Thus the NCWCD market has facilitated water transfers that are beneficial for both municipal and agricultural users and generally for the regional economy, all transactions being on a willing seller-willing buyer basis (Howe and Goemans 2003).

[1] Thanks to Brian Werner of NCWCD for these observations on the changing market scene.

Positive environmental impacts of the NCWCD market take the forms of more prosperous farming operations that can afford conservation practices, particularly regarding soil conservation and the application of fertilizers and other chemicals. Crop farms (as opposed to animal operations) close to urban areas are valued for aesthetic reasons and for increasingly popular farm-to-market horticultural supplies. As noted earlier, Colorado has a very active "instream flow program" under which water rights can be temporarily or permanently devoted to riparian ecosystem and recreational purposes. Thus, the water market has proved to be an *efficient allocation mechanism* in the sense that all trades generate both private and public net benefits. The characteristics of this market appear to be adaptable in other settings in the western US and similar climatic regions.

21.2 Characteristics of the Efficient Market Region

The Northern Colorado Water Conservation District (NCWCD) was established in 1937 to contract with the Federal Government to build a large trans-mountain water transfer project, The Colorado-Big Thompson Project (C-BT) that transfers water from the water plentiful western side of the Rocky Mountains in Colorado to the much drier eastern side of the mountains. NCWCD is responsible for the diversion works of the project and for the allocation of water on the eastern side of the mountains. C-BT is one of hundreds of federal water projects undertaken by the US Bureau of Reclamation under authorization of the 1902 Reclamation Act that was intended to provide subsidized water for the continuing economic development of the western US, especially for irrigated agriculture. The climate conditions of the USA are shown in Fig. 21.1.

The State of Colorado is divided into two distinct regions: the eastern, dry plains starting at roughly 105° west longitude and the western areas that start with the Rocky Mountains and extend through rugged lands to the western border of the State. Rainfall and snow are heavy on the western side of the Rockies, while the eastern slopes of the mountains (the "East Slope") and the plains are semi-arid.

21.2.1 The Northern Colorado Water Conservancy District

The NCWCD is located in the northeastern quadrant of Colorado as shown in Fig. 21.2. The District serves cities all along the eastern side of the mountains, the richest farmlands of Colorado in Larimer and Weld Counties and agricultural lands bordering the South Platte River to the northeastern corner of the State.

NCWCD contains 1.6 million acres (1,000 square miles) in portions of Boulder, Larimer, Weld, Broomfield, Morgan, Logan, Washington and Sedgwick counties. The District was established as the local agency to contract with the federal government to build the Colorado-Big Thompson Project under the federal Reclamation

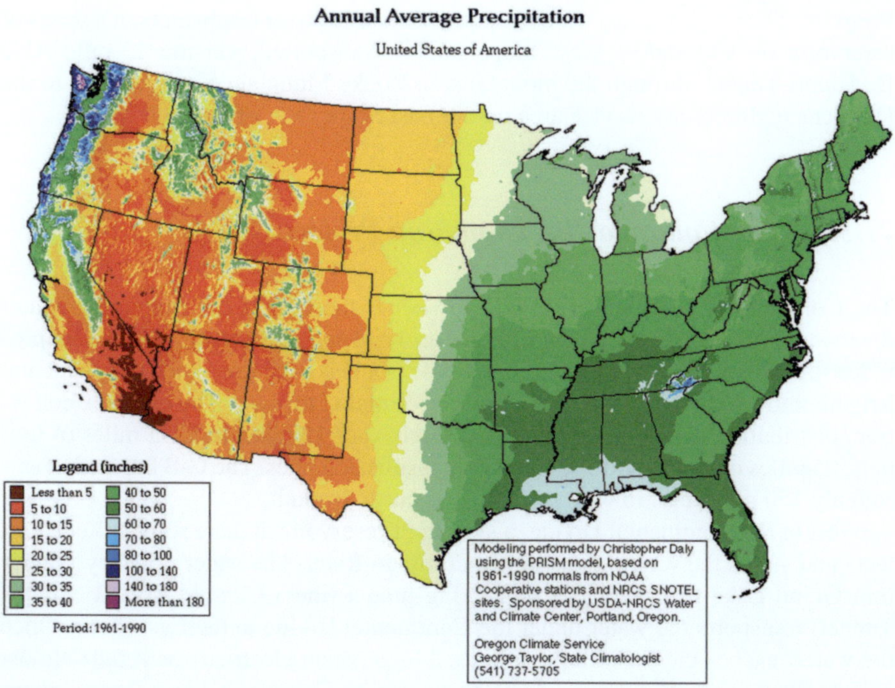

Fig. 21.1 US annual average precipitation (Reproduced from the Website of the US Geological Survey, USGS)

Fig. 21.2 NCWCD and the Colorado-Big Thompson (C-BT) Project (Reproduced from the Website of the NCWCD)

Program. The project stores water from the Colorado River headwaters in a series of reservoirs on Colorado's West Slope and is transported, via the 13-mile Alva B. Adams Tunnel, through the mountains in Rocky Mountain National Park to the District's eight-county service area on the East Slope.

21.2.2 *The Colorado-Big Thompson Project*

The Colorado-Big Thompson (C-BT) Project is the largest trans-mountain water diversion project in Colorado. Built between 1938 and 1957, the C-BT Project provides supplemental water to 33 cities and towns and is used to provide supplemental irrigation to 640,000 acres of northeastern Colorado farmland. The complex collection, distribution and power system is comprised of 12 reservoirs, 35 miles of tunnels, 95 miles of canals and 700 miles of transmission lines. The C-BT system spans roughly 150 miles east to west and 65 from north to south.

West of the Continental Divide, a system of reservoirs at increasing altitude collects and stores the water of the upper Colorado River. The water flows by gravity into Grand Lake from which a pioneering tunnel (the 13.2 mile Alva B. Adams Tunnel) transports the water under the Continental Divide to the East Slope. Once the water reaches the East Slope, it is used to generate electricity as it falls almost half a mile through five power plants on its way to Colorado's Front Range where three major reservoirs store the water. C-BT water is released as needed to supplement native water supplies in the South Platte River basin.

An interesting feature of the C-BT Project is the Green Mountain Reservoir on the western side of the mountains that provides replacement water for the Colorado River Basin to compensate for the water removed from the basin. This replacement water was required to be completed before C-BT began operation in deference to Western Slope interests who had objected to C-BT. This was an innovative form of *compensation to the basin-of-origin*. Compensation to the basin-of-origin is now required for all out-of-basin diversions in Colorado (Howe 2000).

The C-BT Project annually delivers an average of 274,000 acre-feet of water for agricultural, municipal and industrial uses.

21.3 EPI Background: Evolution and Operation of the Allotment Market

21.3.1 *Conditions Leading to the Establishment of NCWCD and C-BT*

The 1927–1937 period was a dry period with severe drought from 1931 to 1935, part of the infamous "dust bowl" of the Great Plains. Flows in the Colorado River (from which C-BT water is diverted) were high from 1896 to 1929, followed by a

38 year dry period from 1930 to 1968, illustrating the decadal variation in climate conditions. The lowest flow on record of only 5.6 million acre-feet occurred in 1934. The US Bureau of Reclamation estimated that 75 % of the 615,000 acres potentially served by C-BT had inadequate (for full yield) water supplies.

Because of these persistent drought conditions, an application was made in August 1933 to the Federal Government for the planning and construction of a supplemental water supply project that would bring water through the mountains to supplement eastern supplies. In addition, an organization to represent the water users of the region and having broad legal powers to contract with the Federal Government was needed. NCWCD was established in 1937. The contract with the Federal Government prescribed the following features for NCWCD:

1. An intended delivery of 310,000 acre-feet annually;
2. A highly subsidized repayment of construction costs;
3. A minimum tax rate on property in the District plus (minimal) annual payments by the water users;
4. District ownership of and arrangements for managing return flows from uses of project water-a key issue.

It was clear that the relative water needs would differ among different types of users and areas. Thus all potential users were allowed to subscribe voluntarily for shares in the District (which are called *allotments*) at very low prices starting in 1939. The 310,000 allotments available[2] were not fully subscribed until 1955. Finally, in 1957 an allotment was legally defined as *a freely transferable* contract between the District and the holder, subject to demonstrated *beneficial use* within the District. Proposed buyers and sellers make a transfer application to the District Board. Beneficial use within a reasonable period must be demonstrated except for municipal users who are allowed to hold "conditional water rights" in anticipation of future growth. This constitutes a deviation from the "no speculation" doctrine of western water law (i.e. water must be put to "beneficial use" by all water users) but realistic use restrictions on volume and time of development have been imposed by the water courts.

21.3.2 Current Operations of the Allotment Market

As water scarcity increases everywhere, flexibility in the allocation of existing supplies becomes increasingly important. In the USA, there is a long history of water marketing, especially in the states of Texas, California, Arizona, Nevada and Colorado. Table 21.1 shows recent evidence of market transfer activity. Some brokers buy allotments at favorable prices, applying the water temporarily to some

---

[2]The anticipated yield of the Project was 310,000 acre-feet, so 310,000 shares (allotments) were made available with the expectation that each allotment would represent 1 acre-foot of water.

Table 21.1 Where are transfers occurring? (Smith 2011)

State	No of transactions	% of the 5-year period total (Period 2005–2010)
Colorado	471	53
California	112	14
Texas	63	7
Nevada	49	6
Arizona	49	6
Total (5 Western States)	744	86

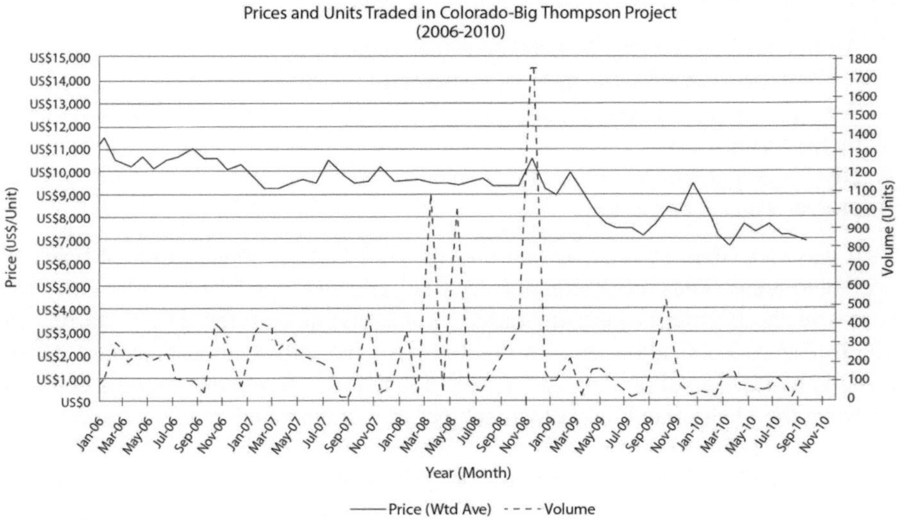

Fig. 21.3 Trends: price and volume in the C-BT market (Smith 2011)

agricultural land until a favorable buyer is located. This "packaging" of allotments is probably beneficial (Howe 2008).

It is clear that Colorado ranks first among the western states. Further, NCWCD's allotment market dominates Colorado transactions. As a result of the active NCWCD market and rapid urban growth, ownership of the District allotments has shifted steadily toward urban users. While ownership has shifted, changes in actual use have been less dramatic since towns typically buy water rights in excess of average needs to protect against drought. In non-drought years, they then rent (lease) substantial amounts of water back to agriculture (permanent sales of allotments and short term leasing are the only types of transactions).

The long term effect of increases in urban and industrial demand has been to drive up the prices of C-BT allotments as shown in Fig. 21.3 which shows the trends in volume of transfers and prices of those transfers since 2006. Volumes and prices are in terms of C-BT allotments. Historically, an allotment has delivered an average of 0.7 acre-feet. The amount delivered depends not only on physical availability but on the "quota" declared annually by the NCWCD Board that allots larger amounts in dry years and less in wet years. For example, the volume traded in November of 2009 was roughly 500 units or 350 acre-feet while prices in that month were in the neighbourhood of US$ 8,000 per unit or roughly US$11,500 per acre-foot in perpetuity.

The large changes in volumes are due to weather conditions and spurts of urban growth. Curiously there has been a downward trend in prices since 2006. This is largely attributable to very effective programs of urban conservation that appear to have permanently reduced urban water use in spite of continued population growth. The City of Denver is currently using 20 % less water than before the drought of the 2000s in spite of a 10 % population growth. Urban use per capita has uniformly fallen throughout Colorado.

21.3.3 Comparative Characteristics of NCWCD Transfers

It is clear that share transfers (permanent) and leases (short term) out of agriculture to urban areas are the predominant type of transfer, but an important feature in the NCWCD market is the high percentage of agriculture-to-agriculture transfers that occur as a result of the fast, low cost transfers. This is critical for irrigated agriculture in semi-arid areas. The size distribution of transfers in NCWCD is exhibited in Fig. 21.4. A striking comparison is that, while the median size of transfer in the South Platte traditional water rights market has been about 367 acre-feet (with a mean of 3,425, not shown), in the NCWCD market over the same period, the median has been only 16.8 acre-feet with a mean of 34 acre-feet (Howe and Goemans 2003; Michelsen 1994).

The differences in the size distributions are attributable to the *low cost and continuity of the NCWCD market*. Cities operating in traditional water markets typically prefer to buy large quantities of agricultural rights in a single transaction

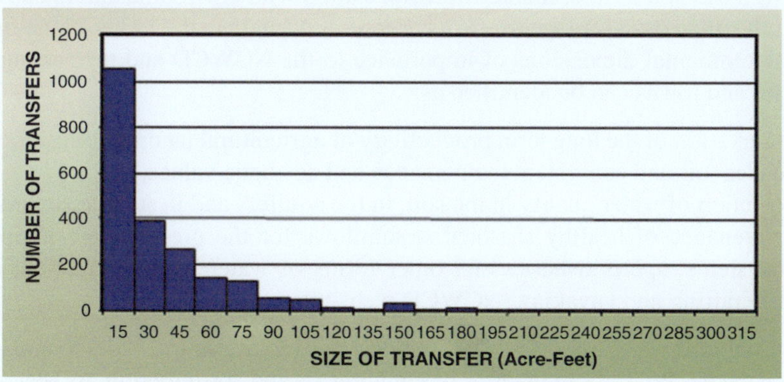

Fig. 21.4 NCWCD distribution of allotment transfers by size. Data period: 1979–1995. Median: 16.8 acre-feet. Mean: 34.0 acre-feet (Reproduced from Howe and Goemans 2003)

because a large part of transaction costs is fixed. In the NCWCD market, however, there is a continuous market in which allotments averaging 0.7 acre-feet/year can usually be purchased at predictable prices, although this situation is changing as more water is transferred to urban and environmental uses. Thus historically, water users have had no need to engage in large, expensive transfers in anticipation of future needs- an important effect of the efficient, low cost NCWCD market.

Further, some studies have shown that, on the average over time, nearly 50 % of the C-BT water available to allotment owners is rented annually, most from cities to agriculture. The volume and direction of rentals are sensitive to weather conditions, with cities withholding water from agriculture and charging somewhat higher prices during drought. Lease prices tend to increase in the late season when farmers often need added water to "finish" a crop and when traditional surface supplies are low. The District favors keeping rental prices low to assist farmers. However, while low rental prices help the farmers who manage to find rental water, it also restricts the supplies that farmers and cities are willing to rent (Goemans and Kroll 2012).

21.4 Assessment Criteria

The EPI in this case study is the efficient water market that has evolved within the administration of the Northern Colorado Conservancy District. The District and the market have evolved together so it is not possible to identify or isolate the environmental, economic or distributional effects of each totally separately. Important lessons would be lost if the institutional lessons from the evolution of the NCWCD were to be omitted.

21.4.1 Environmental Outcomes

The NCWCD and its later market were not started with environmental objectives in mind other than overcoming the effects of serious drought in the 1930s. Nonetheless, the environmental dimensions of importance to the NCWCD and the surrounding counties and towns can be identified as:

1. Preservation of the long term productivity of agricultural lands in terms of crops, broader soil and ecosystem maintenance and aesthetic values;
2. Protection of water quality in the soil, in the aquifers and in surface streams;
3. Maintenance of healthy seasonal streamflows for the preservation of riparian ecosystems, sports fisheries and other forms of water based recreation, especially rafting and kayaking (NCWCD website).

Agricultural water use constitutes over 80 % of total use in Colorado and about 65 % in the NCWCD, both in terms of withdrawals and consumption. As seen in the earlier graphs, while agricultural water use has been declining (urban use expanding), agriculture remains the largest user of NCWCD water. The District has pursued educational and demonstration projects to assist farmers in achieving economic water conservation. These programs are carried out in cooperation with the Agricultural Extension Service and Experiment Stations of the US Department of Agriculture. A major step has been the stimulation of efficient irrigation techniques like the drop line sprinkler. *Adoption of such techniques is stimulated by the active water market that "puts a price on water" and through farmer education.*

As urban use of C-BT water expands, it is increasingly important to motivate economic conservation in the urban setting. Roughly 50 % of urban water use is for the irrigation of lawns, gardens and trees. The major conservation steps encouraged by NCWCD and followed by towns in the District include:

1. Establishment of monthly "water budgets" for residential, commercial industrial and institutional customers;
2. Establishment of increasing block rate structures in conjunction with the water budgets;
3. Issuance of "smart readers" to customers so that the customer can determine current rates of use and cumulative use compared with the budget;
4. Subsidies to installation of water-saving appliances: toilets, washing machines, shower and bath fittings, etc.
5. Educational programs are provided for urban users that center on efficient outdoor use, including demonstration gardens.

These urban conservation programs have resulted in a permanent 30 % reduction in per capita water use in the District's service area. The saved water results in higher stream flows with positive impacts on riparian ecosystems, water related recreation and irrigation water supplies.

The *efficient, continuous market* means that urban areas can acquire water as needed rather than buying large volumes of agricultural water rights that results in drying up large areas.

The *environmental and aesthetic values of agriculture* are increasingly recognized in all areas of public decision-making.

21.4.2 Economic Assessment Criteria: The Economic Efficiency of NCWCD Market Arrangements

The importance of the special provisions governing return flows was not appreciated at the time of project design and construction. Under western US water law, return flows "belong to the stream" and cannot be claimed by the water right holder who made the diversion. Because the Bureau of Reclamation had obtained the needed water rights on the Colorado River and because the water would be new to the South Platte Basin, the contract allowed NCWCD to claim *ownership of all return flows* for recapture and reuse-a feature critical to the subsequent evolution of the NCWCD efficient water market as has been noted above.

The Bureau of Reclamation initially pressured NCWCD to sell the return flows to guarantee further revenues that would help repayment of the construction costs.[3] The District resisted this because it would be impossible to estimate the volume and timing of the return flows with sufficient accuracy to establish clear property rights. The "bookkeeping" would be difficult and subject to challenge.

The most profound effect of the District's refusal to sell return flows (which it owned) was that it left the District free to approve proposed transfers anywhere in the District without recourse to the Water Court procedures that are typically required of water right transfers to guarantee "no injury" to other water users. Only District Board approval is required, subject to Bureau of Reclamation review-usually a formality. While there is no legal obligation to protect return flows, they are largely protected because transfer volumes are limited to the former consumptive use, thus leaving the return flows "in place".

The issue of loses or gains to activities economically linked to Project water users (secondary or indirect effects), e.g. suppliers of agricultural inputs or users of agricultural products, is complicated and has been treated in an extensive literature (Howe and Goemans 2003; Young 1986). The consensus of that literature (in this author's opinion) is that, in a depressed region where there is long term unemployment of resources and capacities, the expansion or contraction of a primary water-using activity (e.g. irrigated crops) can generate "real" (national) economic gains or losses in forward and backward- linked activities by productively employing those resources.

However, in the case of NCWCD, the regional economy is quite prosperous with highly productive irrigated agriculture and expanding urban, industrial and commercial activities. Many water transfers are initiated by changes in land use as urban and commercial activities expand onto farm land. Thus the reduction of agricultural activities does not have negative secondary effects and, indeed, supports the continued expansion of the region's most progressive activities. Thus negative externalities are not a serious issue for NCWCD and the C-BT Project.

Where does this leave us regarding the overall efficiency of the transfer process in NCWCD? The question is whether the advantages of an easy, low cost transfer process are likely to more than offset any net adverse third party effects. Transfers within the agricultural sector are mostly temporary rentals within the same ditch or canal to even out supplies at the end of the crop season. No third party effects are created. When permanent transfers take place within the agricultural sector, it is again likely to be between water users on the same ditch or canal or between users on adjacent ditches, obviating third party effects. Any minor positive and negative effects are likely to be experienced in similar types of agriculture, one offsetting the other. Return flows are likely to return to the same stream (Howe 1987).

[3] This concern about further revenues to help repay construction costs must be understood in light of the depressed economy of the 1930s. While large subsidies were included in the repayment contract (a 50 year repayment period with no interest on the unpaid balance, no adjustments for inflation and 50 % of the costs being repaid in the last 10 years of the repayment period), there was still concern about the District's ability to meet the required payments.

Similarly, if the transfer is from agriculture to urban use, third party return flow effects will be specific to the source and destination locations. However, towns are also increasingly reusing their water supplies thus increasing the net value of ag-to-urban transfers (National Research Council 1992; Oggins and Ingram 1989).

21.4.3 Policy Implementability

This remains an issue. First, the establishment of an efficient market is limited to legal regimes in which water rights are clearly defined and considered to be tradable property, properties of regimes adopting some version of the appropriations doctrine. In the USA and Canada, regions that have used other legal frameworks like the old English riparian doctrine are increasingly changing to more flexible rules (e.g. tradable water extraction permits in the eastern US).

The other issue is the level of transaction costs. In the present case, transaction costs have been kept low because of the return flow arrangements described earlier, i.e. that the C-BT water was imported and NCWCD thus owned the return flows. This relieved NCWCD of "no injury" obligations related to transfers and thus avoided formal court review.

21.4.4 Uncertainty

The uncertainty (more likely, risk) involved in establishing and operating almost any water market stems from climate and hydrology. Most watersheds have long records of streamflow and climate data, these days extended to hundreds of years through dendrochronology. Thus the density functions for historic annual and monthly streamflows are available. A major question facing water planning is the relevance of these historic traces to future conditions under climate change (Oamek et al. 2010; Wensley and Stabler 1998).

The main mechanism for dealing with hydrologic risk is storage. There are limits to the effectiveness of storage in providing reliable supplies. In the case of NCWCD, there are large reservoirs in both West Slope and Eastern Slope regions. This largely eliminates hydrologic variability but weather continues to create some uncertainty on the *demand side*: if there is an extended dry period, demands will increase and the reverse will happen during wet periods. This causes problems of balancing the supply system, i.e. having the water where and when needed.

The conjunctive management of surface and ground waters can be effective in regions with large groundwater stocks in tributary aquifers. During dry periods, the groundwater can be called on to replace surface supplies. While this strategy should be obvious, in some jurisdictions the surface and groundwaters are administered by different agencies and covered by different sets of law (see Howe 2008).

21.5 Conclusions: Lessons Learned

The existence of a flexible water market motivates efficient water use by all users by confronting the users with the real opportunity cost of the water. It can thus overcome the distorting effects of inappropriate pricing policies that are often in place.

The existence of an efficient, continuous water market permits transfers among users on an "as needed" incremental basis rather than infrequent large transfers, thus facilitating transfer funding and easing the indirect economic adjustments that follow from the initiating change in water use.

An efficient rental (lease) market is especially valuable to agriculture in the face of critical demands at different stages of crop growth and variable local supplies. Different water supply agencies (e.g. "ditch companies", conservancy districts, rural water companies) have different sources of supply and may experience different micro-climate effects. Cross-agency balancing of supplies and demands on a quick turn-around basis is possible with the NCWCD type of water market.

Efficient water markets can reduce conflicts that frequently exist between requirements of State water law and putting water to its most valuable uses. Many examples can be found where low-value senior rights call out high value junior rights for extended periods of time (Howe 2008). A water market with low transaction costs has the potential for reducing these conflicts by motivating the shift of low-valued senior rights to higher valued junior rights.

The direct and indirect economic impacts on the transfer area of origin depend on (1) whether the new uses are in the same economic region (usually the same basin) and on (2) the economic vitality of the economy of the area of origin. If water transfers are being induced by the growth of new local economic activity, the transfers reinforce growth. In depressed areas of origin, transfers out of the area reduce activity with no opportunities for investing the water sales proceeds in local activities.

In the case of water transfers out of a depressed region of origin, extra compensation to that region by the buyer is warranted. When C-BT was built, additional reservoir storage (Green Mountain Reservoir) was provided to compensate the Colorado River for reduced streamflows and their effects. Today, urban and commercial buyers frequently negotiate cash payments to local governments in the area of origin to compensate for reduced tax bases.

Cumulative impacts of transfers out of an agricultural region cause increasingly negative impacts, sometimes approaching a "tipping point" at which agriculturally-related businesses begin to fail (Oamek et al. 2010).

Recent experimental research on water markets (Goemans and Kroll 2012) shows that markets for permanent transfers of water rights interact with water rental markets since the two are, to some extent, substitutes. Where efficient, expeditious leasing arrangements are available, a likely result will be that water rights prices are depressed to some extent.

References

Getches, D. H. (1997). *Water law in a nutshell.* St. Paul: West Publishing Company.

Goemans, C., & Kroll, S. (2012). Benefits and costs of introducing leasing to water markets: An experimental analysis. Department of Agricultural Economics, Colorado State University. Fort Collins, Colorado.

Howe, C. W. (1987). Project benefits and costs from national and regional viewpoints: Methodological issues and case study of the Colorado-Big Thompson Project. *Natural Resources Journal, 27*, 5.

Howe, C. W. (2000). Protecting public values in a water market setting: Improving water markets to increase economic efficiency and equity. *University of Denver Water Law Review, 3*(2), 357.

Howe, C. W. (2008). Water law and economics: An assessment of river calls and the South Platte well shutdown. *University of Denver Water Law Review, 12*(1), 181.

Howe, C. W., & Goemans, C. (2003). Water transfers and their impacts: Lessons from three Colorado water markets. *Journal of the American Water Resources Association, 39*(5), 1055–1065.

Howe, C. W., Schurmeir, D. R., & Shaw, W. D. (1982). Innovations in water management: An ex-post analysis of the Colorado-Big Thompson Project and the Northern Colorado Water Conservancy District. University of Colorado, Department of Economics. Unpublished.

Howe, C. W., Boggs, C., & Butler, P. (1990). Transactions costs as determinants of water transfers. *University of Colorado Law Review, 61*(2), 393.

Michelsen, A. M. (1994). Administrative, institutional and structural characteristics of an active water market. *Water Resources Bulletin, 30*(6), 971–982.

National Research Council. (1992). *Water transfers: Efficiency, equity and environment.* Washington, DC: National Academy Press.

Northern Colorado Water Conservancy District: website www.northernwater.org

Oamek, G., Gardner, R., Weber, K., & Howe, C. W. (2010). A proposed method for incorporating rural population-business thresholds ("Tipping points"). In: Honey Creek Resources Inc. Water Transfer Evaluations (A report prepared for the Colorado Water Conservation Board).

Oggins, C. R., & Ingram, H. (1989). *The community consequences of rural to urban water transfers*. Tucson: The Udall Center for Studies in Public Policy.

Smith, R. T. (2011). Strategic Water Group LLC. Talk given at the Congressional Research Service seminar on Water Markets and Transfers, Washington, DC, 17 Feb 2011. Unpublished.

Wensley, M. R., & Stabler, J. C. (1998). Demand-threshold estimation for business activities in rural Saskatchewan. *Journal of Regional Science, 38*(1), 155–177.

Young, R. A. (1986). Why are there so few transactions between water users? *American Journal of Agricultural Economics, 68*, 1143–1151.

Chapter 22
Other Types of Incentives in Water Policy: An Introduction

Alexandros Maziotis and Manuel Lago

Abstract Over the last decades, Cooperative Agreements (CAs) (voluntary, payments for ecosystem services (PES) schemes etc.) have been introduced as supplements to existing command and control regulations, i.e. as part of policy mix, for promoting higher water and environmental efficiency levels than mandated by law. This chapter illustrates the effectiveness and efficiency of CAs among farmers, water companies, authorities and citizens to achieve water policy goals in Europe and beyond. These include voluntary agreements and PES schemes to improve water quality in Dorset (UK), in Evian (France) and in New York (USA) and river restoration in Ebro (Spain). A negotiation agreement to cope with increasing water scarcity by promoting the use of reclaimed water in Tordera and Llobregat (Spain) is also analysed. The economic, environmental and social outcomes from the implementation of these CAs along with their institutional set-up, transactions costs and policy implementability are highlighted. Overall conclusions from the findings of the representative case study areas are finally presented.

Keywords Voluntary agreements • Payments for ecosystem services • Negotiation • Water quality and scarcity • River restoration

22.1 The Role of Other Types of Incentives in Water Policy

Global water and environmental challenges (e.g. water quality, water scarcity, river restoration, greenhouse gas emissions) along with economic development (e.g. population growth, increases in demand) or the need to innovative in terms of new technology (e.g. use of recycled water, clean technology) have persuaded policy makers to search for innovative economic policy instruments. In most Member

---

A. Maziotis (✉)
School of Mechanical, Aerospace and Civil Engineering (MACE),
University of Manchester, Manchester M13 9PL, UK
e-mail: alexandros.maziotis@gmail.com

M. Lago
Ecologic Institute, Pfalzburger Strasse. 43/44, 10717 Berlin, Germany
e-mail: Manuel.lago@ecologic.eu

M. Lago et al. (eds.), *Use of Economic Instruments in Water Policy*,
Global Issues in Water Policy 14, DOI 10.1007/978-3-319-18287-2_22

States (MS) of the European Union (EU), the implementation of water and environmental policy is foreseen through traditional *command and control* policies, however, a closer cooperation among authorities, firms, farmers and citizens are of paramount importance to tackle water and environmental challenges. This implies the use of instruments that include cooperative (CAs) (e.g. voluntary, payments for ecosystems services (PES) schemes) agreements, i.e. negotiated voluntary arrangements between parties to adopt agreed practices often linked to subsidies or offset schemes (Lago and Moller-Gulland 2012; Delacamara et al. 2013).[1] More particularly, there are three main types of voluntary agreements as defined by OECD (1999); *public voluntary agreements*, where the environmental agency defines the rules and conditions of participation; *unilateral commitments* where the agreement is designed by firms and their industry associations and *negotiated agreements* which take the form of formal contract between the environmental agency and industry and are often developed with the expectation that regulators will not introduce more stringent regulation if firms meet pollution targets within a specified time (Borkey et al. 1998; Darnall and Carmin 2005). For the purposes of this book and because of its current relevance as an instrument for water policy in Europe, Voluntary Agreements (VA) have been included as a category in the broad categories of Economic Policy Instruments (EPIs). But it is worth noting that there is an on-going debate in the literature about whether voluntary agreements (VA) can be regarded as a "pure" economic policy instrument or not. Environmental VAs are commonly defined "as an agreement between a government authority and one or more private parties with the aim of achieving environmental objectives or improving environmental performance beyond compliance to regulated obligations. Not all VAs are truly voluntary; some include rewards and/or penalties associated with participating in the agreement or achieving the commitments" (Gupta et al. 2007). Some economists interpret the "Voluntary" nature of the agreements as a version of

[1] In addition to CAs, Chap. 1 introduces another type of instruments, i.e. risk-based mechanisms which rely on the influence of differential insurance premiums and liabilities (compensation) levels (Delacamara et al. 2013). The former refers to insurance schemes against natural and manmade disasters which have recently promoted by the European Commission's Adaptation Strategy to climate change. More particularly, its aim is to improve the market penetration of natural disaster insurance and unleash the full potential of insurance pricing and other financial products for risk-awareness prevention and mitigation and for long-term resilience in investment and business decisions (EC 2013). Example of these schemes is provided by Gomez et al. (2013) where insurance addressing drought risk, i.e. a financial mechanism that covers the loss of or damage to crops caused by insufficient rainfall, was explored in Tagus-Segura (Spain) (Delacamara et al. 2013). Liabilities refer to schemes to prevent and remedy damage to animals, plants, natural habitats and water resources and they are promoted by the Environmental Liability Directive (2004/35/EC). Examples of compensation schemes for environmental damage have been examined for selective case study areas in Europe such as in Sweden (tank collapse and chemical release), Czech Republic (coal mining pollution), UK (effects of abstraction for public water supply on the ecological integrity of river), Germany (compensation in the form of habitat banking, i.e. creation of nature conservation areas from the construction of new infrastructure) (Cole and Kriström 2007). Risk-based mechanisms were not assessed as part of the EPI-WATER project and therefore, no case study areas were included in this chapter.

regulation and therefore, argue that they do not belong to the economic policy instruments category.

Another form of cooperative agreements is PES schemes which are based on voluntary transactions between at least two social actors with the aim of securing the provision of ecosystem services (ES) (e.g. clean water supply, flood risk mitigation, etc.) (Delacamara et al. 2013). Over the last decades, an increasing number of voluntary approaches have been widely implemented in environmental policy such as reduction in CO_2 emissions from energy sector, pollution from the steel sector etc. (OECD 2003; Bryden et al. 2012). Cooperative agreements (e.g., voluntary or PES) have also been carried to improve water quality from pollution by agriculture, highlighting therefore the beneficial interaction between water-related and agricultural policy (Brouwer et al. 2003; Heinz 2008). The benefits of CAs may be significant for both firms and society. Firms could enjoy lower legal costs and increase reputation by improving their environmental performance, whereas societies gain to the extent that firms translate goals into concrete business practices and persuade other firms to follow their example (Gupta et al. 2007). CAs have been introduced as supplements to existing *command and control* regulations, i.e. as part of *policy mix*, e.g. for promoting higher water efficiency levels than mandated by law. The use of cooperative agreements is more and more often seen as an alternative to legislative measures at the EU level but one must not overlook the strengths and merits of the actual existing regulatory system, especially when well devised and effectively enforced (BEUC 2006).

In Europe, the Water Framework Directive (EC/2000/60) and Common Agriculture Policy (CAP) require new approaches in water management (Heinz 2008). WFD's main aims are at achieving good ecological status of water bodies by 2015, tackling water pollution by agriculture and responding to water scarcity and drought risk. Article 4 of the WFD sets the regulations to ensure enhancement and restoration of all surface waters; to guarantee the progressive reduction of pollution of groundwater; and to promote long-term sustainability of water resources (Heinz 2004). To achieve the environmental objectives of the Article 4, each MS needs to adopt the Programme of Measures (Article 11) which will be enhanced by supplementary measures such as negotiated agreements, legislative, economic or financial instruments. These measures should be affordable and should not cause disproportionate costs (Article 4 (5)). Hence, cooperative agreements could assist the implementation of the WFD by allowing for instance farmers and water companies to form an agreement to prevent further pollution. These agreements also provide information on the most cost-effective measures in farming practices (e.g. intercrops to reduce nitrate loads) (Heinz 2004, 2008). Recent publication by the European Commission "A Blueprint to Safeguard Europe's Water Resources" encourages water re-use for irrigation or industrial purposes as an alternative supply option to respond to water scarcity (EC 2012). Cooperative agreements (e.g. water re-use from waste water treatment plants for irrigation or industrial purposes) could be a useful approach to promote a more efficient utilisation of scarce water resources.

Moreover, the Common Agricultural Policy (CAP) is aimed at ensuring the economic sustainability of the agricultural sector and reducing environmental pressures on water bodies. Past reforms of the CAP included cross-compliance direct payments to ensure environmental and agricultural protection e.g. protection of soil and water and avoid the deterioration of habitats and support for rural development policies, whereas the latest reform introduced a green payment to encourage the adoption of agronomic practices by farmers. Therefore, cooperative agreements among relevant parties such as authorities, firms or farmers, could allow knowledge and expertise sharing to avert further pollution and deterioration of water resources. Water companies could engage with farmers to be aware of their efforts to reduce pollution by limiting the excessive use of pesticides and fertilizers and in exchange, farmers could expect compensation payments from water companies and free advisory services (Heinz 2008). In addition to the WFD and CAP, cooperative agreements could assist the implementation of the Nitrates Directive (91/676/EEC). It is a special water-related environmental regulation which requires the identification of water resources that suffer from nitrate pollution by agriculture (Nitrate Vulnerable Zones – NVZs) and the design of action programmes for monitoring the amount of nitrate inputs in these NVZs (Defra 2012). As a result, farmers within those NVZs must comply with certain farming practices for preventing deterioration of water quality and for greater protection of drinking water resources. Cooperative agreements can provide information about best agricultural practices allowing therefore protection against pollution. Finally, the role of cooperative agreements in the form of PES schemes has been promoted and highlighted in other EU legislation and initiatives such as the EU Biodiversity Strategy to 2020 and in the Roadmap for a Resource Efficient Europe. However, clear and transparent definitions and methodologies are still needed at EU level (and national level) to promote the implementation of PES schemes as water-related EPI (Delacamara et al. 2013).

Cooperative (e.g. voluntary, PES) agreements to improve environmental performance have also become popular in countries beyond Europe. In USA, voluntary agreements are defined as programs, codes, agreements, and commitments that encourage organizations to voluntarily reduce their environmental impacts beyond the requirements established by the environmental regulatory system (Carmin et al. 2003; Darnall and Sides 2008). The Environmental Protection Agency (EPA) maintains primary responsibility for setting environmental standards, prescribing the ways in which the regulated community must achieve these standards and imposing penalties (fines) in cases when environmental conditions are violated by companies (Dallar and Carmin 2005). Voluntary agreements had been in place for more than 20 years in USA, around 200 studies exist (Darnall and Sides 2008). As far as PES schemes are concerned, Buric and Gault (2011) and Benett et al. (2013) listed several dozen cases in South America (Brazil, Bolivia, Colombia, Costa Rica, Ecuador, Guatemala, Mexico), Asia (China, India, the Philippines), North America (New York and Santa Fe (USA) and Africa (South Africa, Tanzania, Rwanda) (Delacamara et al. 2013). In Europe the number of CAs (e.g. voluntary or PES) is approximately 500 (the majority of these cases refer to voluntary agreements), with Germany being the country with highest number of such agreements (Heinz et al. 2002; Brouwer et al. 2003; Heinz 2008; Mattheiß et al. 2010).

The chapters in this section of the book discuss the effectiveness and efficiency of cooperative (e.g. voluntary, PES) agreements among farmers, water companies, authorities and municipality to achieve water policy goals in selective case study areas. These include agreements to improve water quality in Dorset (UK), in Evian (France) and in New York (USA), to improve river restoration in Ebro (Spain) and to cope with increasing water scarcity by promoting the use of reclaimed water in Tordera and Llobregat (Spain). The economic, environmental and social outcomes from the implementation of the voluntary agreements along with their institutional set-up, transactions costs and policy implementability are highlighted. Overall conclusions from the findings of the representative case studies are finally presented.

The first chapter in this section comes from the County of Dorset in England and includes a cooperative (voluntary) agreement between a water company (Wessex water) and farmers to reduce water pollution from farming activities. The water company has approached the farmers to cooperate to improve water quality by promoting better practices (catchment approach) instead of opting for other approaches such as water treatment which could be costly. The findings from this study suggest that although the benefits in terms of reduced loads of nutrients in water bodies will become apparent by 2015, the catchment approach proved to be economic efficient (cheaper than alternative solutions) and both farmers and water company were better off (win-win situation). The cooperative agreement between water supplier and farmers was very popular as alternative to regulation and farmers have become keen supporters of the approach, willing its success in order to prove that further regulation is not necessary.

The next chapters discuss cooperative agreements in Evian (France) and in New York (USA), which take the form of payments for ecosystem services (PES) to improve water quality. CA in Evian is developed by the association for the protection of the catchment area of Evian mineral water (APIEME), an association which involves the villages from the spring area that benefit from a government tax on bottled water, the villages from the catchment area, the Evian Company and national public bodies, with local farmers. The French case study illustrates how the Evian Company can maintain a land use and traditional agricultural practices on the catchment area presumed to preserve the quality of the Evian Natural Mineral Water. Although the economic, environmental and distributional effectiveness of this instrument was difficult to quantify with accuracy, it is concluded that this agreement met its ultimate objective, i.e. the environmental protection and sustainable development of the area. Despite the high transaction costs and advanced water regulation and institutions, the involvement of stakeholders and the conduction of a background study to take into account any local particularities and heterogeneous farming were key factors for the successful selection, design and implementation of the agreement. Another example of payments for ecosystem services comes from New York (USA) where the city is paying farmers for services for improving source water quality, i.e. the Watershed Agricultural Programme (WAP). This case study provides an excellent example how the city and farmers are voluntarily working together to protect the quality of a watershed. The study showed that there were substantial benefits in terms of reducing phosphorus loadings to surface waters suggesting that pollution from agriculture is no longer a threat to the city. The economic effectiveness of this agreement was difficult to quantify with accuracy as it was not possible to monetize the value of the water quality benefits that the city received. However, this agreement has certainly been proven to be cheaper than opting for other costly options such as mandatory filtration as required by the Environmental Protection Agency (EPA). Moreover, farmers were better-off but not evenly as larger farmers benefit more than smaller ones. Key successful factor of this agreement was also the importance of well-structured dialogue and negotiations between the city and farmers who were able to work together to identify common ground and solutions to both groups' problems. What remains is for the city to continually monitor and invest in watershed management efforts to control pollutants and excess nutrient loadings as other threats remain (e.g. pollution from exurban development).

The next chapter in this section provides a unique example of voluntary public-private partnerships (the hydropower company (Endesa), the water authorities (Ebro River Basin Authority, ERBA) and the scientific community) for the partial re-naturalization of a significantly modified river in the Lower Ebro (Spain). Changes in the river morphology reduced flood frequency and magnitude, sediment load and altered the river's ecology leading to detrimental effects over many water services such as reduced health and navigation. As a result, macrophytes (visible algae and other flora species) have increased which are detrimental to power generation facilities and their removal through mechanical means is costly. This provides incentives for hydropower companies to cooperate via flushing flows (FF) to improve the ecological potential of the river and control and remove the excess of macrophytes from the river channel. The findings of this study suggest that the benefits in terms of macrophytes removal were substantially high leading to welfare improvement both from a private and social perspective. The voluntary agreement is implemented at an intra-basin level which avoided significant transaction costs and clearly shows that macrophytes removal at a minimum cost has been proved to be the catalyst for agreement and reconciliation of public good concerns and private interests. However, this case is by no means "over" as the progressive drop of macrophytes removal rate may give a chance to a more ambitious agreement (Lago and Moller-Gulland 2012).

The last chapter discusses a voluntary water intra-sectorial transfer (from municipality to irrigators) to promote the use of reclaimed water and decrease the pressure on the local aquifers in Tordera and Llobregat (Spain). This area is characterized by overexploitation of groundwater resources and frequent drought events which could threaten the long-term availability of water resources. To address the growing regional water shortage and pressure on the local aquifers, the Catalonian Water Agency (ACA) considered that a plausible solution would be the use of reclaimed water mainly for irrigation. The findings of this study shows that the reliability of reclaimed water improved the water availability by reducing pumping from groundwater and increased farmers' income by raising the crop yield per hectare. The agreement to promote the use of reclaimed water proved to be the cheapest solution as compared to alternative ones such as sea water desalination and water transfer from other areas leading to a win-win situation both for citizens and irrigators. Key aspects for the success of a water reclaimed agreement were the participation of stakeholders and public, sharing site-specific knowledge and expertise concerning environmental needs and conditions and social awareness and information campaigns for the benefits using reclaimed water to respond to water scarcity risks.

This section illustrates that cooperative agreements (voluntary or PES) are taking place in water policy in several places in Europe and beyond. Even though it would be imprudent to make generalized statements about the advantages of applying cooperative agreements, the following conclusions can be drawn:

- Cooperative agreements (CAs) have been introduced as supplements to existing *command and control* regulations, i.e. as part of *policy mix*.
- CAs target at achieving site-specific objectives in water catchments at minimum cost.
- CAs have met environmental objectives, however, their environmental effectiveness will become apparent in subsequent years.
- The economic benefits of CAs have been proved to be higher than their costs and less costly than alternative solutions.
- Parties involved in the CAs are better-off (win-win situation).
- Voluntary agreements are on their own innovative institutional arrangements. However, Payments for environmental services (PES) are difficult to implement in societies with advanced water regulations and institutions.
- CAs can keep transaction costs at a minimum.
- Trust, knowledge and public & stakeholder participation are key factors for the successful selection, design and implementation of a cooperative agreement.
- Clearly defined targets, robust monitoring system and control of the site-specific objectives are of paramount importance as it may give a chance to more ambitious agreements.

References

Bennett, G., Nathaniel, C., & Hamilton, K. (2013). *Charting new waters: State of watershed payments 2012*. Washington, DC: Forest Trends.

BEUC. (2006). *Voluntary environmental agreements*. The European Consumers' Organisation. Brussels: European Commission.

Borkey, P., Glachant, M., & Leveque, F. (1998). *Voluntary approaches for environmental policy in OECD countries: An assessment*. Paris: CERNA, Centre d'economie Industrielle.

Brouwer, F., Heinz, I., & Zabel, T. (Eds.). (2003). *Governance of water-related conflicts in agriculture – New directions in agrienvironmental and water policies in the EU*. Dordrecht/Boston/London: Kluwer Academic Publishers.

Bryden, A., Petticrew, M., Mays, N., Eastmure, E., & Knai, C. (2012). *Scoping review of evaluations of voluntary agreements between government and business*. London: Policy Research Unit in Policy Innovation Research (PIRU).

Buric, B., & Gault, J. (2011). *Payment for environmental services: First global inventory of schemes provisioning water for cities*. Rome: Natural Resources Management and Environment

Department, Land and Water Division. Food and Agriculture Organisation for the United Nations (FAO).
Carmin, J., Darnall, N., & Mil-Homens, J. (2003). Stakeholder involvement in the design of U.S. environmental initiatives: Does sponsorship matter? *Policy Studies Journal, 31*, 527–543.
Cole, S., & Kriström, B. (2007). *REMEDE Deliverable No. 11: Case study selection report*. REMEDE- Resource Equivalency Methods for Assessing Environmental Damage in the EU.
Darnall, N., & Carmin, J. (2005). Greener and cleaner? The signaling accuracy of U.S. voluntary environmental programs. *Policy Sciences, 38*, 71–90.
Darnall, N., & Sides, S. (2008). Assessing the performance of voluntary environmental programs: Does certification matter? *The Policy Studies Journal, 36*(1), 95–117.
DEFRA. (2012). *Government response to stakeholders' views on the consultation on implementing the nitrates directive*. London: Department for Environment, Food and Rural Affairs.
Delacámara, G., Dworak, T., Gómez, C. M., Lago, M., Maziotis, A., Rouillard, J., & Strosser, P. (2013). *EPI-Water Deliverable 5.3: Guidance on the design and development of economic policy instruments in European water policy*. EPI-Water – Evaluating economic policy instruments for sustainable water management in Europe.
EC. (2012). A blueprint to safeguard Europe's water resources. Communication from the Commission to the European Parliament, the European Economic and Social Committee and the Committee of the Regions, COM(2012) 673 final. Brussels: European Commission.
EC. (2013). An EU strategy on adaptation to climate change – Communication from the Commission to the European Parliament, the Council, the European Economic and Social Committee and the Committee of the Regions. COM(213) final. Brussels: European Commission.
Gómez, C. M., Delacámara, G., Pérez, C. D., Ibáñez, E., & Rodríguez, M. (2013). *EPI-Water Deliverable 4.3: WP4 EX-ANTE case studies droughts and water scarcity – Tagus and Segura SE (Central Spain & Portugal) interconnected river basins*. EPI-Water – Evaluating economic policy instruments for sustainable water management in Europe.
Gupta, S., Tirpak, D. A., Burger, N., Gupta, J., Höhne, N., Boncheva, A. I., Kanoan, G. M., Kolstad, C., Kruger, J. A., Michaelowa, A., Murase, S., Pershing, J., Saijo, T., & Sari, A. (2007). Policies, instruments and co-operative arrangements. In B. Metz, O. R. Davidson, P. R. Bosch, R. Dave, L. A. Meyer (Eds.), *Climate change 2007: Mitigation*. Contribution of Working Group III to the Fourth Assessment Report of the Intergovernmental Panel on Climate Change. Cambridge/New York: Cambridge University Press.
Heinz, I. (2004). Sustainable farming as a result of negotiations: An analysis at European level. Paper presented at the CIGR International Conference, Beijing, China.
Heinz, I. (2008). Co-operative agreements and the EU water framework directive in conjunction with the common agricultural policy. *Hydrology and Earth System Sciences, 12*, 715–726.
Heinz, I., Andrews, K., Brouwer, F., & Zabel, T. (2002). Voluntary arrangements to cope diffuse pollution from agriculture and their role in European water policy. *Water Science Technology, 46*(6–7), 27–34.
Lago, M., & Moller-Gulland, J. (2012). *EPI-Water Deliverable 3.2: WP3 EX-POST case studies comparative analysis report*. EPI-Water – Evaluating economic policy instruments for sustainable water management in Europe.
Mattheiß, V., Strosser, P., & Carrasco, J. M. (2010). *Background notes on financing water resources management*. Report prepared for OECD, Paris.
OECD. (1999). *Voluntary approaches for environmental policy: An assessment*. Paris: OECD.
OECD. (2003). *Voluntary approaches for environmental policy: Effectiveness, efficiency and usage in policy mixes*. Paris: OECD.

Chapter 23
Cooperative Agreements Between Water Supply Companies and Farmers in Dorset (E)

Christophe Viavattene, Simon McCarthy, Colin Green, and Joanna Pardoe

Abstract This case study, located in the English County of Dorset, is a classic example of a water company (Wessex Water in England) facing increasing nitrate groundwater contamination. The pollution is mainly the result of farming activities. Potential "cheap" solutions such as blending the water from different sources are increasingly difficult to undertake due to the extent and increase in contamination. As a result the water company has two options: the treatment option or a catchment management approach. In this case to avoid the high operational and maintenance and construction costs of the treatment option Wessex Water has approached the farmers in order to cooperate to improve the water quality by promoting better practices. The cooperation started in 2005 involves information and education support but also phased incentive payments. This chapter illustrates the effectiveness and the continuity of such cooperation and the findings are expected to be of great interest to highlight the pro and con of a cooperative agreement as experienced in England to improve the water quality.

Keywords Cooperative agreement • Nitrate • Ground water contamination • Water utility • Farmers

23.1 Introduction

The Frome and the Piddle are two catchments located in the County of Dorset in England. The geology under these catchments includes Cretaceous Chalk which provides excellent conditions for high quality water aquifers suitable for domestic supply. However, nitrate pollution primarily threatens the quality of water. In 2005 Wessex Water Utilities decided to apply a catchment approach within three geographically bordering pilot catchments (Frome, Piddle and Wey river catchments) to improve the situation on eight water supply sources classified as 'endangered water

C. Viavattene (✉) • S. McCarthy • C. Green • J. Pardoe
Flood Hazard Research Centre, Middlesex University,
North London Business Park, London N11 1NP, UK
e-mail: c.viavattene@mdx.ac.uk; s.mccarthy@mdx.ac.uk;
c.green@mdx.ac.uk; j.pardoe@mdx.ac.uk

M. Lago et al. (eds.), *Use of Economic Instruments in Water Policy*,
Global Issues in Water Policy 14, DOI 10.1007/978-3-319-18287-2_23

bodies'. For Wessex Water Utilities the primary policy aims are environmental and health related, i.e. the respect of Water Framework Directive standards and the provision of good water quality. The objectives are an inversion of the Nitrate (N) trend or at least a stabilisation of the N level in the long-term and the reduction of the seasonal pollution peak to an acceptable level in order to protect the public water supply sources (WAgriCO 2008a). However, due to the complexity of the system response, recognising the long-term response of groundwater to changes on the land surface, the objectives have to be redefined in agronomic terms, i.e. the objective becomes the reduction of N quantity in the soil by the adoption of better practices by a maximum of designated farmers. A cooperation scheme started in 2005 was initiated as part of a Life Project (2005–2010) called WAgrico (LIFE05 ENV/D/000182). This gave the advantages, such as: removing some institutional barriers; encouraging stakeholders and farmers' participation by providing financial support in the form of grants. It has also allowed setting up a scientific monitoring process (sampling collection, farmers' surveys, and modelling support) and as a consequence providing substantial information for ex-post assessment. Whereas nowadays catchment based approaches are promoted and recognized as necessary for a sustainable water management strategy (DEFRA 2011), the Wessex Water Utilities in England was a pioneer in the country. Engaging farmers and other stakeholders is a long and evolving process and several years are necessary before any lessons might be learned. As such investigating how the scheme has evolved since 2005 and what were the current social, economic and environmental outcomes drove our interest on this particular case study to highlight the advantages and disadvantages of a voluntary agreement instrument. The assessment described in this case study is based on the available literature, on the surveys conducted by Wessex Water and on interviews with the people involved in this project.

23.2 Setting the Scene

Wessex Water supplies drinking water of approximately 370 Ml/day to a population of 1.2 million. The predominant sources of this water accounting for 80 % of Wessex Waters' domestic supply are the aquifers underlying two catchments, the Frome (198 km^2) and the Piddle (107 km^2). These catchments are located in Dorset, part of the South West region of England. Dorset is typical of the South West of England in that it is a predominantly a rural region, where agriculture occupies the majority of the land (79 %). This includes 39 % arable, 34 % grassland and 6 % rough grazing (WagriCo 2008c). The remainder of the area comprises forested land (11 %), urban (9 %) and water and wetland (1 %). N pollution originates from intensive farming practices which have developed since 1975. The average long-term trends of N indicated a potential to exceed the drinking water limit of 11.3 mgN/l by 2015, the situation varying from one source to another. A command and control policy is in place (e.g. Nitrate Vulnerable Zone for the UK) in addition to the regional farming economic sector decreasing in recent years but this has not reduced the problem.

Potential "cheap" solutions such as blending the water from different sources are now increasingly difficult to realise due to the extent and increase in contamination. Rather than treating the water to remove the pollutants, the water company has opted for a catchment management approach involving a cooperative agreement (CA). Cooperative agreements are defined as voluntary agreements between farmers and water supply companies. Such agreements have to meet the following four key requirements (Heinz et al. 2002):

- It is established on a voluntary basis between farmers and at least one water supplier and relying on the self-interest of the parties involved
- It is based on self-regulation among the key actors
- It includes an important role of the water supplier, either in the negotiation process and/or in the provision of financial resources
- It is targeted to a specific area (e.g. water catchment area; groundwater protection zone)

However, they may have different aims. Brouwer et al. (2003) identified three aims associated with CAs regarding the pollution situation: remedial statutory (drinking water standard is exceeded), preventative statutory (drinking water standard is at risk of being exceeded in the future) or discretionary (no risk but a desire to obtain the purest water). These three situations and the capacity of the natural system to respond to changes are very important to consider as they may impact on the negotiation process. It is worth mentioning as an illustration that in the case of N the pollution may be of two types: long-term pollution of the groundwater taking effect over a couple of years or decades and the yearly seasonal peak of pollution. In this case study given the extent of agricultural land in the catchment area combined with the shallow layer of land between the surface and the aquifer, both situations exists and, where aquifers are particularly close to the surface, transmission of nitrates can be rapid.

Such approaches were already applied in other parts of Europe however, remained limited at the time (Brouwer et al. 2003; Barraque and Viavattene 2009). Whereas few CAs exist in the UK, France and the Netherlands, in contrast, 435 CAs could be identified in Germany. The authors provide many explanations for this unbalanced occurrence of CAs within the EU, such as the different sized populations and shares of groundwater in water abstraction, the assignment of statutory groundwater protection zones, difficulties in enforcing compulsory rules, and the willingness of water companies and consumers to pay the costs necessary to stimulate farmers to change their production methods. In the UK most common agreements with farmers were not formed with water suppliers but with nature conservation organisations. The main barrier here has been the legislative and regulation system associated with the economic regulation system. The Water Services Regulation Authority taking care that the cost of reducing pollution was not passed on to the consumers through higher water prices under the guise of the polluter payer principle. Cases of bargaining with farmers and drinking utilities existed in the UK but most of them were still exploratory. The Wessex Water case study was one of them supported by a Life Project (2005–2010) called WAgriCo (LIFE05 ENV/D/000182).

23.3 The Cooperative Agreement in Action

The approach had two phases: the preparation of the agreement and the agreement itself. Farms presenting high risks of N loss (based on hydrology and the farms' activities characteristics) were usually targeted for more efficiency. The preparation phase aimed to identify these farmers in the areas presenting the highest risks and to identify the potential solutions for the farmers to reduce the loss of N, the practicality of solutions and their costs. On this basis some farmers agreed to adopt certain practices in exchange for a grant. As much as possible the implementation of primary measures was promoted as *simple and flexible measures are essential for acceptance* (*under voluntary measures*). However, it was recognized that the N loss could only be reduced by 5–15 % with such measures and that, beyond this, drastic management may be required. The cooperative agreement allowed for the selected measures to be reviewed each year based on the field N samples. After 2010 the approach has been maintained by Wessex Water. However, the uses of a legal agreement and of the grants have stopped. The cooperation is now limited to verbal agreements mainly for exchanging free advice with farmers granting access to their land for N sampling. In some discreet cases financial exchanges are realised. The catchment officers play a central role in the management of the cooperative process.

23.3.1 The EPI Contribution

23.3.1.1 Environmental Outcomes

During 2005–2010, 45 farms of 74 farms agreed to participate in a preliminary assessment and 28 farms agreed to a farm-gate nutrient assessment. Following these assessments a set of measures were proposed to the farmers, i.e. fertiliser recommendation, manure management plans and farms waste audits, use of cover crops, fertilizer best recommendation, moving application of slurries and poultry manure and the calculation of N efficiency. Fifty-two farms agreed to participate and received a grant in exchange for adopting some of these practices. Preferred measures were fertiliser recommendations and manure management plans. However, it was highlighted that the farmers were already using these existing recommendations as some regulations (e.g. Nitrate Vulnerable Zones[1]) were already in place. As such the EPI aims to optimise these practices. Fertilizer calibration and N efficiency calculation were also appreciated by the farmers. Cover crops had a good uptake considering that this approach is not applicable on every field. However, many of the farmers have indicated that they would not grow cover crops unless they were paid to do so as these practices present some inconveniences (such as weed growth

[1] Restrictions on fertiliser use are placed on Nitrate Vulnerable Zones as designated areas to reduce nitrates levels in water bodies.

and then spray off action). After 2010, the participation of farmers has been maintained without the grants. The surface area covered by the farmers engaged with Wessex Water represents 80–100 % of the medium and high risk catchments. The rate of uptake of the different measures may have changed but no information was available on this.

How did these changes in individual behaviour translate into lower pressures on water? The reduction of the pressures on the water can be measured in two different ways at the soil interface: by sampling the quantity of N in the soil after the harvest and by sampling the concentration of N in the leaching water. Following the change of practices a reduction of 55 % of the Soil Mineral Nitrogen (SMN) values is observed on average for the different crops. The quantities of SMN after the EPI are more or less similar as the one observed on Nitrate Vulnerable Zones at a national scale. In 2009 low yields and higher SMN values were observed apart from 2009 (Fig. 23.1) probably as the consequence of a difficult year with high rainfall observed for the third successive year (DEFRA 2010a).

The hydro and geologic survey and modelling on the catchment have highlighted that there were no general trends indicating future increase of N concentration. The phenomena of a plateau in levels are mainly observed (WagriCo 2008b; DeVial 2008; PHCI 2014). The current samples tend to confirm this assumption: no particular change has been observed in the current groundwater concentration since the EPI implementation, but effects on the background concentration were not expected in the short-term. However, the current management reduces the amplitude of the short-term pressure (peak of nitrates). Therefore the EPI maintains the provision of good water quality.

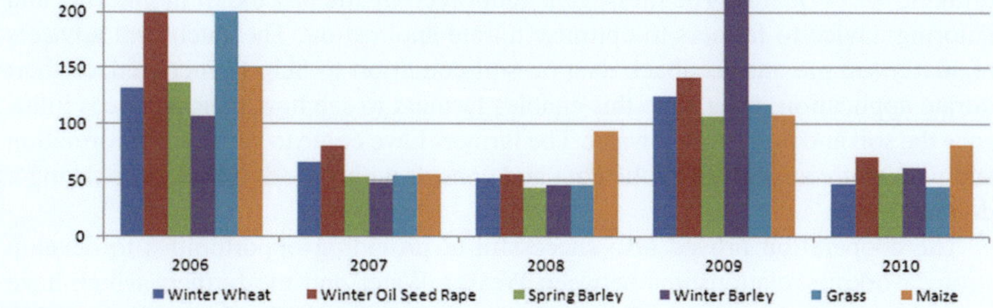

Fig. 23.1 Winter SMN values for different crops (Data source: Wessex water 2011)

23.3.1.2 Economic Outcomes

The comparison of alternative approaches to reduce groundwater contamination is often a difficult and tentative exercise. The catchment management approach (advice only) annual cost was estimated at 8 % of the annual costs of a treatment alternative approach, knowing that the annual treatment costs per catchment depend on the water quantity to be treated and range between 0.2 and 1.5 £millions (DeVial 2008). However, in terms of effectiveness, the treatment options guarantee a good drinking water quality as soon as it is operational. Due to the lack of response of the hydro system and the uncertainties associated with climates, agricultural practices and future land use changes, catchment management measures tends to reduce the risk but do not ensure a constant water quality in both the short and long-term.

23.3.1.3 Enhanced Distributional Impacts

Interviews were conducted with key stakeholders to understand the enhanced distributional impacts associated with the cooperative agreement. Due to the project limits only representatives from Wessex Water Manager and the catchment advisers were interviewed. Ideally, with further time and resources, farmers directly involved would have been consulted to provide a robust assessment from their perspective.

The economic impacts were limited. In many ways the measures were deemed advantageous to the farmers. For example, the involvement of catchment advisers provided free information and expertise that would allow them to tailor their fertiliser application so as to ensure less wastage, thus potentially saving the farmers' money. The catchment managers and advisers were keen to ensure that they did not impact the farmers' businesses negatively. They understand, for example, that if they are encouraging farmers to apply less fertiliser, they need to ensure that the yields are not subsequently reduced.

The EPI facilitates a subtle form of knowledge transfer and education. Educating farmers on the benefits of reducing their nitrate application has been central to changing attitudes and practices to reduce nitrate pollution in drinking water supplies. The interviewees commented that through their contact with farmers they had been able to improve farmers' understanding of how their practices influence local drinking water quality. The message is reinforced by the process of monitoring and tailoring advice to farmers to optimise nitrate application. The catchment advisers regularly sample and feedback data on soil condition to help farmers reduce their nitrate application. Over time this enables farmers to see how their practices influence the soil and thus ground water. The farmers have come to value this information and now many actively seek the data and consult with the advisers before taking a decision.

The cooperation proved also successful at providing opportunities to develop strong working relationships between Wessex Water and the farmers which have proved mutually beneficial. The catchment advisers visit the farms on a weekly basis maintaining regular contact. Wessex Water feels this is vital to remaining in close cooperation with farmers and reminding them of the importance of the work. The catchment advisers emphasise the importance of time in building relationships with the farmers. They report negative experiences when first approaching farmers as the farmers resent the intrusion into their activities, but the interviewees comment that this can improve over time with effort to develop a connection. Furthermore, they highlight that such voluntary agreements improve the farmer's impressions of Wessex Water and result in more positive feelings compared to the regulatory approach.

The cooperative agreement and catchment management approach is dependent on Wessex Water developing and maintaining these strong connections and relationships with farmers. However, the farmers themselves often have social connections between each other. This can be problematic where a varied and tailored approach is taken between farms. If one farm receives one set of benefits, other farmers may find out and expect the same. As such it is important for Wessex Water to ensure that they take a discreet or consistent approach.

23.3.2 The EPI Setting Up

23.3.2.1 Institutions

The existing pattern of rules means that only the water supply companies have any power to directly seek to encourage farmers to reduce nutrient flows to groundwater, although the various agri-environment programmes may have an indirect effect. However, the incentive for the water company to intervene is determined by the formula used in the quinnenial price review. The Water Services Regulatory Authority (OFWAT) is the economic regulator of the English and Welsh privatised water and sewerage industry. As part of its duties, OFWAT is responsible for approving water pricing tariffs and reviews these on a five yearly basis. This formula has two components:

Well run companies are entitled to earn a fair return upon their regulatory asset capital; over time, the price regulator has sought to drive the return down to a fair return but the current allowed return is arguably generous. Thus, the price rules encourage them to adopt capital intensive strategies.

The formula is based upon rpi − x + k where rpi is the rate of inflation, x is the anticipated improvement in operating efficiency, and k is the allowance for the capital investment required for the agreed programme of improvements over the next 5 years.

The first element of the formula encourages the companies to make capital investment s whilst the x factor in the overall formula encourages them to drive down operating costs (and shift from operating costs to capital costs). Hence, there are strong disincentives to the water companies in adopting the catchment management approach as the costs involved are all operating costs and add nothing to their regulatory asset capital. This may be one reason why the catchment approach has not been more widely adopted by water companies and why payments to farmers have now been discontinued. The incoming government has promised a White Paper on water management and all parties are lobbying hard for the White Paper to set out a framework for promoting a transition to sustainable water management. A change to the price formula is strongly considered as a precondition to promote a shift to sustainable urban water management. Such change may require primary legislation.

The failure of the approach either to be replicated by other water companies or compensation payments to continue to be made by Wessex Water can reasonably be associated with the much wider failure to develop the integrated institutional framework to deliver sustainable water management in England. It is one failure amongst many, there being neither an integrated approach nor a framework of powers and associated rules to deliver such an integrated approach (Green and Anton 2012; Green 2010).

23.3.2.2 Transaction Costs and Design

In its simple form the stakeholders involved in the approach are the water company and the farmers. The catchment advisors of Wessex Water represent the key actors. They have a clear role: to monitor, to control, to advise and to report to the farmers and the water company. The main transaction costs are the costs of the catchment officers working with the farmers to better tailor their practices, i.e. around £40,000/year (2008 values).

23.3.2.3 Implementation

The EPI, cooperative agreements, is used by Wessex Water as a highly flexible instrument which is tailored to individual farms and farmers via interactions with the catchment advisers from Wessex Water. The most successful strategies for engaging the farmers have been "softly softy" approaches whereby catchment advisors persevere in establishing contact with farmers to develop working relationships with them. Through ongoing contact with the farmers, the advisers monitor the measures through sampling and data collection, the results of which are fed back to the farmers along with tailored advice. In some cases it is necessary to take a slightly different approach of actually paying farmers in particularly high risk locations where there is a significant danger of exceeding pollution limits. In these cases farmers may be paid directly to not apply fertilisers or pesticides. This method is controversial, even within Wessex Water and as such much thought is given as to whether this approach is appropriate for a particular farm. There is concern that other farmers may expect payment if they become aware that some receive payment. With regards to targets and deadlines, these were particularly flexible for the farmers. Wessex Water has targets to ensure that their water supplies don't exceed the limits for nitrates. The farmers are made aware of the N limits, however, there is very little emphasis on specific targets for the farmers and no deadlines are provided as this is seen as an ongoing, long-term process of engagement. Deadlines and targets have not been necessary to see success in the approach.

The cooperative agreements were accepted by the farmers and catchment advisers as a novel but sustainable means to achieve the goal. They were very popular as alternatives to regulation and some farmers have become keen advocates of the approach, willing its success in order to prove that 'red tape' and further regulation is not necessary.

In terms of compliance, the cooperative agreements are fully voluntary. The fact that those farmers participating in the EPI remain involved even after several years demonstrates the success of the EPI approach from the farmers' perspective. The EPI's safeguarding mechanisms can be considered as the work in kind that the catchment advisers carry out (i.e. soil sampling) which offset the negative impacts such as having to spend more time in discussion with the catchment advisers.

Fully embedded into the EPI are mechanisms for monitoring the effectiveness of the approach. These mechanisms are monitoring nitrates through regular soil sampling and monitoring boreholes and nitrates in water storage sites.

In terms of barriers to the achievement of the objectives of the EPI in this case study there is no clear evidence of policies that provide such obstacles. For policies that Wessex Water could take advantage of, there are regulatory policies such as Nitrate Vulnerable Zones which restrict farmer's fertiliser use. However, the catchment advisers are reluctant to use this policy as they believe they will have greater success in achieving continued compliance by working with farmers in a voluntary approach. The Wessex Water approach supports 'Catchment Sensitive Farming' under the England Catchment Sensitive Farming Delivery Initiative (ECSFDI). The ECSFDI encourages voluntary action to achieve the goals of the Water Framework Directive by managing land and optimising fertiliser use to reduce pollution. The mechanisms for achieving this are monitoring, evaluation and advice to farmers. These mechanisms are the same as those used by Wessex Water, however, Wessex Water's catchment advisors are at an advantage over the ECSFDI advisers as they cover smaller areas and therefore have greater contact with farmers, fostering cooperative relationships. The Wessex Water approach, therefore, has a strong synergy with ECSFDI.

23.4 Conclusion

This case study is focusing on a specific economic policy instrument called cooperative agreements involving farmers and a Water Company. The main aim of the EPI is for the water company to provide good water quality to its customers by maintaining an acceptable level of N in their water sources by a catchment management approach rather than by using expensive treatment plants. The long-term objective is to maintain or reduce the N trends in the groundwater in different catchments. A short-term objective is to reduce the risk of N peaks in the Spring. The catchment approach is currently limited to a recommendation approach for an optimal use of the fertilizers and for the adoption of mitigation measures such as grass cover when necessary. No change in cropping patterns or on the type of crops such as conversion to grassland and extensification are promoted. This method differs from other UK approaches as the catchment officers are working closely with the farmers to better tailor with them their use of fertilizers in accordance with the national recommendations (DEFRA 2009, 2010b). The cooperation started in 2005. It is therefore too early to judge the effectiveness of the instrument in terms of change in the groundwater quality.[2]

EPIs are primarily designed to change the behaviour of individuals. On the issue of diffuse pollution of groundwater resources the uptake of measures is therefore a good indicator of the effectiveness of the EPI. Farmers' participation in the different catchments can be considered as a success; between 80 % and 100 % of the catchments at medium and high risks are now engaged with Wessex Water. The level of cooperation has also been maintained after the suspension of grants indicating a strong cooperation between both parties. The measures proposed in the cooperation are not too restrictive for the farmers which may also explain the high participation. The current Soil Mineral Nitrogen values sampled in the field following the establishment of EPI indicates similar values as the ones observed on average at a national scale for the Nitrogen Vulnerable Zones, stressing good farming practices and appropriate fertilizer use. It is difficult to conclude empirically to what proportion the EPI contributes by itself to these good practices. However, in principle, the close Winter monitoring (SMN values and nitrates leaching pots) on various fields and their use as a risk assessment tool seems very appropriate to discuss with the farmers the options to reduce the risks of nitrate leaching and to find common solutions. The use of compensation can be discussed as part of these solutions.

The annual cost of the catchment management approach is very low compare to the treatment costs options, circa 8 %. The costs mainly include the catchment officer costs and the sampling costs. These costs per farm are 20 times higher than those observed for standard catchment management approaches. The needs of secondary measures such as grassland conversion will also increase the costs of the approach if compensation was paid by the water company.

The EPI has mainly a high impact on social capital: trust, social connection and the relationship between the farmers and the water company are enhanced as well as their common and local knowledge on water catchment management and the diffuse pollution process.

The system of economic regulation of the water industry is still the most significant barrier to the development of such an approach. The approach works in harmony with environmental policy such as the England Catchment Sensitive Farming Delivery Initiative and the Nitrate Vulnerable Zone by supporting similar objectives. The cooperation approach allows a focused, tailored and adaptive approach on specific areas which could not be achieved by national approach. If required, the EPI could also be used to support farmers in entering schemes promoting greater environmental benefits such as the Entry Level Stewardship or the High Level Stewardship schemes.

[2] The ex-post assessment was conducted in 2011.

References

Barraque, B., & Viavattene, C. (2009) Eau des Villes et Eau des Champs vers des accords cooperatives entre services publics et agriculteurs? Economie Rurale “La gestion de l’eau en France” 310.

Brouwer, F., Heinz, I., & Zabel, T. (2003). *Governance of water-related conflicts in agriculture: New directions in agri-environmental and water policies in the EU* (Environment and policy, Vol. 37). Dordrecht: Kluwer Academic Publishers. 225 p.

DEFRA. (2009). Protecting our water, soil and air: A code of good agriculture practice for farmers, growers and land managers. ISBN 978 0 11 243284 5.

DEFRA. (2010a). Agriculture in the south west of England 2009/2010. http://farmbusinesssurvey.co.uk/regional/commentary/2009/southwest.pdf. Accessed Oct 2014.

DEFRA. (2010b). Fertiliser manual (RB2009). ISBN 978 0 11 243286 9.

DEFRA. (2011). Water for life – Policy paper. 105 p. ISBN 9780101823029.

DeVial, L. (2008). Bargaining with farmers to protect drinking water resources. Powerpoint presentation – Paris preliminary seminar to the Fontainebleau Seminar: Bargaining with farmers to protect drinking water resources?

Green, C. H. (2010) The transition to sustainable urban water management: The London case study. SWITCH EU project http://www.switchurbanwater.eu. Accessed Oct 2014.

Green, C. H., & Anton, B. (2012). Why is Germany thirty years ahead of England? *International Journal of Water, 6*, 195–214. ISSN 1465-6620.

Heinz, I., Brouwer, F., Andrews, K., Zabel, T. (2002). Co-operative agreements in agriculture as an instrument to improve the economic and environmental effectiveness of the European Union water policy. Research Project Final report, funded by the EU-Commission, Contract no. ENV4-CT98-0782.

PHCI (Poole Harbour Catchment Initiative). (2014). Catchment plan update – May 2014. http://www.wessexwater.co.uk/environment/twocol.aspx?id=11790&linkidentifier=id&itemid=11790. Accessed Oct 2014.

WAgriCo. (2008a). Technical final report, LIFE05 ENV/D/000182 WAgriCo project.

WAgriCo. (2008b). Results from monitoring and an evaluation of mitigation Methods, Annex 51-Uk Interim report. LIFE05 ENV/D/000182 WAgriCo project.

WAgriCo. (2008c). Diffuse pollution control in England, Annex 7-Uk Interim report. LIFE05 ENV/D/000182 WAgriCo project.

Wessex Water. (2011a). Wessexwater a YTL company. Internal document.

Wessex Water. (2011b). Catchment management: managing water-managing land. http://www.wessexwater.co.uk/workarea/linkit.aspx?linkidentifier=id&itemid=7201. Accessed Oct 2014.

Chapter 24
Financial Compensation for Environmental Services: The Case of the Evian Natural Mineral Water (France)

Pierre Defrance

Abstract The Evian bottled Natural Mineral Water Company in France initiated in the late 1980s a promising multisectorial water protection policy aiming at maintaining the Evian Natural Mineral Water (NMW) quality by promoting a sustainable development of its catchment area. The assessment illustrated in this chapter focuses on the payment for ecosystem services (PES) scheme developed by the association for the protection of the catchment area of Evian mineral water (APIEME) with local farmers. It demonstrates how the Evian Company can maintain a land use and traditional agricultural practices on the catchment area presumed to preserve the quality of the Evian Natural Mineral Water, without buying any land around the catchment area, by financing agricultural related projects. It also demonstrates that the financial dimension of PES schemes may not be the most important one to explain their success. Defining precisely what is the issue, gathering all stakeholders, sharing knowledge and building trust are all important components of a successful PES, even if they are creating a system defined by high transaction costs. Lessons learned from the Evian case study should help designing and implementing PES schemes in Europe and contribute to the development of preventive policies.

Keywords Water quality • Payment for environmental services • Natural mineral water • Transaction costs

24.1 Introduction

Evian (owned by Danone group) is one of the major brands of bottled Natural Mineral Water (NMW) in the world. Its water comes from several sources in the French Alps, around the city of *Evian-les-Bains*. The French legislation for NMW is very strict: the purity, composition, temperature and other essential characteristics

P. Defrance (✉)
AequitO, 32 chemin Commins, La Montagne, 97417 Saint-Denis, France

ACTeon, 5 place sainte Catherine, 68000 Colmar, France
e-mail: defrancep@gmail.com

M. Lago et al. (eds.), *Use of Economic Instruments in Water Policy*,
Global Issues in Water Policy 14, DOI 10.1007/978-3-319-18287-2_24

of natural mineral water must remain stable. In addition to the geological natural protection, water from Evian, as a Natural Mineral Water, is defined by its groundwater origin, its purity and the stability of its mineral content and the absence of any chemical treatment, and of any additive, disinfectant or preservative. The right to use the "Natural Mineral Water" label would be lost if mineral concentration was to change.

Twenty years ago, two long-term evolutions could have affected the Evian Natural Mineral Water (NMW) and the agricultural area related to it: (i) the evolution of agricultural practices from traditional dairy farming to more intensive agricultural practices, and (ii) the drive to open up the area by improving links to other regions in France and Switzerland.

A few managers of Evian looked at these evolutions seriously, even though the NMW was not reported to be threatened by any kind of pollution at that time. Learning the lessons from what happened to another NMW company recognized at an international level (Vittel, Nestlé Waters), they initiated in the late 1980s a promising multisectorial water protection policy tackling wastewater collection and treatment, town and country planning, wetland protection, tourism, biodiversity and agriculture.

This policy mix (regulatory approach and economic instruments) relies on the association for the protection of the catchment area of Evian mineral water (APIEME), an association which comprises the villages from the spring area that benefit from a government tax on bottled water, the villages from the catchment area, the Evian Company and national public bodies. Its objective is to protect the Evian Natural Mineral Water (NMW) by promoting a sustainable development of its catchment area.

The APIEME "agricultural economic instrument" policy which can be classified as a scheme of payment for ecosystem services (voluntary agreement between farmers and one industry), is part of the policy mix. This instrument is oriented towards the development of a modern environmentally friendly agriculture focusing on dairy production linked to cheese making under the protected designation of origin (PDO). Basically, the Evian Company helps financing projects to maintain a land use on the catchment area presumed to preserve the quality of the Evian Natural Mineral Water. For each project, an agreement was signed by the APIEME and the project owner designed by the Gavot Plateau farmers' association (SICA). For instance, subsidies were targeting small to medium-size farms, helping them to follow the European sanitary norms evolution and to favour close loops and a higher income.

The economic policy instrument developed by the Evian Company through the APIEME to preserve the Evian NMW quality can be referred to as one of the rare schemes for environmental services in France. While the institutional context (haziness of the definition, lack of guidance) and high transaction costs are among the major barriers to PES schemes development in Europe, the example of Evian reveals both can be seen as opportunities. This first assessment of the Evian case study also contributes to the definition of preconditions for the implementation of such EPI.

24.2 Setting the Scene: Challenges, Opportunities and EPIs

The city of Evian-les-Bains is located on the banks of Lake Leman in the north-east of the Rhône-Méditerranée and Corse River basin district (Haute Savoie, French department in the Rhône-Alpes Region). The Evian bottling plant is located in *Amphion-les-Bains* (*Publier* commune), next to *Evian-les Bains*. It constitutes one of the most important plants of its kind in the world, producing six million bottles per day (2014). 2,200 million of Evian NMW bottles are thus consumed in France and also, for more than half of the volumes, worldwide in about 140 countries. In France, more than 1,800 jobs are directly linked to the Evian Natural Mineral Water (over the 10,000 jobs that are linked to Natural Mineral Water in France) and indirect jobs would be three times more (around 30,000 jobs in France).[1]

The catchment area is located on the Gavot Plateau, at an elevation ranging from 800 to 1,200 m and exhibits a middle mountain climate. In turn, the spring area is located at an elevation around 400 m and benefits from a more temperate climate influenced by the Lake Geneva. Due to a particular geological configuration, the water of Evian is well protected in a confined (artesian) aquifer. Rain- and snowmelt-water infiltrates on the 35 km^2 catchment area and flows to the spring through, first a multilayer quite low hydraulic conductivity system, during more than 20 years, giving to the water of Evian its particular composition, and second, in the last part of the NMW transit, through high permeability sands. In addition to the natural geological protection, the Natural Mineral Water also benefits from two kind of protection: (i) legal protection (the "Declaration of Public Interest" – DIP) that is mostly conceived to maintain the integrity of the impermeable cover of the aquifer, and (ii) technical protection (design and protection of the spring catchwork such as using stainless steel pipes).

Consequently there is no qualitative issue for this resource: concentration of nitrate is stable around 3.7 mg/l while the maximum allowed nitrate concentration in France is 10 mg/l for infants, 15 mg/l for mineral water and 50 mg/l for tap water; and no traces of pesticides were ever found (concentration are below the analytical detection thresholds); more generally, no traces of organics, mineral or biological contaminants were ever reported.

However, the aforementioned threefold protection does not protect the catchment area whereas the high quality of the NMW was interpreted amongst others as the result of harmless traditional agricultural practices. The main economic activity in the catchment area is agriculture (that represent 60 % of the total land use, among which 51 % of meadows and around 9 % of crops), represented by dairy cow breading for a typical local protected designation of origin (PDO) cheese production.[2] Fifty-five farms, mainly small to medium-size farms, are located on this area covering 2,100 ha of farm land (Buric et al. 2011).

[1] CSEM, 2007, 'L'eau minérale naturelle: Un produit naturel et protégé, une industrie responsable, un emballage recyclable'. Livre Blanc

[2] *Abondance* and *Reblochon*, two brands among the most famous of the French cheeses.

But, in the late 1980s, direct subsidy from the European Common Agricultural Policy (CAP) did not benefit small and large-size farms in the same way and small scale hill farming might not have been profitable enough to keep their traditional practices (Bazin 1994): regrouping of land would have become an option. With the removal of hedges and agricultural intensification, farmers would have increased their production of maize (instead of meadows) and used more fertilizers and pesticides. The change of agricultural practices and urban development might have become possible threats to the hydrological balance conservation of the site.

The Evian bottling company directors thus decided to develop a new water protection policy based on win-win actions, downstream-upstream economical redistribution processes and voluntary agreements. The policy was launched more than 20 years ago (in 1992) when the association for the protection of the APIEME was created. The villages from the spring area (one-third, less than 5 % of their legal tax revenues[3]) and The Evian Company (two-third) finance this association that works as a "*democratic water parliament*". This association plays the role of an intermediary, funding collective projects aiming at maintaining and developing modern environmentally friendly agriculture. It is translated in the ground by limiting the number of dairy cows grazing on the impluvium and which are only fed by local pasture.

24.3 The Payment for Environmental Services Scheme in Action

24.3.1 The EPI Contribution

24.3.1.1 Environmental Outcomes

Evian's preservation policy can clearly be classified as a preventive policy. The EPI aims at changing farming practices and reducing pressures on the catchment area. This makes conclusion on the EPI efficiency difficult.

The concentrations of pollutants have not changed when we analyse the thousands of tests that are carried out in line with European and French legislations associated to NMW. Otherwise, Evian would have lost the NMW label. Without the status the Company would have lost the high quality premium of NMW. In addition, the impacts of changes of agricultural practices early 1990s would only start being measured today or in a few years in terms of water quality changes considering the 20-year transit time of the infiltration.

[3] *The villages from the spring area benefit from an old French regulation that institutes a specific tax on bottled natural mineral waters such as Evian NMW. Thus the Evian Company gives money to these cities for each bottle of Evian sold while the villages of the catchment area do not receive anything.*

But the EPI clearly contributes to maintain a specific land use and traditional agricultural practices. Interviews with experts and the diagnostic of the French National Institute for Agricultural Research (INRA) (Christofini et al. 1994) indicated that, in the absence of EPI, agriculture may have continued its intensification and specialisation. The surface of maize in the impluvium area would thus have increased, and the number of farmers would have decreased. In addition, milk production would have partly switched from products of quality (milk used for the production of PDO cheese) to industrial production (selling milk to cooperatives located in the plain).

When the APIEME realized an inventory in 2002 with approximately the same methodology and typology than INRA, there were 71 farms (livestock farming, mostly dairying, including part-time farmers) in the area instead of 100 in 1992 (also including part-time farmers), that is a decrease of almost 30 % within 10 years. But the number of professional farmers remained almost unchanged during the last 20 years (Buric et al. 2011). The dairy farms' production was dedicated to the production of PDO cheese (*Reblochon* and *Abondance*), that is around 7.7 million litres of milk (about 770 tonnes of cheese) per year instead of 7.2 million litres of milk in 1993. In general, the dairy production is based on maize, which is considered as an important factor of nitrates increase (Perrot-Maître and Davis 2001). But the situation is different for the Gavot Plateau thanks to the EPI. Developing modern environmentally friendly agriculture associated to the PDO system allowed to limit the increase of maize surface and even reduce maize surface in the catchment area (from 3.8 % of the total agricultural surface to 2.6 %) at least for the 10-year period considered. As an efficient preventive tool, the PES scheme leads to a reduction of pressure and is crucial to lift the long-term threat that agricultural intensification is posing to the NMW preservation.

24.3.1.2 Economic Outcomes

When facing a situation of pressure or pollution, natural mineral water bottlers have five alternative options (Déprés et al. 2008):

1. Doing nothing and relying only on natural protection;
2. Forcing polluters to change their practices relying on legal or regulatory actions;
3. Relocating their activity by choosing new and non-contaminated resources;
4. Buying all lands around their catchment area;
5. Achieving a contractual arrangement or a voluntary agreement with polluters.

When facing this choice in the early 1990s, the Evian bottling Company (Danone Waters) was in the comfortable situation to have time: the water resource was not reported to be threatened, except by some chloride ion (Cl^-) but such pressures were considered to be very limited.

At that time, no specific study was undertaken to quantitatively define the least-cost alternative or to compare costs to benefits in order to support decision-making. However, the evolution in land use became rapidly obvious to the managers of

Evian and they identified the need to reconcile the development of villages from the source and impluvium areas by integrating them into the decision-making process. The means chosen was to design win-win actions based on voluntary agreement and downstream-upstream financial redistribution that would maximise economic, environmental and social benefits (option 5).

The cost of this payment for environmental scheme is estimated to around EUR 85,000 per year and EUR 35 per hectare (projects dedicated to agriculture represent 13 % of the EUR 700,000 annual budget of the APIEME). Budget forecast defined in the agreement signed by the parties in line with recommendations made by INRA in 1994 is the following:

- To comply with standards of livestock buildings (impermeabilisation and coverage with a roof of manure farm dunghills and increase storage facilities) and to comply with standards of dairy farms: both subsidies were designed for a 6-year period from 1996 to 2001 and the total was constrained to a maximum EUR 33,500 yearly contribution from the APIEME;
- To renovate and establish cooperative dairies for cheese production: these subsidies were designed for a 15-year period from 1995 to 2009 and they were constrained to a maximum EUR 61,000 yearly contribution from the APIEME;
- To prevent any leakage of the pesticides or fertilizer spread on the few maize plots of the plateau, technical studies implemented with the farmers allowed elaborating an adapted methodology. The resulting protocol does not ban pesticide use and helps farmers to adopt environmentally friendly practices (shallow ploughing between the maize rows and light herbicide application on the rows). A new manure management plan was also designed in order to avoid the excess of fertilizer on specific plots these projects were designed for 5-year period from 1995 to 1999 and they were constrained to a maximum EUR 24,500 yearly contribution from the APIEME. They favoured milk processing operations and closed loops in order to maintain traditional farms and increase farmers' incomes;
- Technical support from the Chamber of agriculture with experimental sites: the APIEME contributed up to EUR 10,500 yearly to this action.
- In addition to these actions, a charter of good practices was developed with the contribution of INRA, the SICA, farmers and the APIEME. Some of these subsidies were depending on the signature of this charter.

The budget parties agreed on is around EUR 1.3 million. But the effective total budget allocated to actions aiming at developing a modern environmentally friendly agriculture is even higher (more than EUR 1.5 million). Most of the contribution comes from the Evian Company (more than two-thirds). Thus the Evian Company and the villages located in the spring area, which are the beneficiaries of the EPI, support most of the cost of its design and implementation.[4]

[4]With the exception of a little part of the total budget of the APIEME that comes from subscriptions. It represents EUR 35,000 over the 19-year period of implementation, that is around EUR 170 per year per village.

Looking back, other options that were identified as feasible in theory can be considered as too risky or very expensive compared to the voluntary agreement. For instance, buying all (or part of) the lands of the impluvium area could have been an option but it was not realistic at that time for economic, legal and social reasons. The price of land in this area is quite high due to the proximity to the Leman Lake and Switzerland. In addition to this financial barrier, national laws prevent the purchase of agricultural lands for non-agricultural uses. And finally, this policy might have led to social tension weakening the relationship between Evian Company and the local stakeholders.

24.3.1.3 Distributional Effects and Social Equity

Three types of co-benefits can be identified. Regarding how the EPI impacts on farmers' activity (costs, profits, incomes), the first basic impact to be considered should be the redistributional effect of the functioning of the APIEME. Subsidies granted to farmers by the APIEME through the EPI come from the global budget of the APIEME, which is funded by the Evian Company (two-third) and the remaining by the villages located in the spring area. Thus money is redistributed from downstream (the beneficiaries of the EPI) to upstream (the famers who contributes to maintain the quality of NMW).

This redistribution of money through the EPI compensates additional efforts farmers have to make (increase of production costs), for instance by reducing their use of pesticides. The EPI also helps small farmers to face additional expenses associated with new regulations (around EUR 300 per dairy cow for 16 farms for complying with standards of livestock buildings).

In addition, the political voice of famers has been heard through meeting during the design (surveys amongst other), implementation and operation of the EPI. They have greater say since the creation of the SICA and thus thanks to the implementation of the EPI.

Finally, villages located in the catchment area and the Evian Company have both beneficiated from the creation of the APIEME: at local scale, villages have improved their access to the decision-making process; the Evian Company found a new space for discussion at local scale and reinforced its legitimacy at international scale.

24.3.2 The EPI Setting Up

24.3.2.1 Institutional Set-Up

The most embedded institutions relevant for the EPI are found at local level because both the environmental asset (quality of NMW) and the EPI (voluntary agreement between farmers and one industry) are very specific. First, the quality and properties of Evian NMW used to be "miraculous" and "timeless" for consumers and the general public. People generally do not know where the drinking water they receive at their tap is coming from and get a poor understanding of groundwater functioning or problems (Rinaudo 2008). The case of NMW reinforces this common perception as water emerges from the underground and people can only see the source.

In addition, the relationship between the Evian Company and the farmers was complex in the 1990s because most of the part-time farmers used to be employed by the company or had someone in their family employed by Evian. Thus, people living in the villages located in the catchment area were connected to the people living in the villages located in the source area and to the company. The company was used to negotiate with farmers in particular during the locally well-known strikes. But this link was becoming weak in the 1990s as more and more people living in the villages located in the catchment area found jobs in Switzerland and got disconnected from the company. This situation might have made negotiations more complicated because of a loss of reciprocal knowledge, trust and understanding.

The implementation of the payment for environmental services scheme also benefited from three types of intermediaries. First, the SICA created in 1993 actively contributed to the partnership between farmers, the APIEME and the Evian Company. One of the members of the SICA in particular played an important role in the process. While he was experiencing the intensification and specialization of agriculture in the Gavot Plateau and in its own farm, he decided to shift back to traditional farming and to promote products of quality (milk used for the production of PDO cheese and tourism). He fully contributed to the success of the EPI as he became the president of the SICA.

Then, the research team from INRA who helped to switch from "ready-to-use" solutions at plot of land scale to solutions compatible with the maintaining of a traditional and sustainable agricultural based on quality products.

And finally, the APIEME, as a neutral organisation, gave space to discussion and negotiation and become one of the most important preconditions to the success of the EPI. The idea of including the villages of the spring area (as beneficiaries) also increases the fairness of the instrument and made easier negotiation and agreement on the design of the EPI.

24.3.2.2 Transaction Costs and Design

A specific attention was paid to transaction costs as they are considered to be the main barrier to the development of payment for environmental schemes. Transaction costs occur during the formulation, the design and the development of the EPI as well as during the implementation and operation of the EPI. In the case of the Evian NMW, transactions costs were relatively high and concentrated during the first years.

The choices of the EPI and its design have not been guided by any models or tools. However two types of studies were undertaken in order to help decision-making. The first type of studies was related to the understanding of the hydrogeological functioning of the system, i.e. understanding where the natural mineral water comes from and how it infiltrates. These studies were essential to give a space to the idea of protecting the water resource at source by defining and delineating the catchment area. But they are not specific to the design of the EPI and are not considered in the analysis of transaction costs. These studies were implemented at least since the 1960s and are still ongoing.

Second, a partnership was developed between the Evian bottling Water Company and a research team from INRA starting in 1990 and ending in 1997.[5] The objective was to get a better understanding of the catchment area in terms of ecological functioning and human activities. The partnership played a strong role in determining the preventive approach and actions as Evian did not have competencies in agriculture and did not know (i) which where the most relevant levies to maintain a traditional agriculture in the area and (ii) how to reduce pressures. Their conclusions indicate the need for a water preservation policy pointing to the fact that pressures existed and were increasing on the catchment area. However, the risks for the NMW were unknown, in particular because of stocking and denitrification phenomena occurring in soil and wetlands. An interesting part of this partnership was dedicated to make a diagnostic of current activities in the catchment area identifying potential pressures, in particular coming from agricultural practices (Christofini et al. 1994). This diagnostic lasted 2 years including a survey of farmers which aim was to develop a typology of farms based on practices and impacts on water quality. It played also a mediating role ensuring mutual comprehension and allowing negotiations between the Evian Company and farmers.

Thanks to the recommendations of INRA, it did not take much time to select the EPI (define which will be the projects funded) and implement it. Based on the diagnostic and their experience, the farmers created the SICA to harmonise their requests and put forward feasible measures. They were negotiated and accepted by the APIEME. As such, both the INRA and the APIEME helped reducing transaction costs during the design and the implementation of the EPI.

Since 2006, from three to six meetings are organised each year, gathering one representative from the Evian Company, representatives from the SICA and the Chamber of agriculture and two representatives of farmers for each villages located in the impluvium area. These meetings aim at discussing progress, barriers and future initiatives of the EPI. Before 2006, similar meetings were organised but in a less structured and regular way. In addition, one of the representatives of Danone Waters is partly dedicated to the EPI through the APIEME, but the sharing has not been estimated between the contribution to the EPI and the functioning of the APIEME.

The TCs associated to the monitoring and the enforcement (ex-post TCs) are quite low because most of the subsidies are distributed in exchange of invoice. However, transactions costs associated to the charter of good practices are not well

[5] None of these related studies was published. Most of the information was confidential in the 1990s. One reason for this was the risk of misunderstanding. Communicating about a water quality preservation policy could have been counterproductive in that context. It was thus focused on experts or stakeholders capable of understanding these issues.

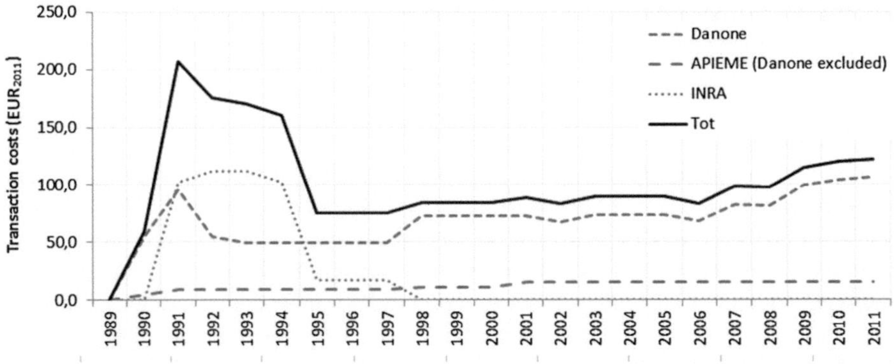

Fig. 24.1 Evolution of transaction costs related to the APIEME creation and implementation (Source: Own author's elaboration)

defined, but they are probably low, potentially at the expense of the effectiveness of the EPI. Other monitoring costs are partly shared with other actions of the APIEME, reducing their importance for the EPI (Fig. 24.1).

At the end, transaction costs were estimated at around EUR 100,000 per year in average, and more than EUR 150,000 in average during the first 5 years of the process. The values are quite uncertain because part of the costs considered can also be attributed to the whole policy mix and a few others are not considered in the analysis: regular meetings between the SICA, the Chamber of agriculture, representatives of the farmers and the APIEME as well as the creation of the SICA in 1993 should be considered as TC and added.

However, the estimated transaction costs are relatively high in comparison to the cost of the EPI (between EUR 85,000 and EUR 200,000 per year depending on the period considered). But this is a condition for the success of the payment for environmental services scheme anticipated by the Evian Company before it implemented the EPI: it was necessary to give time and space for negotiation to get a compromise between the expectations of the Evian Company and the requests of farmers.

24.3.2.3 Implementation

The EPI has been designed to be very flexible. First of all, the diagnostic (made by INRA) contributed to take into consideration local particularities, heterogeneous farming and a diversity of pressures. A list of actions was thus developed considering three types of farms and various scales of action. In addition, both the schedule and the funding were discussed and negotiated during the process: the period of certain subcontracts was extended allowing more farmers to beneficiate from the financial facilities aiming at complying with the standards of livestock buildings; the budget of the APIEME allocated to agriculture increased from around EUR 85,000 per year to EUR 200,000 to finance new relevant projects; and in some specific cases, the Evian Company added money when the annual budget of the APIEME was insufficient compared to the needs. The flexibility of the EPI thus contributed to make implementation easier.

The contribution of stakeholders also played an important role during the design and implementation of the EPI. As the instrument relies on voluntary agreements, farmers have been interviewed during the early stage of the design phase to identify which actions would be relevant. In addition to this, discussions and consultations were organized with farmers (the SICA) to negotiate the financial conditions of the contribution of the APIEME and define collective projects without individual contracting. From 2005 to now, regular meetings (from three to six times a year) are organized by the APIEME gathering the SICA, the Chamber of agriculture and two representatives of farmers from each villages of the catchment area. Other stakeholder representatives were consulted through the APIEME (villages and the Chamber of agriculture) and decisions were finally taken within the association chaired by the mayor of one of the villages located in the catchment area.

However, the creation of the APIEME was the initiative of the Evian Company. It can be considered as the most important driving force behind the whole process and in particular the EPI. Preserving the quality of the Evian NMW is a priority for the parent company. Evian's investment in terms of time, money, ideas and technical support seems to be one of the key of the success of the EPI and more generally the success of the APIEME and its policy mix. The effort made to understand farmers and their traditional agriculture and to establish a dialogue with all stakeholders were at least as important as financial contribution and technical support.

24.4 Conclusions

In the Evian Natural Mineral Water case study, the financial dimension may not be the most important one to explain the success of the EPI as it remains relatively low in comparison to potential benefits (for the Evian Company, for villages located in the spring area and for the villages located in the catchment area). Gathering all stakeholders and sharing knowledge and point of views to define and fund collective projects ahead of its time has to be considered as the main reason to both the preservation of the stability of the Evian NMW and the development of a modern environmentally friendly agriculture. Even though environmental, economic and social outcomes were not quantified with accuracy, the EPI seems to send right and coherent incentives to stakeholders with preliminary results showing that the situation evolves in the right direction (a sustainable development of the catchment area contributing to protect the NMW).

Estimated transaction costs are relatively high in comparison to the cost of the EPI, both ex-ante fixed costs and ex-post variable costs. But it appears surprisingly to be a condition for the success of the EPI anticipated by the Evian Company before its implementation. First, the partnership developed between the Evian Company and INRA in 1990 contributes to get a better understanding of the catchment area in terms of ecological functioning and the diversity of practices and potential pressures. Thus it played a strong role in determining the preventing approach and actions as Evian did not have competencies in agriculture. The diagnostic helped to reduce asymmetric information while the results were shared with farmers. Involving INRA in the process finally contributes to reinforce reciprocal trust between the Evian Company and farmers.

Second, the creation of the APIEME allowed parties to build shared ownership on the issues and to take part in the decision-making. It also gave space to discussion and negotiation by externalizing the initiative. In addition, the creation of the SICA helps harmonizing the request of the farmers and contributes to reduce TCs, while the Chamber of agriculture provides technical support.

Finally, the delivery mechanism the Evian Company chooses through the APIEME both contributes to the high level of transaction costs and helped reducing them. Indeed, the EPI allows flexibility (extension of the subsidies' duration) and requires regular meetings with stakeholders. But, it also prevents from conflicts and complex legal procedures – both associated with high transaction costs – by trying to reach compromises between the expectations of the Evian Company and the requests of farmers. The EPI has thus been welcomed by most of the stakeholders.

These three dimensions (financial, technical and social) and their relative influence over the process were also described as key factors to explain the success of the PES scheme used by Vittel (Nestlé Waters) to protect its mineral water (Perrot-Maître 2006). The water protection policy developed by the Evian Company is also in line with the final recommendation of Perrot-Maître (2006) by not focusing on one particular polluter but by taking a multisectorial approach. All potential sources of pollution or positive land use (and land cover) are taken into account by the APIEME through a coherent water protection policy mix.

Acknowledgements This chapter is the result of discussions between all partners in the EPI-WATER consortium. It benefited from specific contributions of Pedro Andrés Garzon Delvaux and Verena Mattheiß, ACTeon. The author would like to thank Patrick Lachassagne and Jean-Jacques Beley from Danone Water France and Cathy Le Hec from the APIEME who shared most of the information.

References

Bazin, G. (1994). Aides directes et disparités de revenus agricoles'. *Économie Rurale, 220–221*, 192–196.

Buric, B., Gault, J., & Bertoye, F. (2011). Payment for environmental services: First global inventory of schemes provisioning water for cities. FAO, Working paper.

Christofini, B., Dorioz, J. M., Ferhi, A., Roybin D., Bibet, P., Trévisan, D., & Vansteelan, J. Y. (1994). '*Activités agricoles, milieux physiques et qualité des eaux dans l'impluvium d'Evian*'. Final report, Evian/INRA contract, 52 p. (not published).

Déprés, C., Grolleau, G., & Mzoughi, N. (2008). Contracting for environmental property rights: The case of Vittel. *Economica, 75*, 412–434. doi:10.1111/j.1468-0335.2007.00620.x.

Perrot-Maître, D. (2006). *The Vittel payments for ecosystem services: A "perfect" PES case?* London: International Institute for Environment and Development.

Perrot-Maître, D., & Davis, P. (2001). *Case studies of markets and innovative financial mechanisms for water services from forests*. Washington: Forest Trends.

Rinaudo, J. D. (2008). Assessing the benefits of groundwater protection – A case study in the Rhine district (IVM website, 2008), http://www.ivm.vu.nl/en/. Accessed 31 Oct 2014.

Chapter 25
New York City's Watershed Agricultural Program

Carolyn Kousky

Abstract For over a century, New York City's unfiltered drinking water was characterized as the "champagne of tap water." In the 1980s, the quality of New York City's water was declining and the Environmental Protection Agency considered mandating filtration. The city looked for a way around this expense and began watershed management as an alternative. This chapter explores one component of the city's management approach, the Watershed Agricultural Program. A farmer-run, non-profit institution was established to develop and implement best management practices (BMPs) on farms whose owners voluntarily participate. The city finances the operating costs of the WAC and covers the costs to farmers of adopting BMPs. This case study demonstrates the viability of watershed management to protect source water quality and the potential of voluntary agreements to produce meaningful changes in agricultural practices. It also demonstrates the challenges of a city trying to influence land use outside its jurisdiction, the challenges of avoiding filtration in developed watersheds, and the role of regulations as forcing functions.

Keywords Water quality • New York City • Watershed agricultural program • Filtration avoidance

25.1 Introduction

New York City gets its drinking water from three watersheds that are grouped into two systems—the Cronton system and the Catskill-Delaware (Cat-Del) system. For over a century, New York City did not have to filter its water, as these watersheds provided it with what was characterized as the "champagne of tap water." In the 1980s, however, it became clear that the quality of New York City's water was declining, due in large measure to exurban development and an intensification of farming. This led the Environmental Protection Agency (EPA) to consider mandating filtration under the Surface Water Treatment Rule (SWTR) of the Safe Drinking

C. Kousky (✉)
Resources for the Future, 1616 P Street NW, Washington, DC 20036, USA
e-mail: kousky@rff.org

M. Lago et al. (eds.), *Use of Economic Instruments in Water Policy*,
Global Issues in Water Policy 14, DOI 10.1007/978-3-319-18287-2_25

Water Act. New York City preemptively decided to filter the Cronton system since the watershed was already highly developed. Building a filtration plant for the Cat-Del system (90 % of the city's water by volume) was estimated to cost US$4–8 billion in 1990, and today is estimated to cost roughly US$8–12 billion in up-front capital costs and US$350 million annually in operating costs (Appleton 2006). The city looked for a way around this huge expense and began to explore watershed management.

Following years of negotiation between the city, farmers in the watershed, the watershed towns, and the EPA, a Memorandum of Agreement (MOA) was signed in 1997. Under the agreement, New York City is financing a watershed agricultural program, purchasing critical lands, to some extent regulating land uses, and investing to upgrade infrastructure, such as septic systems and wastewater treatment plants. This is costing the city substantially less, around US$1.5 billion to date. In July, 2007 the city was granted 10 more years of filtration avoidance from the EPA as the program was improving water quality.

This chapter focuses specifically on the watershed agricultural program (WAP). Under an agreement with farmers, a farmer-run institution, the Watershed Agricultural Council (WAC), was established to develop and implement best management practices (BMPs) on farms whose owners voluntarily participate. The city is financing the operating costs and covering all the costs to farmers of adopting BMPs. In one sense, then, the WAP is an example of "payments for ecosystem services": the city is paying farmers for the service of improved source water quality. To be clear, however, the city's payments are not contingent upon clean water outcomes, but are payments for certain outputs shown to be correlated with water quality. This program has been analyzed by several academics and widely discussed in "gray" literature for its innovation and success.

The WAP program has saved the city billions of dollars in terms of the avoided costs of building a water filtration plant. It has also helped to preserve agriculture in the Cat-Del watershed. Farmers were having a difficult time making a viable living and development pressure was forcing some farmers to sell their land to developers. New York City's commitment to environmentally-friendly agriculture has helped the local economy. Water quality has improved and the city has been able to avoid filtration. That said, the issue cannot be considered "solved," as water quality threats remain, particularly from exurban development, which is not addressed in the farm program.

This case study is worth close examination for a couple reasons. First, at the time the policy was adopted, the idea of watershed management for water quality improvements and voluntary agreements to manage non-point source water pollution were considered unlikely to work in practice (Appleton 2006). The WAP proved that voluntary programs can work—at least when everyone has something to gain from the program and supports its overarching mission. The WAP, in conjunction with the city's other efforts in the watershed, has also demonstrated that watershed management can cost-effectively produce high-quality drinking water. This can be done in a working agricultural landscape, and thus, the program has also shown that the economic viability of farming and environmental protection need not be at odds.

That said, however, the New York City approach has not been widely replicated. It appears that there were a unique set of circumstances in New York City that may not be easily copied. For example, it is unlikely the city's approach would have worked in a watershed that was much more developed. The only other cities in the USA that have been able to avoid filtration through watershed management have a substantial portion or all of their watersheds in public ownership. Finally, it is unlikely New York City would have invested the time or resources in watershed management if not facing the unthinkably high costs of filtration if it failed to do so. A regulatory stick was needed.

25.2 Setting the Scene: Challenges, Opportunities, and EPIs

The New York City Department of Environmental Protection (DEP) oversees the delivery of safe drinking water to over nine million people in New York City and surrounding areas. New York City's drinking water is supplied by a system that includes 19 reservoirs and 3 controlled lakes (New York City Department of Environmental Protection 2011). Water from these storage areas is taken via aqueducts to terminal reservoirs, from which it can be piped into the city's distribution system. The watershed supplying the city with its drinking water is almost 3,219 km^2, located northwest of the city. It is divided informally into two sources. The Cronton River watershed, which provides 10 % of the city's drinking water, has experienced significant amounts of suburban development, and the decision was made early to filter its water. The Cat-Del system, a combination of the Catskill and Delaware watersheds, has received a filtration avoidance determination from EPA. This case study focuses on a program implemented by the city in the Cat-Del watershed.

The Cat-Del system is 404,686 ha. The Catskill part of the Cat-Del system is largely forested, with some farming and vacation homes (Appleton 2006). The Delaware River basin has rolling hills, some forested areas, and a significant amount of dairy farming (Appleton 2006). Around 20 % of the land area in the Cat-Del watershed is in the New York State Catskill Forest Preserve. Close to three-quarters of the watershed is forested, 85 % of which is privately owned (Watershed Agricultural Council 2012). Farming, centered in the valleys, is the second largest land use, making pollutant loadings from agriculture key to watershed management for potable drinking water. There are 40 towns with some land area in the Cat-Del watershed (National Research Council 2000). Urban extent, while small, still harms water quality through contaminated runoff, leaking septic systems, and wastewater treatment plants that have not been upgraded to high standards for discharging clean water. The city reservoirs have faced problems with eutrophication. Nitrogen and phosphorus can both lead to eutrophication, but phosphorous is of more concern because it is the limiting factor for algae growth (National Research Council 2000).

At the time of the 1997 agreement, New York City owned about 7 % of the watershed, state and conservation organizations owned another 20 %, and the rest was in private hands (Platt et al. 2000). Much of the watershed has steep slopes that are not well-suited for development (Hoffman 2008). Still, farming has occurred in this region for over 200 years (Bryant et al. 2008). While not the majority of jobs, farming is a large part of the community of the watershed (Isakson 2002). In 2000, there were 351 large farms (defined as having a gross annual salary of at least US$10,000 in operation in the Cat-Del watershed, 90 % of them being dairy farms with between 50 and 200 animals) (National Research Council 2000). Only 39 % of farmers in watershed counties, however, claim farming as their principal occupation (Isakson 2002).

The watershed is not an affluent region. A survey of watershed farmers found that of those with a gross annual income over US$10,000, a quarter earns less than US$20,000, although around 40 % do report earning over US$150,000 (Isakson 2002). Over half the jobs in the Cat-Del watershed are relatively low-wage service industries, with the average annual wage in 2003 being just under US$26,000 (Hoffman 2008). The Cat-Del watershed has a population that varies seasonally between 50,000 and 200,000 (National Research Council 2000). The Catskill system is estimated to have a population density of 24 people per square kilometer and the Delaware system is estimated to have a population density of 17 per square kilometer (Pires 2004). While rural and not heavily populated, the entire New York City watershed still has the highest population density of any unfiltered watershed in the USA (Finnegan 1997).

Drinking water in the USA is regulated through the Safe Drinking Water Act. In 1986, the EPA issued the Surface Water Treatment Rule (SWTR) in response to an amendment to that act. The SWTR sets forth requirements that water supply systems must meet in order to obtain a Filtration Avoidance Determination (FAD), which allows them to forgo filtration of drinking water. These include: monitoring of fecal coliform and total coliform; meeting certain testing requirements and concentration levels in the source water; providing adequate disinfection; meeting site specific criteria for the presence of certain viruses, total coliforms, and disinfectant byproducts; meeting certain turbidity levels; developing and implementing a watershed control program; undertaking annual third-party inspections; and ensuring that the system is never the source of a waterborne disease outbreak.

In the USA, only a few major cities have been able to avoid filtration of their drinking water. These include San Francisco, California; Seattle, Washington; and Portland, Oregon where 100 % of the land is in public ownership, and Boston, Massachusetts where 53 % is in public ownership (United Nations Development Programme et al. 2000). In contrast, at the time of the MOA, New York City owned less than 7 % of the land (and most of this was the land under the reservoirs), and 20 % was owned by the state. New York City thus faced a much greater challenge of having to find a way to manage land uses on private lands.

25.3 The Watershed Agricultural Program in Action

When the EPA began to consider filtration for New York City, the DEP looked for a way to avoid the enormous expense. Filtering the larger system would have easily doubled water and sewer rates for New York residents (Appleton 2006). As a quirk of history, New York City has the authority to directly regulate the watersheds from which it obtains its drinking water, subject to oversight from the New York State Department of Health (Finnegan 1997). In 1990, the DEP released draft watershed regulations and in its 1993 report to the EPA for a FAD, New York City based its watershed protection approach on land acquisition. Watershed residents reacted with hostility to both regulations and land acquisition. They were concerned about a curtailing of economic development, a drop in property values, and a decrease in tax revenues for local governments (Platt et al. 2000). It was clear that the city would be unable to move forward with regulations or the particular approach to land acquisition it initially developed.

This case focuses on how the city went about addressing pollution from agriculture. Soon after the release of the 1990 regulations, a local farmer invited the DEP to his land to demonstrate how economically destructive the city's regulations would be to farmers; DEP Commissioner Al Appleton accepted and from this visit realized that enforcing stringent regulations on farmers was not going to be the answer to the city's problem (Appleton 2006). In addition, it would be impossible for the city to monitor non-point source pollution from farms, and so without landowner buy-in, the regulations would not be sufficient to protect water quality.

Following the DEP visit, the Deputy Commissioner of the New York State Department of Agriculture suggested that the farmers and the city begin a process of mutual education (Appleton 2006). An Ad Hoc Taskforce on Agriculture was created, chaired by the DEP Commissioner and facilitated by Dennis Rapp from the New York State Department of Agriculture (Isakson 2002). At the end of 1991, the Taskforce produced an agreement between farmers and the city. Under the agreement, the farmers would be held harmless from regulations, except for willful polluters, and in exchange, a new locally controlled non-profit organization was established, the WAC, to implement Whole Farm Plans (WFPs) on watershed farms financed by the city.

This approach is a variant of payments for ecosystem services policies, where the beneficiary (here, New York City) of an ecosystem service (here, source water quality) pays providers (here, farmers) for producing that service. The New York City case is slightly different than a pure payments scheme in two respects. First, the city is not paying directly for the service of water quality, but for actions it believes to be contributors to water quality. Second, the payments are part of a larger institutional structure providing assistance to the farmers.

Whole Farm Planning is not a concept unique to the New York City policy. The idea behind it is to assess farm operator goals and conditions, as well as all off-farm impacts from farming activities, and then develop a holistic plan to improve environmental impacts through the adoption of BMPs, while safeguarding the farmer's goals (Ervin and Smith 1996). Plans are tailored to individual farms. Some examples of BMPs include stream bank fencing, developing a nutrient management plan, improving manure storage, developing animal trails, precision feeding, and installing a trough or tank.

The WAC is farmer-run, operated with financing from the city. Cornell University provides research support. A sticking point in the agreement was whether participation by the farmers would be voluntary. The DEP was concerned that voluntary programs had historically been a failure. The DEP Commissioner and a Delaware County farmer finally came to a resolution: the program would be voluntary for individual farmers, but the WAC would guarantee a participation rate of 85 % within 5 years, and if not attained, the city could revert back to traditional regulation (Appleton 2006). This agreement was signed in 1991 and in the same year the city received its first FAD from the EPA, granted for 2 years. It received another 3 year avoidance at the end of 1993. The WAP first focused on establishing WFPs on large farms and in 2009, New York City extended the WAP to small farms, as well.

New York City received another 5 year FAD in 2002 and a 10-year FAD in 2007. Both required updating and improving its watershed protection efforts. The 2002 FAD included commitment to build an ultraviolet light disinfection plant for the Cat-Del system. The 2007 FAD included waterfowl management, land acquisition, land management, a watershed forestry program, stream management, riparian buffer protection, wetlands protection, Croton watershed management, Kensico water quality control, turbidity control, infrastructure upgrades (e.g. for septic systems, wastewater treatment plants), as well as on-going efforts of the WAP discussed in this case study (New York City Department of Environmental Protection 2011). The continued issuance of the FADs from the EPA is a clear indication that the approach to watershed protection taken by the city is working at meeting drinking water quality standards.

25.3.1 The EPI Contribution

25.3.1.1 Environmental Outcomes

Perhaps surprisingly, New York City established no explicit environmental goals for the WAP. Instead, all success metrics were based on observable actions, such as number of farms enrolled. These outputs need not be perfectly correlated with the outcome of surface water quality. On output metrics, the program has been a huge success. Since 1992, the WAP has established WFPs on over 416 farms (New York City Department of Environmental Protection 2011). Around 95 % of all the large commercial farms in the Cat-Del watershed have WFPs (New York City Department of Environmental Protection 2011). In 2010, the WAP met a goal of the FAD to have 90 % of participating large farms meet "substantially implemented status." In addition, the WAP has secured over 7,284 ha of conservation easements on watershed farms (New York City Department of Environmental Protection 2011).

In a review, the National Research Council (2000) noted that these output metrics, however, do not give an indication of the total impact on water quality, and that monitoring is needed to achieve this. The report recommended that phosphorus load reduction goals be established and farm-scale monitoring undertaken. At the time the city began its watershed management efforts, all the reservoirs were receiving excess phosphorus and phosphorus levels were increasing, with loads exceeding an amount that causes eutrophication (National Research Council 2000). Even without farm specific monitoring of soil and water quality, planning teams do monitor all farms for maintenance of the BMPs (National Research Council 2000).

At a system level, the DEP continually monitors the state of its water. The Kensico Reservior, which is an endpoint for water from the Cat-Del system, has consistently met all turbidity and fecal coliform standards established under the SWTR. Moving further back in the system, the Cannonsville Reservoir has not been listed as phosphorus-restricted[1] since 2002 and this is attributed to a combination of the WAP and an upgrading of wastewater treatment plants and septic systems (Bureau of Water Supply 2006). More recently, research based on monitoring in the Cannonsville Reservoir conservatively indicates, after accounting for reductions from other sources, that the WAP decreased dissolved phosphorus by 50 % and decreased total phosphorus by 17 % when the period 2000–2004 is compared with the period 1992–1999 (Bryant et al. 2008).

In addition, over the years there has been research quantifying the impact of BMPs on water quality (e.g. James et al. 2007; Bishop et al. 2007). Research done in the 1970s and 1980s in the Cat-Del watershed focused on the source of phosphorus loadings and the impact of BMPs (National Research Council 2000). Since then, USDA Agricultural Research Service scientists and Cornell University scientists have been working together to document the impacts of selected BPMs on reducing phosphorus loadings to surface waters (Bryant et al. 2008). Cornell scientists have developed models that can be used as planning tools for the WAP (National Research Council 2000). A total of 10 % of the WAC budget in the first phase was devoted to research and monitoring (Willett and Porter 2001).

25.3.1.2 Economic Outcomes

The costs of filtration, as mentioned in the introduction, would be large: US$8–12 billion in up-front costs and US$350 million in annual operations. In the 1980s and early 1990s, due to an increase in capital expenditures, water and sewer rates were increasing at around 15 % a year and there was huge outcry (Appleton 2006). Another period of large increases was not going to be tenable and the DEP thus chose to pursue watershed management. This was the least cost solution to the drinking water quality regulations the city faced from EPA. Initially, the city wanted to directly regulate land use in the watershed. This would have been cheaper for the city, but was not politically feasible, due to intense outrage from the regulated communities. New York City first applied for a filtration waiver in 1991. Between that first application and 2010, the city spent over US$1.5 billion on source water protection in the Catskill and Delaware watersheds, but note these also included costs of other activities beyond the WAC not addressed in this case study (New York City

[1] Reservoirs that do not meet state phosphorus guidance values are labeled phosphorus-restricted and stricter regulations apply.

Department of Environmental Protection 2010). This is substantially less than the cost of a filtration plant.

As far as I am aware, no analysis has been done to monetize the value of the water quality benefits that New York City receives or the willingness-to-pay of water users in the city for source water quality. Since the city faced a regulatory trade-off between filtration and watershed management, it was clear from a cost-effectiveness standpoint that watershed management was preferred, but it is impossible to say if, at the margin, an extra dollar from the city spent on water quality improvements is worth the cost.

25.3.1.3 Distributional Effects and Social Equity

Socio-economic data is not collected at a watershed level, making inferences difficult at that scale. Around the year 2000, it was estimated that roughly 20 % of watershed residents lived in poverty (Isakson 2002). In the early 1990s, 30 % of the population of Delaware County, a large portion of which is in the watershed, was on welfare and rural poverty was widespread (Appleton 2006). It is unclear how much New York City's program has impacted such measures in the watershed. One study by Hoffman (2008) isolated employers in the Cat-Del watershed and analyzed trends between 1990 and 2003. Hoffman found that New York City's watershed protection programs had not reduced investment in the watershed and might have generated net gains for employers. Of course, this analysis is on the entire suite of activities being undertaken by the city.

The WAP programs are designed to not impose any costs on participating farmers, as the installation of all BMPs is paid for by the city. The farmers also receive numerous co-benefits from participation in the WAP. Many of the BMPs improve herd health; for example, providing cattle with troughs to drink from instead of streams limits their exposure to infections (James et al. 2005). New barnyards have anecdotally reduced hoof problems and rates of mastitis in cow udders (Isakson 2002). Other BPMs save money—for instance, increasing the efficiency of nutrient use allows farmers to buy less fertilizer. Some also save the farmer time, and often time from the particularly unpleasant task of handling manure (Appleton 2002). By participating in the WAP, farmers have also been able to preserve their autonomy, build social capital, and have a voice in the future direction of the watershed (Isakson 2002). Finally, the WAP helps them avoid the costs of other regulations. Not only are they exempted from New York City's regulations, but they are better positioned to meet federal regulations, such as a 2009 Animal Feeding Law requiring WFPs (Isakson 2002). Farmers that participate in the WAP are also eligible for other federal programs that can be financially beneficial, such as the Conservation Reserve Enhancement Program.[2]

[2] There are capital costs to joining the federal program and the program covers 40 % of those costs. All costs are paid in full by the city for WAP participants so when they join, they get to pocket the 40 % cost share as a bonus payment (Isakson 2002).

It appears that overall the WAP has improved the livelihood of participating watershed farmers. The only possible negative impact is a concern by some that the WAP might lower real estate values, but local real estate agents say there has been no impact on resale values (Isakson 2002). Despite clear benefits overall, these benefits are not distributed evenly among farmers. The largest farms benefit the most from WAP programs. This is partly due to the initial focus on large farms, and also because funding is dependent on the size of the farm and the level of production of key pollutants (Isakson 2002). This can be seen in farmer surveys. Two thirds of farmers earning over US$100,000 said the program helped their economic well-being but for farmers earning under US$20,000, just over a quarter claimed economic benefits from the program (Isakson 2002). Those with low incomes from farming, however, may have supplemental income and so farm income does not necessarily correlate highly with total household income. The WAP program is also targeted at dairy farmers, which some other farmers resent (Isakson 2002). The actual number of farmers that have experienced significant benefits may thus be a more limited subset of watershed farmers.

While benefitting farmers economically, it also seems that the magnitude of benefits may not be enough to actually preserve farming as a viable profession. In the Cannonsville Reservoir watershed, for example, there has been a decline in agriculture (Bryant et al. 2008). The WAC also notes in its Strategic Plan for 2011–2014 that farming is declining in the region due to diminishing returns from agriculture, with the number and size of farms falling (Watershed Agricultural Council 2011). The WAP program is focused on improving water quality, not maintaining farming per se, although the city has noted that exurban development could be worse for water quality.

25.3.2 The EPI Design

25.3.2.1 Institutional Set-Up

The WAC was a newly created non-profit institution. It was an innovative solution created by Commissioner Appleton and the farmers (Appleton 2002). The farmers insisted that the program be farmer-run so that they could maintain autonomy. The board of directors is composed almost exclusively of local farmers, with one representative from the DEP. The WAC has a dual mission of improving surface water quality and supporting the economic viability of farms. Participation in the WAC's programs is voluntary for farmers in the watershed, but the WAC agreed to ensure a participation rate of at least 85 %, which they have exceeded. The WAC attributes the success to "kitchen meetings," where a farmer already participating invites his friends and neighbors to have WAC staff explain the benefits of enrolling (almost 100 % of farmers who attended such kitchen meetings ultimately enrolled); farmers trust each other and distrust city officials (Isakson 2002). As mentioned below, farmers were willing to join largely to improve their farm, be better stewards, and to be held harmless from city regulations (Isakson 2002). While New York City provides the majority of the funding, the WAC also receives some federal money and technical assistance from Cornell University, County Soil and Water Conservation Districts, and the USDA Natural Resource Conservation Service. Their website lists dozens of organizations they partner with to achieve their goals.

25.3.2.2 Transaction Costs and Design

The transaction costs of establishing the New York City WAP program include the costs of initial negotiations with the farmers. Estimates of the man-hours for these negotiations are unavailable, but anecdotal evidence suggests they were significant. The DEP took trips to the watershed to meet with farmers and observe how their proposed regulations may impact the economics of farming operations. The entire negation period occurred over about 3 years (Appleton 2002). The costs of the WAP also include the costs of directly implementing the BMPs, the costs of developing the whole farm plans, the administration of the WAP, and the research that supports its activities. Farms are monitored annually for compliance. The WAC currently employees 19 people in its main office (although a few of these work on the forestry program, not discussed in this case study). The 2010 Annual Report for the WAC shows that program administration accounted for just under 12 % of total expenditures the previous fiscal year. The city has found these costs, along with those of its other watershed programs, to be less than filtration.

25.3.2.3 Implementation

The WAC was the outcome of negotiations between the city and the farmers. There was long-standing animosity towards New York City from watershed residents who saw the history of the establishment of New York City's drinking water system as one in which the city repeatedly hurt, abused, and alienated them (Finnegan 1997). All participants note that a key to coming to an agreement was developing a clear vision that all parties could agree on. This was that drinking water protection and economic returns from farming could be consistent goals (Smith and Porter 2010). Participants agreed on this through the sessions of mutual education undertaken by the city and watershed farmers. The discussions made each side realize the legitimate concerns of the other and the outcome was the starting point that agriculture could be "watershed friendly" (Appleton 2006). Also essential to implementation of the WAP was the fact that pollution reduction would be tailored to each farm to not jeopardize other management goals.

25.4 Conclusions

The case of New York City's WAP has been told many times, in part because it overturned two commonly held assumptions about environmental policy in the USA: that voluntary programs do not work and that watershed management could not generate high-quality drinking water (Appleton 2006). There is not a counterfactual as to the state of water quality without the city's investments, but the research to date suggests their actions are having a positive impact on water quality. The New York City case is often cited as proof in the concept of ecosystem services—that natural systems do generate economic value and that with proper policies, this can be captured, harmonizing ecological and economic objectives.

The process New York City went through to achieve this holds lessons for other localities seeking to harness the economic benefits provided by natural systems. First, some forms of agriculture and the provision of potable drinking water need not be at odds with each other. In working landscapes, it is possible to identify areas of overlap between the goals of environmental protection and economic development. As Al Appleton, the DEP Commissioner at the time reflected, "the ecosystem must be seen as including both its natural and human resources. One cannot be sacrificed to the other," (Appleton 2002). Instead of imposing one-size-fits-all regulations, if water quality improvements can be tailored to the farm, they can be made compatible with management goals.

That said, the second lesson from New York City is that cities cannot expect watershed residents to bear the costs of maintaining source water quality. On-farm investments or changes in agricultural practices required to reduce pollutant loadings are often expensive and especially in communities where farmers are struggling economically, the beneficiaries of these changes must be prepared to help cover the costs. As this case demonstrates, however, these costs may sometimes be much less than the cost of alternatives.

The WAP program also highlights the importance of well-structured dialogue and negotiations. The city and farmers were able to work together to identify common ground and solutions to both groups' problems. This is not easy, however, and should not be under-appreciated. In the end, however, it produced a program where, for farmers struggling economically, not only are they relieved of having to bear the costs of improving the city's drinking water (something argued as highly inequitable), but they are receiving aid for improvements that provide them with other benefits.

A key contextual reason New York City was able to engage in watershed management, however, was because the watersheds from which it obtains 90 % of its drinking water had not yet been highly developed. Perhaps one reason there has not been much emulation of the city's approach is because localities that obtain water from already developed watersheds have no other option but to filter their water.

Once they are expending the money to do this, they may see little added value to also investing in watershed protection. It was EPA's threat of forcing filtration, and thus a huge cost on the city, that pushed it into watershed management.

Finally, as is often the case with innovative policies, key individuals proved decisive in establishing the WAP. These individuals are often referred to in academic literature as "public entrepreneurs." For instance, Al Appleton, the DEP Commissioner, recounts how it was a conversation between him and a watershed farmer, who was also a leader in the community, which found the compromise of making participation in the WAP voluntary for the farmers, while requiring a minimum level of participation.

This case is by no means "over" (see Soll 2013 for an overview of the history and most recent policies). New York City will need to continually monitor and invest in watershed management efforts to control pollutants and excess nutrient loadings. While the WAP is fairly well established, such that agriculture is no longer a leading threat to the city's water, other threats remain. In particular, exurban development continues to be a problem, particularly as economic hardship continues in the area and people sell land to developers building second homes. This often creates more pollutant problems than agriculture. It may be more cost-effective for the city to buy out some farms to preserve as open space and prevent exurban development, but such land acquisition has met with objections from watershed communities.

References

Appleton, A. F. (2002). *How New York City used an ecosystem services strategy carried out through an urban-rural partnership to preserve the pristine quality of its drinking water and save billions of dollars*. New York City: Forest Trends.

Appleton, A. F. (2006). Regional landscape preservation and the New York City watershed protection program: Some reflections. In K. S. Goldfeld (Ed.), *The race for space: The politics and economics of state open space programs* (pp. 39–76). Princeton: The Policy Research Institute for the Region, Woodrow Wilson School of Public and International Affairs, Princeton University.

Bishop, P. L., Lojpersberger, J. L., et al. (2007). New York City watershed: An eleven-year study of the effectiveness of agricultural BMPs in reducing farm pollutant losses. *NWQEP Notes, 125*(June), 1–13.

Bryant, R. B., Veith, T. L., et al. (2008). Cannonsville reservoir and town brook watersheds: Documenting conservation efforts to protect New York City's drinking water. *Journal of Soil and Water Conservation, 63*(6), 339–344.

Bureau of Water Supply. (2006). *2006 long-term watershed protection program*. New York City: New York City Department of Environmental Protection.

Ervin, D. E., & Smith, K. R. (1996). *What it takes to "get to yes" for whole farm planning policy*. Greenbelt: Henry A. Wallace Institute for Alternative Agriculture.

Finnegan, M. C. (1997). New York City's watershed agreement: A lesson in sharing responsibility. *Pace Environmental Law Review, 14*, 577–644. Paper 315.

Hoffman, J. (2008). Watershed shift: Collaboration and employers in the New York City Catskill/Delaware watershed from 1990–2003. *Ecological Economics, 68*, 141–161.

Isakson, R. S. (2002). *Payments for environmental services in the catskills: A socio-economic analysis of the agricultural strategy in New York City's watershed management plan*. Payment for Environmental Services in the Americas Project sponsored by the FORD Foundation and conducted by Fundación PRISMA.

James, E., Kleinman, P., et al. (2005). Keep your cows out of streams and ponds. *Hoards Dairyman, 150*, 472.

James, E., Kleinman, P., et al. (2007). Phosphorus contributions from pastured dairy cattle to streams of the Cannonsville watershed, New York. *Journal of Soil and Water Conservation, 62*(1), 40–47.

National Research Council. (2000). *Watershed management for potable water supply*. Washington, DC: National Academy Press.

New York City Department of Environmental Protection. (2010). *Filtration avoidance annual report for the period January 1 through December 31, 2009*. New York City: New York City Department of Environmental Protection.

New York City Department of Environmental Protection. (2011). *2009 watershed water quality annual report; updated May 2011*. New York City: New York City Department of Environmental Protection.

Pires, M. (2004). Watershed protection for a world city: The case of New York. *Land Use Policy, 21*, 161–175.

Platt, R. H., Barton, P. K., et al. (2000). A full, clean glass? Managing New York City's Watersheds. *Environment, 42*, 8–20. June.

Smith, L. E. D., & Porter, K. S. (2010). Management of catchments for the protection of water resources: Drawing on the New York City watershed experience. *Regional Environmental Change, 10*, 311–326.

Soll, D. (2013). *Empire of water: An environmental and political history of The New York City water supply*. Ithaca: Cornell University Press.

United Nations Environment Programme et al. (2000). *World resources 2000–2001: People and ecosystems: The fraying web of life*. Washington, DC: World Resources Institute.

Watershed Agricultural Council. (2011). *Strategic plan 2011–2014*. Walton: Watershed Agricultural Council.

Watershed Agricultural Council. (2012). *Croton & Catskill/Delaware watersheds*. http://www.nycwatershed.org/aw_watersheds.html. Accessed 4 Jan 2012.

Willett, I. R., & Porter, K. S. (2001). *Watershed management for water quality improvement: The role of agricultural research*. Canberra: Australian Centre for International Agricultural Research.

Chapter 26
Voluntary Agreement for River Regime Restoration Services in the Ebro River Basin (Spain)

Carlos M. Gómez, Gonzalo Delacámara, C. Dionisio Pérez-Blanco, and Marta Rodríguez

Abstract The construction of the Mequinenza and Ribarroja dams back in the 1960s modified the hydrology and changed the physical and environmental conditions of the Lower Ebro River in northeastern Spain. These conditions have stirred the uncontrolled proliferation of macrophytes, which have become a relevant concern in the area since 2000. Among other environmental and economic impacts, macrophytes threaten hydroelectric power infrastructures, increasing operating costs and reducing the productivity of power-generating plants. Macrophyte blooms thus became the catalyser for collaboration between the hydropower operator and the Ebro River Basin Authority, within a larger consortium with academic experts on floods and sediment flows, to deliver controlled water floods (flushing flows). The economic instrument assessed in this chapter consists of the voluntary acceptance, based upon public and private incentives, to deliver a set of pulses or artificial floods designed ad hoc for the partial restoration of the river regime in the Lower Ebro. Since 2003 and with the exception of 2004 and 2005 (dry years) and also 2008 and the spring of 2009 (natural floods), flushing flows have been regularly performed twice a year (at the end of spring and autumn) and have resulted in macrophyte removal rates as high as 95 % in areas close to the dam.

C.M. Gómez
Departamento de Economía, Universidad de Alcalá and IMDEA-Water Foundation, c/ Plaza de la Victoria s/n, 28803 Alcalá de Henares (Madrid), Spain
e-mail: mario.gomez@uah.es

G. Delacámara • M. Rodríguez
Madrid Institute for Advanced Studies in Water (IMDEA-Water Foundation), Punto Net 4, floor 1, Edificio ZYE, Parque Científico-Tecnológico de la Universidad de Alcalá, 28805 Alcalá de Henares (Madrid), Spain
e-mail: gonzalo.delacamara@imdea.org; marta.rodriguez@imdea.org

C.D. Pérez-Blanco (✉)
Centro Euro-Mediterraneo sui Cambiamenti Climatici, Fondazione Eni Enrico Mattei (FEEM), Isola di San Giorgio Maggiore, 30124 Venice, Italy
e-mail: dionisio.perez@feem.it

M. Lago et al. (eds.), *Use of Economic Instruments in Water Policy*, Global Issues in Water Policy 14, DOI 10.1007/978-3-319-18287-2_26

Keywords Voluntary agreements • Flushing flows • River restoration • Hydropower • Lower Ebro River

26.1 Introduction

The dams of Mequinenza (1,533 hm^3, 7,540 ha of reservoir, 79 m high) and Ribarroja (218 hm^3, 2,029 ha), built back in the 1960s (1966 and 1969, respectively), altered the hydrology of the Lower Ebro River in northeastern Spain (85,914 km^2, 426 m^3/s of average flow). Although the river still experiences natural floods, its physical and environmental conditions have changed within the last decades. Changes in the flow regime (particularly through reduced flood magnitude) and diminished sediment supply have resulted in a series of morphological effects, including re-vegetation of formerly active areas of the river channel (Batalla et al. 2006), local incision and riverbed armouring (Vericat et al. 2006). These conditions have stirred the massive colonisation of macrophytes, which have become a relevant concern in the area since 2000.

Macrophytes are visible algae and other flora species that are rooted in shallow waters with vegetative parts emerging above the water surface (Haslam and Wolseley 2014). In lakes, they are considered as eco-indicators; in heavily engineered rivers, its presence is an evidence of degradation, rather than of good ecological status, and may result in negative impacts on a number of stakeholders (Gómez et al. 2014). In the case of hydropower operators, macrophytes threaten river infrastructures, increasing operating costs and reducing the productivity of power stations. Macrophyte proliferation started after the modification of the flow regime that followed the construction of the dam complex (comprising the above-mentioned Mequinenza and Ribarroja dams and a smaller one, Flix: 11 hm^3), and experiences periodic blooms during intense droughts and low flow conditions for long periods of time (Montesinos et al. 2009).

Following two dry years corresponding to one of the most remarkable macrophyte blooms ever (ERBA 2013), the hydropower operator (ENDESA) and the Ebro River Basin Authority (ERBA) initiated a collaboration to deliver controlled water floods (flushing flows). Co-ordination within a larger consortium including academic experts on floods and sediment flows started in 2002 and flushing flows were implemented from 2003 on. Since then and with the exception of 2004 and 2005 (dry years), and also 2008 and the spring of 2009 (natural floods), flushing flows have been regularly performed twice a year (at the end of spring and autumn) and have resulted in macrophyte removal rates as high as 95 % in areas close to the dam (Batalla and Vericat 2009).

Macrophyte removal was not the main objective, indeed, or at least it was not so from a public perspective; yet, it proved to be the catalyst for agreement and reconciliation of public good concerns (river restoration) and private interests. Initially, the private interest of the hydropower operator claimed the attention to mainly focus on the capacity of the artificial floods to remove the macrophytes in the vicinity of the power generation facilities (which are actually located far away from the river mouth). The good news is that the hydropower operator was willing to consider water flow patterns that were not only designed to maximize financial profits within the range of prevailing regulations but also to deliver some improvements in the ecology of the river system, paving the way for a collaborative agreement facilitated by the remarkable research effort made in the area.

26.2 Setting the Scene: Challenges, Opportunities and Responses via Voluntary Agreements

The Lower Ebro River in northeastern Spain is located between the Mequinenza-Ribarroja-Flix Dam Complex (MRFDC) and the Ebro River outlet to the Mediterranean Sea (see Fig. 26.1). Administratively, the catchment roughly corresponds to the *vegueria* (county) of Tierras del Ebro in southernmost Catalonia (NUTS2). Tierras del Ebro is an agricultural (64.32 % of land use and 9.3 % of 2010 GDP) and depopulated area (188,878 inhabitants in 2013, 3,340.87 km^2 and a

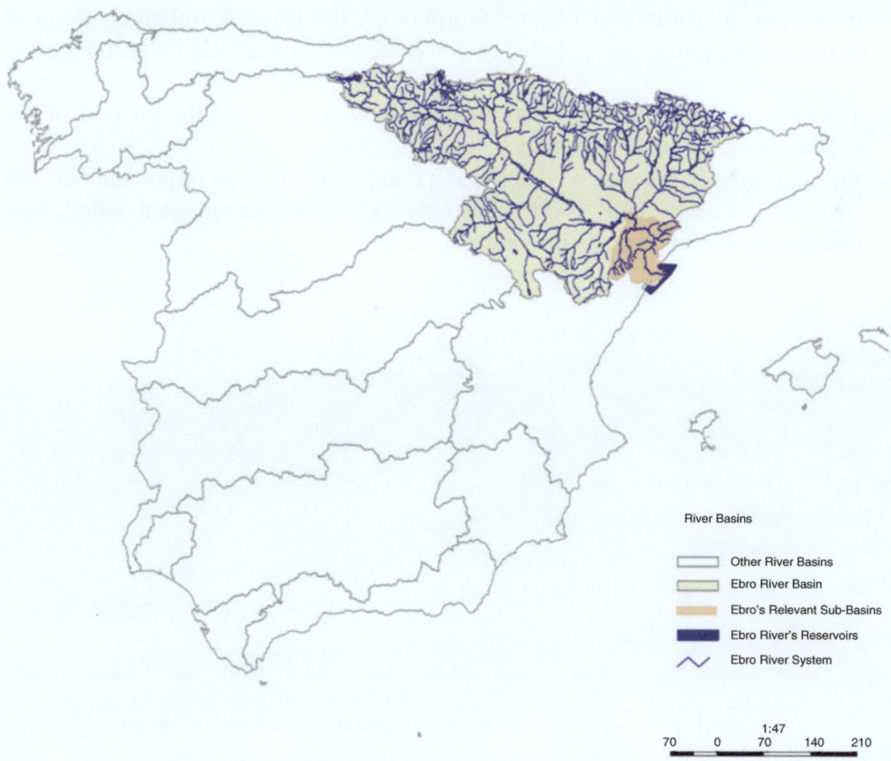

Fig. 26.1 Lower Ebro River in NE Spain (Source: Own elaboration from IGN (2014))

population density of 56.5 inhabitants/km^2 well below the Spanish average of 93 inhabitants/km^2) (Generalitat de Catalunya 2011; IDESCAT n.d.; INE 2014).

Despite the growing trends observed in water demand and the average reduction of 25 % in runoff due to upstream afforestation processes, water is still relatively abundant and droughts are rare (ERBA 2008, 2013). The main environmental concern in the area is related to the impoverished ecological status that resulted from the alteration of the river's hydrology (see Fig. 26.2) and, subsequently, the channel morphology after the construction of the MRFDC. Anytime any day the water flow in the Lower Ebro depends on decisions made by the hydropower utility, which tries to make the best out of its power generation capacity, and every month total water flowing down the river increasingly depends on the needs of the irrigation sector than on the priorities of the hydropower sector. Literature provides abundant evidence on how these private decisions may largely differ from wider societal goals (i.e., environmental) (Glenn et al. 2008; Lessard et al. 2013; Truong 2012).

The MRFDC comprises the dams of Mequinenza, Ribarroja, and Flix privately managed by the hydropower operator ENDESA S.A. (shared in 92 % by the Italian company ENEL Energy Europe S.L.). The most important ecological impacts of the MRFDC include (Batalla et al. 2006; Batalla and Vericat 2009; Vericat et al. 2006):

- The attenuation of flood frequency and magnitude, which are the energy source for keeping an active river channel morphology (for example, relatively frequent floods, with a return period between 2 and 25 years, have been reduced by 25 % in average);
- The reduction of the river's sediment load, which implies the erosion of the coarser fractions in the channel;
- The alteration of the river's ecology, as a compound effect of impoundment, low frequency of bed moving floods, slow moving waters, fine sediment deficit, high temperatures, and excess nutrient load.

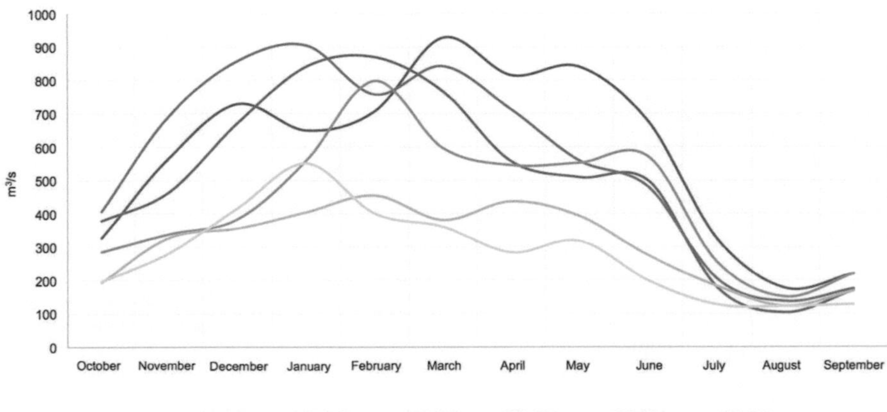

Fig. 26.2 Evolution of the monthly river flow in the Ebro (at Tortosa, river mouth) (1912–2004) (Source: Own elaboration from ERBA historical data)

Although the river still experiences natural floods and the impact of regulation is much smaller than that found in commensurable large rivers such as the Sacramento and the San Joaquin in California (Kondolf and Batalla 2005), and even in some of its main tributaries (Ollero 2010), the river's physical and environmental conditions have remarkably changed in the last decades. This new set of environmental conditions, together with similar changes in the upstream main tributaries, seems to explain the uncontrolled proliferation of macrophytes in the Lower Ebro River channel (Montesinos et al. 2009).

Macrophyte sprawl has taken place in formerly active channel areas in the Lower Ebro, and has caused a number of problems for a wide range of stakeholders, including irrigation pumping stations, hydropower plants and a nuclear power plant (Ascó),[1] (Batalla et al. 2008). Besides, evidence shows that macrophytes have been creating problems in water intakes and navigation (ERBA 2008). Macrophytes are also seen as the main cause of a plague of black flies (*Simulium spp*.), which became a major public-health threat, especially during the summer, since they transmit diseases such as *onchocerciasis* (river blindness) (Gómez et al. 2011; WHO 2014). Competition for space and resources resulting from the stabilisation of dense macrophyte stands affects the biology of the river ecosystem in many different ways (Batalla and Vericat 2009).

Experimental flushing flow releases (Batalla and Vericat 2009) have been undertaken with the main aim of controlling macrophyte biomass growth downstream the MRFDC. The design of these artificial floods was based on the sediment entrainment method (Kondolf and Wilcock 1996), that mobilises an active layer equal to the maximum root depth of algae, and has been continually informed by sediment attributes and macrophyte removal at representative sites throughout the river channel.

It is important to note that the design of flushing flows is constrained by a number of factors such as the operation of the hydropower dam system (water storage and power output), water availability in the second reservoir (Ribarroja, from where the flushing flows are released), and the risk of flood in riparian human settlements (Batalla et al. 2008). Higher demand of electricity in winter limits the opportunity for flushing flows, and this explains the time sequence of these artificial floods (see above).

The challenge thus becomes how to combine or merge private and public interests within the above-mentioned constraints so as to achieve a meaningful impact of flushing flows on macrophytes removal and on the overall ecological status of the Lower Ebro within a reasonable cost. The economic instrument assessed in this chapter consists of the voluntary acceptance, based upon public and private incentives, to deliver a set of pulses or artificial floods (flushing flows) designed ad hoc for the partial restoration of the river regime in the Lower Ebro.

Even a mild alteration in the river hydrology would imply changes in the overall amount of water delivered and in the river regime throughout time with major

[1] For instance, a power decrease at Ascó II (one of the reactors of the a downstream nuclear power station), distorted measures at Ascó and Tortosa hydrographical gauging stations (ERBA 2010).

consequences on the value of water for power generation. Taking this into account and considering that new operation rules would mean a shift in prevailing water use rights, the EPI was designed as a reciprocated collaboration scheme. Indeed, since its early stages it involved the possibility of a side payment to the energy operators both to compensate for additional opportunity costs and for the delivery of additional ecosystem services resulting from the river restoration scheme.

26.3 Voluntary Agreements (VAs) in Action

Flushing flows are delivered by the private operator of the three hydropower dams (ENDESA S.A.). Besides the hydropower operator and the basin district authority, since 2002 the development of this instrument involves the scientific community. In 2008, a working group on macrophytes in the Ebro River Basin was created (ERBA 2013). ENDESA's Sustainability Report (ENDESA 2013) records design, monitoring, and implementation studies in the period 2005–2010. The outcomes of this collaboration can be synthesised as follows:

- Flushing flows are overall an effective way of removing macrophytes (although maybe insufficient to control their development) and are far from being incompatible with hydropower production. As a matter of fact they can actually yield positive impacts for energy generation and water pumping for irrigation by mitigating the clogging of water intakes (critical, for instance, for the cooling of the two nuclear fission reactors at Ascó) (1,032.5 and 1,027.2 MW respectively).
- Socio-economic benefits from the restoration of the natural river regime stem from pest prevention cost savings, significant improvements in the efficient use of water, maintenance of water infrastructures, risk abatement, and natural river habitat enhancement (both in-stream and riparian). In addition, opportunity costs of flushing flows consist of production losses in the economic uses of water diverted for river restoration (specifically in hydropower generation). The balance between opportunity costs and environmental benefits determines the economic feasibility of these experimental floods, which need to be regularly re-assessed and re-designed.

26.3.1 Contribution of This Voluntary Agreement

26.3.1.1 Environmental Outcomes

Since 2002, a series of controlled floods have been regularly implemented in the Lower Ebro. At the outset, as above, this was only for experimental purposes, supported by an ambitious research program to design floods and to monitor and maximize its effectiveness; more recently as part of the Ebro River Basin management planning process (ERBA 2013). These efforts were integrated in the design of the river plan and finished with the agreement to deliver two controlled floods every year, deliberately defined to maximize macrophyte removal rates and implying the delivery of more than 36 hm^3 along 16 h in each controlled flood.

The efficiency of flushing flows in macrophytes removal depends on the amount of macrophytes, distance to the dams, natural flow variability and macrophyte life cycle. For example, removal rates are considerably higher during autumn than during spring, when macrophytes are growing and stalks are stronger (according to the macrophytes life cycle, macrophytes mass reaches its peak in summer: URS-España 2010). Also, effects are better in the vicinity of big dams and hastily decrease with only marginal changes in the river estuary (Batalla et al. 2008).

Artificial floods have proved themselves a useful means to maintain the river ecosystem, with the highest macrophyte concentration after years where flushing flows were not implemented (ERBA 2008). However, removal rates have been reduced both in intensity and extension since 2002, demonstrating that the design of the flushing flows being assessed are not enough to keep macrophytes under control in the long term.

Paradoxically, while the success in improving the chemical status of the river within these last 10 years is a fact, this seems to have driven an increase in the potential for the proliferation of macrophytes and boosted its rate of renewal after every controlled flood (URS-España 2010). New research efforts are currently being undertaken to shed light on the limits of better-designed or more regular controlled floods. The provisional balance, according to the experts involved in the field, indicates that designed floods help in river restoration but are not sufficient to offset a number of hydromorphological changes affecting the Lower Ebro. To deliver its expected outcome artificial floods should be part of a strategy involving at least better-designed environmental flows in order to make the ecology of the river less appropriate for typical lake standing species of flora.

On the other hand, flushing flows are tested means to enhance biological productivity of the physical habitat, to entrain and convey sediments for the restoration of the river channel, to remove pollution loads and to improve water quality, to control salt intrusion and to supply sediments to the delta and transitional waters (ecotones) (Batalla and Vericat 2009).

26.3.1.2 Economic Outcomes

The costs of environmental restoration not based on actual social willingness to pay (WTP) can be rather justified on the basis of the precautionary principle in cases when expenditure is aimed at avoiding irreversible effects on natural assets. On the other hand, when this expenditure maintains or increases the supply of goods and services above safe minimum standards for habitat preservation, expenditure is not justified without social profitability (Bishop 1978). In the Lower Ebro, river alteration is actually relatively low and there is no irreversibility; hence, social profitability is required.

The implementation of flushing flows has contributed to improve the ecological status of the river at a reasonable cost, especially if compared to the costs of removing macrophytes using exclusively labour and physical capital (i.e. mechanical prune, scrape or twirl). Flushing flows reduce financial revenue of the hydropower company, but different environmental valuation studies show that expected social welfare gains are significantly higher.

Flushing flows in the Lower Ebro have an estimated cost of EUR 109,000 a year, compared to the estimated daily revenue of the company of EUR 250,000 (thus, losses mean only 0.16 % of the average yearly revenue) (Gómez et al. 2014). Losses stem from the impact of the drop in the stock on hydropower generation during the absorption period[2] and the regulation of the production timing during the flushing flows, which prevents the company from adapting the production to those moments of the day when energy prices are at their highest value; these two effects overcome the positive effect of the increase in power output during the flood.

As above, artificial floods require 36 hm^3 along 16 h, which implies a cost of EUR 76,000 in the autumn flood and EUR 33,000 in the spring flood. On average, it can be shown that the cost per cubic meter delivered is around EUR 0.002 for the autumn flood and half that cost for the spring one. The differential is explained by higher energy prices in autumn than in spring. Energy prices in the model are average values, so that actual costs can be lower or higher than expected, as hydropower generation is a very volatile market with daily price variations as large as 64 % (Spanish Electricity Market Operator 2014). The value displayed of EUR 109,000 is the long-term annual average.

The expected reduction in the energy output is equivalent to 0.06 % of the hydroelectricity produced by the system in an average year. The implementation of the river restoration programme does not seem to be in conflict with the potential role of hydropower as a clean energy source and is unlikely to generate relevant externalities (e.g. an increase in greenhouse gas emissions). The opportunity cost of the periodical release of flushing floods by reservoir operating rule curves also seems to be lower than any other alternative of obtaining water from other sources (such as agricultural water use, urban uses or reclaimed water reuse, desalination and so forth).

For the measurement of environmental benefits, several methodologies can be applied, such as contingent valuation, travel costs, hedonic prices, and choice experiments (environmental valuation) (Mehrnaz 2013). However, their cost (time, money) would be too high for our purpose here and there still would be doubts about the convenience to use these valuation techniques for such an analysis.

---

[2] The flushing flows alter the decision making process of the hydropower operator, moving away the observed stocks and released water flows from the optimal path (baseline). As a response to the lower water stock in the dam resulting from flushing flows, the hydropower operator will release less water than he would otherwise do in the baseline scenario without flushing flows. This will happen until the amount of water stored in this alternative scenario finally converges to the amount of water stored in the baseline scenario. This time span is known as the absorption period.

The findings from a series of studies by Berrens et al. 1998; Brown and Duffield 1995; González-Cabán and Loomis 1997; Johnston et al. 2005; Loomis et al. 2000; Loomis 1996, 1998; Magat 2000 showed that WTP for flushing flows in areas that resemble the Lower Ebro ranged from US$6.7/year per person to US$377/year per person (roughly EUR_{2014} 5.32–299). Even from a narrow perspective and considering that ecosystem services were to be enjoyed and afforded only by the local population of 191,568 inhabitants (which is not necessarily the case), the average cost would be only EUR 0.52/year/person. Should river restoration measures be paid by the million people living in areas close to the Ebro River, the cost would fall to EUR 0.1/year per person; EUR 0.01/year per person if taking the whole river basin as a reference (ten million inhabitants).

Provided flushing flows are implemented by using sound economic criteria their opportunity cost is lower even in 1 or 2 orders of magnitude than people's WTP to secure the benefits of river restoration programs. In spite of the variability in the flushing flood opportunity cost, due to the uncertain behaviour of water streams and stocks in Mediterranean rivers, this cost is lower than the benefits associated to the river restoration programs as measured by individual willingness to pay. Depending on the size of the program beneficiaries, the opportunity cost may vary within the above-mentioned range, whereas the willingness to secure the benefits of river restoration programs can be as high as US$21 per person-month as reported for example by Loomis et al. (2000). This information might be considered sufficient to judge that the agreement would be compatible with a cost-benefit decision rule, and no ad-hoc valuation exercise would be required.

26.3.1.3 Distributional Effects and Social Equity

The economic instrument assessed is a voluntary agreement between the hydropower operator and the water authority on behalf of the public interest, and equity issues at stake are not especially remarkable. No distributional consequences can be directly associated to the introduction of this instrument. As a matter of fact, a financial compensation (due to profit loss for the hydropower company), which was part of the negotiations between the energy company and the river authorities, was not actually paid after all; incentives for an agreement were clear even in its absence, which is a very insightful lesson from this EPI: the critical part is the scheme of incentives (both for the power-generating companies and the river basin authority) rather than whatever monetary compensation. Noteworthy, payments for environmental services are difficult to implement in societies with advanced water regulations and institutions, especially in EU countries where water resources are not private assets and where private (use) rights can only be issued under certain conditions. Side payments for good practices are not easy to accommodate within existing regulations and will require important legal amendments besides other transaction costs. Difficulties in implementing payments for environmental services presumably reduce the scope for VAs of the kind illustrated by this example.

No significant deprivation of water uses or other equity concerns were at stake. We see the potential for some conflict of uses in the Lower Ebro, however, although nothing has been raised so far either as part of the analysis or the periodic stakeholder consultation, (in a basin district which is renowned for its public participation processes in Spanish basins) (Gómez et al. 2011).

It is important to note that restoration programs (the restoration of natural river regimes compatible with private hydropower generation) may imply a reduction in the water flow to be turbined (or the operational rules involved in such a procedure). Batalla and Vericat (2009) insist that flushing flows are far from being incompatible with hydropower production and can actually create positive impacts by decreasing of clogging of water intakes in the downstream nuclear power plants and irrigation pumping stations, therefore inducing positive externalities. Yet, flushing flows designed for the Lower Ebro have been partially insufficient to avoid clogging, and complementary actions such as mechanical extraction of macrophytes has been in place, at least in areas upstream the intakes.

As in Sect. 26.2, a significant public-health risk, linked to a plague of black flies due to macrophyte accumulation is to be taken into account, though. Black flies nourish by feeding on the blood of mammals, including humans. In several reaches of the Lower Ebro there has been evidence of black fly plagues, which actually became a common nuisance for the local population and visitors. The public-health threat is due to the fact that black flies spread several diseases (although the incidence is especially higher in Africa and South America). Intense feeding is said to cause a fever, with headache, nausea, high temperature, swollen lymph nodes, and aching joints, besides some sort of allergic reactions. There are records that in 2010, 4,500 people were seen in primary health care centres, which implied a cost of *circa* EUR 45,000. This adds to around EUR 300,000 allocated by the Regional Government to the prevention and minimization of effects linked to black flies (Gómez et al. 2011).

26.3.2 Institutional Issues

26.3.2.1 Transaction Costs and Design

The transaction costs linked to this EPI consist both of the costs of arranging the agreement *ex ante* and monitoring and enforcing it *ex post*. It should be clear (and this is a good example) that transaction costs are not to be avoided (they are indeed critical to the success of this EPI), but rather to be minimized.

No specific definition or estimation of *ex ante* transactions costs has been identified in the literature review for this case study. However, we are aware that there were significant *ex post* monitoring costs of the agreement, mostly based upon research projects, which, in most cases, were funded through competitive research programmes at a national level (National Plan of R&D). Research investment totalled EUR 543,768 in the period 2001–2014 (Gómez et al. 2011).

26.3.2.2 Implementability

Adverse effects on channel geomorphology such as armouring and incision would have persisted had non-appropriate decisions regarding frequency and magnitude of flushing flows had been made disregarding the driving controls of the natural flood regime (Batalla and Vericat 2009). Gravel injection to minimize incision during floods should be considered to mitigate this problem. Likewise, possible aggravation in reaches where flow competences is lower should also be examined. As a matter of fact, there is a clear need for the co-ordination of different water policy goals regarding the river restoration and also sectoral interests at stake if considering this restoration program as a whole, and not just in terms of macrophyte removal. The co-ordination level of significant stakeholders is remarkable so far, at least since the creation of a task force in 2008. The Ebro River Basin Authority has a long tradition of public participation and accountability, ranking high in the *Transparency International* index[3] (Transparency International – Spain 2013). Its ability to engage stakeholders proved as a catalyst for the success of this EPI.

In the Ebro River Basin Management Plan 2010–2015 (ERBA 2013) a specific Action Plan to tackle macrophyte massive growth was included. A significant bias towards algae removal (and not other river restoration measures) has been observed within the past few years, and therefore much emphasis is placed in this chapter in analysing implementability concerns in this regard, since the public good variables at stake might have been fading.

As in other countries, water in Spain belongs to the public domain (BOE 2001, sec. 2) and is subject to state planning (BOE 2001, sec. 3). State functions are in turn subject to policy principles, including, inter alia, the economy of water (BOE 2001, sec. 14.1). River Basin Authorities are responsible for the administration, management and control of water resources, the preparation of water plans, the operation of common works, and the preparation, construction and exploitation of water projects (BOE 2001, sec. 23). All these legal provisions frame the voluntary agreement assessed in this chapter, but have not posed implementability challenges from a policy perspective, since they are linked to the overall integrated water management approach and have not been an obstacle, rather the opposite, to the EPI implementation. The fact that macrophyte removal became a spur for the agreement does not necessarily show a bias towards the interest of the hydropower company but rather a practical means to find common ground for such a cooperation agreement.

Furthermore, this EPI shows that water uses can also provide important benefits downstream and on the catchment. This EPI, in fact, provides evidence in an area, which has not been sufficiently developed in the literature, providing information on the current and potential future contribution of the hydropower sector not only to renewable energy targets or greenhouse gas emission abatement, but also to the regeneration of the river regime.

[3] INTRAG (Index of Transparency in Water Management, Índice de Transparencia en la Gestión del Agua in Spanish).

26.4 Conclusions

As they have only been recently developed and mostly by practitioners (Glachant 2005), the literature on VAs is still limited but it is growing rapidly. Specifically, the experience on VAs for the delivery of artificial floods in the Lower Ebro is a unique example of public-private partnerships for the partial re-naturalization of a modified river. It has helped build a transparent bargaining scheme supported by long-term focused research enabling a better understanding of the river ecology and contributing to a better design of restoration alternatives.

This case also shows how the public interest in restoring water ecosystems can make use of the potential gains for water users to build a self-enforcing cooperation agreement and may serve to deeply change the reactive attitude from many private firms into a proactive one. Businesses engaging in the agreement do not only enjoy certain financial benefits but can also integrate these actions into their corporate social responsibility strategy (ENDESA 2013). Building cooperative agreements is only feasible when private interest is somehow compatible with the actual purposes of water policy, such as the recovery of some ecological potential of the river system.

Moreover, in this kind of cooperation setting, when the voluntary participation of critical water users is key, the emphasis can easily be placed on the design of alternatives with a better potential to contribute to the objectives of private partners (e.g. the removal of macrophytes in the closer areas of the power generation plants at the least opportunity cost in terms of power output and foregone turnover), rather than those objectives of water policy (e.g. maximizing the social benefits of river regime restoration along the whole river).

The effective contribution of the agreed flushing floods may depend on the previous set-up of many other measures designed to recover the ecological potential of the river, such as a properly defined and effectively enforced environmental flows, which are not already in place and that cannot be expected just as the result of an agreement with water users. In fact, VAs are possible regarding particular measures that are easy to define and to observe, but the recovery of water ecosystems usually involves many different measures that may need to be coordinated.

References

Batalla, R., & Vericat, D. (2009). Hydrological and sediment transport dynamics of flushing flows: Implications for management in large Mediterranean rivers. *River Research and Applications, 25*(3), 297–314. doi:10.1002/rra.1160.

Batalla, R. J., Vericat, D., & Martínez, T. I. (2006). River-channel changes downstream from dams in the lower Ebro River. *Zeitschrift fur Geomorphologie Supplementband, 143*, 1–14.

Batalla, R. J., Vericat, D., & Palau, A. (2008). Efectos de las presas en la dinámica geomorfológica del tramo bajo del Ebro.: crecidas controladas. *Ing Agua, 15*, 243–255.

Berrens, R. P., Bohara, A. K., Jenkins-Smith, H., Silva, C. L., Ganderton, P., & Brookshire, D. (1998). A joint investigation of public support and public values: Case of instream flows in New Mexico. *Ecological Economics, 27*, 189–203. doi:10.1016/S0921-8009(97)00169-9.

Bishop, R. (1978). Endangered species and uncertainty: The economics of a safe mínimum standard. *American Journal of Agricultural Economics, 60*(1), 10–17.

BOE. (2001). Real Decreto Legislativo 1/2001, de 20 de julio, por el que se aprueba el texto refundido de la Ley de Aguas.

Brown, T. C., & Duffield, J. W. (1995). Testing part-whole valuation effects in contingent valuation of instream flow protection. *Water Resources Research, 31*, 2341–2351. doi:10.1029/95WR01744.

ENDESA. (2013). Memoria de Sostenibilidad (report no. 2012). ENDESA S.A.

ERBA. (2008). Esquema provisional de Temas importantes de la demarcación hidrográfica del Ebro (No. Report). Ebro River Basin Authority.

ERBA. (2010). *La Confederación Hidrográfica autoriza un nuevo desembalse controlado de Flix para frenar la proliferación de algas del Bajo Ebro*. [Press note, November 3rd, 2010]. Zaragoza: ERBA (CHE/Confederación Hidrográfica del Ebro).

ERBA. (2013). Propuesta de proyecto de Plan Hidrológico de la cuenca del Ebro 2010–2015 informada por el Consejo del Agua de la Cuenca del Ebro y el Comité de Autoridades Competentes en julio de 2013 (River Basin Management Plan – Draft). Ebro River Basin Authority.

Generalitat de Catalunya. (2011). La Reserva de la Biosfera de las Tierras del Ebro (Report). Generalitat de Catalunya.

Glachant, J. M., & Finon, D. (2005). A competitive fringe in the shadow of a state owned incumbent: The case of France. *The Energy Journal, 26*(Special issue), 181–204.

Glenn, E. P., Hucklebridge, K., Hinojosa-Huerta, O., Nagler, P. L., & Pitt, J. (2008). Reconciling environmental and flood control goals on an Arid-Zone river: Case study of the limitrophe region of the lower Colorado river in the United States and Mexico. *Environmental Management, 41*, 322–335. doi:10.1007/s00267-007-9056-4.

Gómez, C. M., Delacámara, G., Pérez-Blanco, C. D., & Rodríguez, M. (2011). Lower Ebro (Spain): Voluntary agreement for river regime restoration services (EX-POST case studies no. deliverable no.: D3.1 – review reports), WP3.

Gómez, C. M., Pérez-Blanco, C. D., Batalla, R. J., & Part, A. (2014). Tradeoffs in river restoration: Flushing flows vs. hydropower generation in the lower Ebro river, Spain. *Journal of Hydrology, 518*, 130–139. doi:10.1016/j.jhydrol.2013.08.029. Creating Partnerships Between Hydrology and Social Science: A Priority for Progress.

González-Cabán, A., & Loomis, J. (1997). Economic benefits of maintaining ecological integrity of Río Mameyes, in Puerto Rico. *Ecological Economics, 21*, 63–75. doi:10.1016/S0921-8009(96)00093-6.

Haslam, S. M., & Wolseley, P. A. (2014). *River plants of western Europe*. New York: Cambridge University Press.

IDESCAT. (n.d.). Población extranjera. Por comarcas 2013 [WWW Document]. Ist. Estad. Catalunya IDESCAT database. http://www.idescat.cat/es/. Accessed 31 Oct 2014.

IGN. (2014). Instituto Geográfico Nacional – database [WWW Document]. http://centrodedescargas.cnig.es/CentroDescargas/index.jsp

INE. (2014). Entorno físico y medio ambiente (territorio Población, superficie y densidad por CCAA y provincias) [WWW Document]. INEBase. http://www.ine.es/inebmenu/indice.htm. Accessed 31 Oct 2014.

Johnston, R. J., Besedin, E. Y., Iovanna, R., Miller, C. J., Wardwell, R. F., & Ranson, M. H. (2005). Systematic variation in willingness to pay for aquatic resource improvements and implications for benefit transfer: A meta-analysis. *Canadian Journal of Agricultural Economics Canada Agroeconomie, 53*, 221–248. doi:10.1111/j.1744-7976.2005.04018.x.

Kondolf, M. G., & Batalla, R. J. (2005). Chapter 11: Hydrological effects of dams and water diversions on rivers of Mediterranean-climate regions: Examples from California. In C. Garcia & R. J. Batalla (Eds.), *Developments in earth surface processes, catchment dynamics and river processes Mediterranean and other climate regions* (pp. 197–211). Amsterdam: Elsevier.

Kondolf, G. M., & Wilcock, P. R. (1996). The flushing flow problem: Defining and evaluating objectives. *Water Resources Research, 32*, 2589–2599. doi:10.1029/96WR00898.

Lessard, J., Hicks, D. M., Snelder, T. H., Arscott, D. B., Larned, S. T., Booker, D., & Suren, A. M. (2013). Dam design can impede adaptive management of environmental flows: A case study from the Opuha Dam, New Zealand. *Environmental Management, 51*, 459–473. doi:10.1007/s00267-012-9971-x.

Loomis, J. B. (1996). Measuring the economic benefits of removing dams and restoring the Elwha river: Results of a contingent valuation survey. *Water Resources Research, 32*, 441–447. doi:10.1029/95WR03243.

Loomis, J. B. (1998). Estimating the public's values for instream flow: Economic techniques and dollar values. *Journal of the American Water Research Association, 34*, 1007–1014. doi:10.1111/j.1752-1688.1998.tb04149.x.

Loomis, J., Kent, P., Strange, L., Fausch, K., & Covich, A. (2000). Measuring the total economic value of restoring ecosystem services in an impaired river basin: Results from a contingent valuation survey. *Ecological Economics, 33*, 103–117. doi:10.1016/S0921-8009(99)00131-7.

Magat, W. A. (2000). An iterative choice approach to valuing clean lakes, rivers, and streams. *Journal of Risk and Uncertainty, 21*, 7–43.

Mehrnaz, R. (2013). *Freshwater values framework. A review of water valuation methods utilised within total economic valuation (Auckland Council working report No. WR2013/001).* Auckland: Auckland Council, Research, Investigations and Monitoring Unit.

Montesinos, S., Bea, M., & Losada, J. A. (2009). Determinación de macrófitos en el río Ebro entre Flix y Mora D'Ebre. In S. Montesinos & L. Fernández (Eds.) *Teledetección: Agua y desarrollo sostenible. XIII Congreso de la Asociación Española de Teledetección*, Calatayud, 23–26 de septiembre de 2009 (pp. 137–140). Zaragoza: Universidad de Zaragoza.

Ollero, A. (2010). Channel changes and floodplain management in the meandering middle Ebro river, Spain. *Geomorphology, 117*, 247–260. doi:10.1016/j.geomorph.2009.01.015. Introduction to Management of Large Rivers.

Spanish Electricity Market Operator. (2014). Red Eléctrica Española Database [WWW Document]. Red Eléctrica Esp. Database. http://www.ree.es/sistema_electrico/informeSEE.asp

Transparency International – Spain. (2013). Índice de Transparencia en la Gestión del Agua (INTRAG) 2013 [WWW Document]. Rank. Transparencia Glob. http://www.transparencia.org.es/INTRAG/INTRAG_2013/%C3%8Dndice_INTRAG_2013.htm

Truong, C. (2012). An analysis of storage capacity reallocation impacts on the irrigation sector. *Environmental and Resource Economics, 51*, 141–159.

URS-España. (2010). *Asistencia técnica para el control de macrófitos: Mejora de la gestión de los embalses del Bajo Ebro. (Report No. Final).* Barcelona: United Research Services España S.L.

Vericat, D., Batalla, R., & Garcia, C. (2006). Breakup and reestablishment of the armour layer in a large gravel-bed river below dams: The lower Ebro. *Geomorphology, 76*, 122–136. doi:10.1016/j.geomorph.2005.10.005.

WHO. (2014). Onchocerciasis. Fact sheet N°374 [Updated March 2014] [WWW Document]. World Health Organ. – Media Cent. http://www.who.int/mediacentre/factsheets/fs374/en/. Accessed 30 Oct 2014.

Chapter 27
Voluntary Agreements to Promote the Use of Reclaimed Water at Tordera River Basin

Francesc Hernández-Sancho, María Molinos-Senante, and Ramón Sala-Garrido

Abstract The voluntary agreement to promote the use of reclaimed water is an economic policy instrument (EPI) which focuses on improving water management by using reclaimed water. Following a win-win strategy, this EPI was implemented in the Tordera river basin (Spain), an area with endemic water scarcity problems and high competition among users for water resources. The assessment of the EPI suggests that significant positive outcomes have achieved from an environmental and economic point of view. Thus, the demand of freshwater has decreased and the availability of water is guaranteed even during summer period allowing therefore for the maintenance of economic activities (agriculture and golf course) and for reducing overexploitation of aquifers. The social acceptance of the use of reclaimed water and the institutional framework (regulation and previous experiences) were two key factors for the success of this EPI.

Keywords Voluntary agreements • Reclaimed water • Win-win strategy • Water scarcity

---

F. Hernández-Sancho (✉)
Department of Applied Economics II, Faculty of Economics, University of Valencia, 46022 Valencia, Spain
e-mail: Francesc.Hernandez@uv.es

M. Molinos-Senante (✉)
Departamento de Ingeniería Hidráulica y Ambiental, Pontificia Universidad Católica de Chile, 4860 Santiago, Chile
e-mail: mmolinos@uc.cl

R. Sala-Garrido
Department of Mathematics for Economics, Faculty of Economics, University of Valencia, 46022 Valencia, Spain
e-mail: Ramon.Sala@uv.es

M. Lago et al. (eds.), *Use of Economic Instruments in Water Policy*,
Global Issues in Water Policy 14, DOI 10.1007/978-3-319-18287-2_27

27.1 Introduction

The voluntary agreement to promote the use of reclaimed water is an EPI focused on the management of water resources. It is considered to result in a win-win strategy since all parties involved in the process improve their situation in economic terms or water availability.

On the one hand, the instrument has an economic objective which is to ensure the availability of water including in the summer and so maintain economic activity. On the other hand, the environmental objective is to reduce the overexploitation of local aquifers by reducing freshwater and energy consumption.

The instrument was implemented in an area within the Tordera River Basin (TRB) in the southern boundary of the Costa Brava in the North-Eastern Spain. This is an area with endemic water scarcity problems and competition for water access. In the last 20 years, due to the development of golf courses water shortage problems have accentuated.

Water has been reused for golf course irrigation in the Costa Brava since 1985 and the social acceptance is better than in areas where recycled water hasn't been used for so long. The water administration believed that a good alternative to address the growing regional water shortage and pressures on the local aquifers was to promote the use of reclaimed water further through a voluntary agreement.

The philosophy of the EPI is clearly shown in the case of the Mas Pijoan Farm which reached an agreement with the managers of a nearby golf course. Until 2003, the farm worked on 35 ha that was irrigated from the local aquifer (FAO 2010). The yield of the wells at the beginning of the summer could reach 150 m^3/ha, but would decrease during the season to 20 m^3/ha, thus water could not be guaranteed at crucial crop growing stages. In this context, the Mas Pijoan Farm found that connecting to the reclaimed water pipeline of the Costa Brava Golf Course was a reasonable solution. The golf course irrigation is in operation from 9 pm to 7 am every day, and the water is supplied to agriculture during the rest of the day. The cost of the connection to the existing pipeline, the storage pond was partly funded (70 %) by the European Agricultural Fund for Rural Development (EAFRD). At the same time, the farm owner signed a 25-year service contract to share the use and operation and maintenance cost of the reclaimed water pipeline of the Golf course property (FAO 2010). The arrangement has provided reliability and flexibility to both users. This example shows the importance of a voluntary agreement between parties in order to promote the use of reclaimed water following a win-win strategy.

27.2 Setting the Scene: Challenges, Opportunities and EPIs

The case study focuses on an area within the delta of the river Tordera which is located in north-eastern Spain. The TRB belongs to the Catalonia Basin District and covers an area of 894 km^2.

The predominant climate is Mediterranean, with a high concentration of rainfall in spring and autumn, with summer being the dry season. It is important to note that during the period 1988–2007 six periods of drought warnings have forced the adoption of exceptional measures to guarantee water supply.

The main use of water in the basin is urban supply and this represents approximately the 77 % (including industry services) while the remaining 23 % is used for agricultural purposes. Groundwater presents serious problems of contamination; as well as overexploitation which result in marine intrusion and the salinization of the water. Total underground resources exploited by the system are estimated approximately at 42.6 hm^3/year (FAO 2010).

In the study area there are three wastewater treatment plants (WWTPs), two with tertiary treatment –Blanes and Tordera – and one with secondary treatment (Castell-Platjad'Aro). Effluent from the Blanes plant (around 3.5 hm^3/year) is used mainly for reaching the aquifer, though a few farmers also use it for irrigation. The Tordera WWTP, produces around 1 hm^3/year of reclaimed effluent which is discharged into the Tordera river. The effluent of Castell-Platja d'Aro WWTP (5.5 hm^3/year) is treated to secondary and tertiary levels, and is used for golf course watering, groundwater recharge, and agricultural irrigation, with the residue discharged into the sea.

The area of the Tordera Delta is characterized by the fact that irrigation water is taken entirely from groundwater, with no surface supply. Because of the low rainfall, during summer the freshwater availability decreases significantly. Thus, water cannot be guaranteed at crucial crop growing stages. Moreover, the area is characterized by a high level of tourism activity with a significant number of golf courses. Taking into account that in Catalonia, there is a prohibition on the use of groundwater for golf course irrigation, competition for water is always high.

To address the growing regional water shortage and pressure on the local aquifers, the Catalonian Water Agency (ACA) considered that a plausible solution would be the use of reclaimed water mainly for irrigation. Because golf courses shifted in 1998 to the use of reclaimed water it was considered that a voluntary agreement between farmers and municipalities was an appropriate instrument for promoting the use of reclaimed water.

The main challenges in the Tordera area were on the one hand, to ensure the availability of water even in the summer season, resulting in smaller pumping costs and an increased irrigated agricultural area, and on the other hand, to reduce overexploitation of local aquifers by reducing water and energy consumption. This case study provides evidence that negotiation enables agreements that benefits different parties following a win-win strategy.

27.3 The Voluntary Agreement in Action

The EPI focused on improving water management by using reclaimed water. This EPI involved three main parts; namely golf courses, farmers and water authority. The negotiation process was performed between farmers and golf courses, while water authority introduced monetary incentives to promote the use of reclaimed water.

27.3.1 The EPI Contribution

27.3.1.1 Environmental Outcomes

The assessment of the environmental outcomes focuses on the changes in water supply for urban and farmer use as a consequence of the use of regenerated water.

The voluntary agreement to promote the use of reclaimed water in the Tordera Delta resulted in significant environmental benefits due to the release of freshwater. By restoring aquifers and recharging the Tordera river, the water availability in the watershed has increased and damage to the aquatic environment has diminished. It is well know that reused water involves significant environmental benefits at river basin level.

As a result of the implementation of the EPI, the municipalities of Plata d'Aro and Santa Cristina d'Aro have started to use reclaimed water for garden irrigation. Hence, as it is shown in Fig. 27.1, that despite increases in population growth during the years 2003–2010 (by 44 % for Santa Cristina d'Aro and by 32 % for Platja d'Aro), urban water demand has decreased significantly in Santa Cristina d'Aro (by 50 %) and slightly in Platja d'Aro (by 10 %).

This reduction in the urban water demand is not exclusively attributable to the introduction of the EPI, but we must bear in mind that, in the municipality of Santa Cristina d'Aro, the average annual decrease in water demand has been 216,300 m^3/year and the volume of water reused for garden irrigation was around 126,000 m^3/year. In the case of Platja d'Aro, average urban demand reduction has been 227,400 m^3/year, while the volume of water regenerated for municipal use was 162,000 m^3/year (FAO 2010).

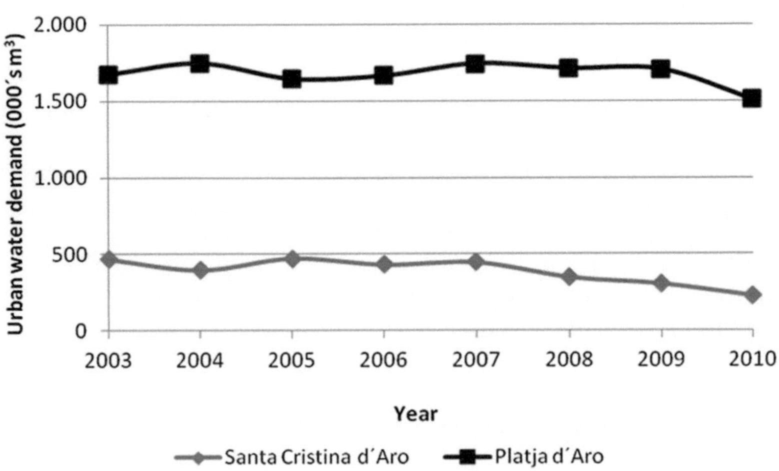

Fig. 27.1 Urban water demand in two municipalities of the Tordera River Basin (000's m^3), 2003–2010 (Source: Own elaboration from Consorci Costa Brava (CCB) 2011)

Table 27.1 Comparison between with and without use of reclaimed water at Mas Pijoan Ranch

	Situation in 2003	Situation in 2006	Change 2003–2006 (%)
Total irrigated land (ha)	35	41.6	18.9
Land irrigated with reclaimed water (ha)	0	25	–
Land irrigated with well water (ha)	35	9	–74.3
Land irrigated with mixed water (ha)	0	7.6	–
Crop water requirement (m³/ha)	5,000	5,000	0.0
Well water used (m³/year)	175,000	71,240	–59.3
Reclaimed water used (m³/year)	0	136,760	–
Crop yield (kg/ha)	50,000	70,000	40
Irrigation cost (EUR/m³)	0.075	0.115	53.3

Source: FAO (2010)

One of the most successful water reuse plans in the Tordera Delta has been carried out by the Mas Pijoan Farm. The use of 0.137 hm³/year of reclaimed water has involved changes in irrigation practices in comparison with the situation when no reclaimed water was used (See Table 27.1). The recycled water accomplished the thresholds established by the Spanish Royal Decree 1620/2007. Moreover, nitrogen and phosphorus were not removed from wastewater which involved a significant reduction in the use of fertilizers and an increase in the crop yield (Sala and Serra 2004).

As Table 27.1 shows, the situation in 2006 differs significantly from the past in terms of cultivated land, crop yield, water use, and irrigation costs. The farmer irrigated in 2006 part of the land with reclaimed water, part with well water, and mixed water which involves both reclaimed and well water. Specifically, 25 ha were irrigated by reclaimed and 7.6 ha with mixed water, whereas the farmland irrigated with well water decreased from 35 ha to 9 ha. The extraction of well water was reduced by 59.3 % in 2006 due to the use of reclaimed water. The reliability of reclaimed water improved the water availability and raised the crop yield per hectare by 40 %. It should be noted that crop yield depends on several factors and therefore the increase in the crop yield from 2003 to 2006 cannot be attributed only to the reliability of reclaimed water.

The decrease in the use of local underground resources has entailed other environmental outcomes. Energy savings, associated with the reduction of groundwater pumping, could be quantified approximately by 30,000 kWh/year. According to the Spanish national electrical production grid (Spanish Ministry of Industry and Tourism 2009), this figure means a saving of 10,800 kg of CO_2 equivalent per year. It has been verified that the level of the Tordera aquifer has increased by approximately 10 m in few wells. This improvement mainly refers to the coastal points and during the summer months, namely when there is the greatest increase in water consumption. In recent years, the level of the aquifer has not been below zero meters above the sea level and so avoiding the aquifer salinization and contributing to the preservation of this strategic reserve (Muñoz and Sala 2007).

Because reclaimed water contents nutrients, no chemical fertilizers are needed in the fields (25 ha) that are irrigated with reclaimed water. In this sense, the implementation of the EPI has enabled a reduction of 14,500 kg/year of chemical fertilizer use (CCB 2011) and this means significant energy savings and increased availability of non-renewable resources such as phosphorus.

Because all treated wastewater was previously discharged into the sea, the use of reclaimed water for irrigation purposes, and for recharging the Tordera aquifer has not affected the river regimes. The chemical parameters of the water and the biological and morphological indicators of the river remained unchanged (Sala 2010).

27.3.1.2 Economic Outcomes

In the area of study, guarantee urban water supply is a challenge faced by water authorities. Hence, the main alternative to the EPI was the implementation of a seawater desalination project. By dividing the total cost of reclaiming and distributing the water (EUR 1,313,281/year) with the volume of wastewater recycled (3,475,691 m^3/year), the unit cost is approximately at EUR 0.38/m^3. This unit cost may also be compared with the average cost of seawater desalination ranging from EUR 0.45/m^3 to over EUR 1.0/m^3 (Water Reuse Association 2012). Obviously, the reuse of treated wastewater in irrigated agriculture is a preferable alternative to desalination projects.

The illustrative cost of the reclaimed effluent in the Tordera Delta (EUR 0.38/m^3) is much higher than the cost of pumping groundwater (EUR 0.11/m^3) (CCB 2011). There is no present source of cross-subsidy from farmers, where urban and recreational users could in principle afford the economic tariff. However, these users only account for a minor part of consumption. Users accept to use regenerated water since this source ensures the availability of water in summer time, i.e. in a typical water scarcity period.

Governmental subsidies are provided by the European Commission (European Fund for Rural Development). Such subsidies support the financing of water reclamation projects, even though they do not influence the economic efficiency of water transfer projects. However, they facilitate the implementation of such projects and can essentially influence the readiness of farmers to join in the use of reclaimed water. EC supported such projects since they improved significantly the environmental conditions of the water river basin.

As a first step to assess the economic outcomes of the EPI, the water reuse project at Mas Pijoan Ranch was evaluated. Firstly, the cost-effectiveness of the project was evaluated. Due to the conversion of reclaimed water and the expanded farmland, the additional yields lead to an increase in sales revenue of the amount of EUR 174,300/year, which means an increase of 66.4 % with respect to the situation without the use of reclaimed water. Groundwater pumping cost reductions result in savings of EUR 7,782/year (59.3 %). The cost of fertilizing decreased by EUR 3,588/year (52.6 %), albeit the cultivated farmland has been expanded (FAO 2010). However, the use of reclaimed water leads to a cost of EUR 34,529/year for carrying the water to the fields.

In comparison with the past, the farmer's income has been increased by EUR 185,670/year (76.5 %) due to cost savings in pumping and fertilization and to the increase in the irrigated land. Due to the high increase in income, the principle of cost recovery is 100 % fulfilled since the users of the reclaimed water pay the entire cost of the treatment (Sala 2009).

Furthermore, the investment of one euro in the use of reclaimed water yields an income increase in agriculture of EUR 11.80, if farmers do not pay the conveyance cost. If they pay these costs, the income return factor is EUR 10.80. This result is in line with the expectations from the theoretical point of view (Sala 2009).

The cost of water reclamation in Platja d'Aro WWTP can be summarised as follows. The investment cost of the tertiary treatment is EUR 1.2 million. Hence, by considering that the lifetime of the project is 25 years and the rate of interest if 6 %, the investment cost is EUR 93,840/year. When taking into account that the volume of tertiary treated water is 990,489 m^3/year and the cost is EUR 0.05/m^3, then the operation and maintenance cost is EUR 49,524/year. By considering investment and operating costs, the total annual cost of tertiary treatment at Platja d'Aro WWTP is EUR 143,364/year (ACA 2007, 2009).

The improvement in economic efficiency of irrigation due to the use of reclaimed water from Platja d'Aro WWTP can be summarised as follows. The use of reclaimed water leads to economic net benefits of EUR 169,890/year, if only the transport cost is considered. If the tertiary treatment cost of EUR 143,364/year is included in the cost of reclaimed water, then the economic net benefit amounts to only EUR 26,526/year (ACA 2009). Nevertheless, the reclaimed water cost is lower than the added value in agriculture.

The strategy for Platja d'Aro was to increase the reclaimed water production reaching 20,000 m^3/day with similar water quality i.e. using only a recycling scheme, and build new pumping stations, pipelines, and water reservoirs (Borràs 2002). The construction costs of these facilities to provide reclaimed water is shared proportionally with each of the users as shown in Table 27.2.

Of the total investment cost of around EUR 7.7 million, 16 % is required for the enlargement of tertiary treatment, 48 % for the pipelines, and 33 % for storage facilities.

The conversion from groundwater to reclaimed water in irrigated agriculture has led to several benefits for other water users and the aquatic environment. In particular, the reduction in the use of groundwater has avoided the construction of a new pipeline to carry water from Ter River to meet the increasing water demand in the Costa Brava. An investment cost of EUR 27 million has therefore been saved (Borràs et al. 2007).

The economic net benefit resulting from the use of reclaimed water at Tordera Delta area has been estimated by taking into account the total benefit and total cost. To quantify the total economic benefit the following items have been considered: cost savings for farmers (fertilization, water extraction, resource development); and increased sales revenues in agriculture. The total economic cost includes the wastewater treatment, the conveyance, and storage of reclaimed water and the application of reclaimed water.

Table 27.2 Investment cost of reclaimed water use at Platja d'Aro area

	Request water (m^3/year)	Tertiary treatment (EUR)	Pipe lines (EUR)	Pumping (EUR)	Storage (EUR)	Amount by user (EUR)
Municipality						
Platja d'Aro	162,000	78,192	618,387	39,194	148,065	883,837
Santa Cristina d'Aro	126,000	60,816	371,127	32,727	118,227	582,898
Golf						
Pitch & Putt Platja d'Aro	24,000	11,584	91,613	5,806	21,935	130,939
Golf d'Aro[a]	210,000	–	–	–	–	0
Finca Lara	30,000	21,471	25,342	978	27,907	90,178
Pitch & Putt Mas Torrelles	28,000	13,515	82,473	7,273	26,273	129,533
Golf Costa Brava[a]	250,000	–	–	–	–	0
Pitch & Putt La Llave	116,200	56,086	181,320	3,790	108,093	349,289
Farmers						
Mas Pijoan Ranch	136,760	–	124,052	25,125	125,253	–
Plots near the WWTP[a]	91,250	–	–	–	–	0
Farmers in Soilius[a]	171,500	–	–	–	–	0
Farmers in Llagostera	1,000,000	482,664	1,579,684	127,616	2,100,000	4,289,964
ACA						
Ecological water flow	1,000,000	482,664	715,684	32,616	–	1,230,964
Cost of each action (EUR)		1,200,000	3,687,100	250,000	2,550,500	7,687,600
Total request water (m^3/year)	**3,345,710**					

Source: FAO (2010)
[a]Old user

The total cost of the use of reclaimed water in the Tordera Delta, is around EUR 1.3 million/year, of which 27.7 % and 72.3 % is for tertiary treatments and conveyance, respectively. The total benefits amount to around EUR 3.3 million/year, of which 26.7 % are due to increases in crop sales and 69.1 % due to cost savings in transmitting distant resources. The economic net benefit of the use of reclaimed water in Tordera Delta is estimated to be approximately EUR 2 million/year (FAO 2010).

The implementation of the EPI has not led to a cost saving for water users since according the economic feasibility study, the unit cost to produce regenerated water is approximated EUR 0.38/m^3, while the cost of pumping groundwater is only EUR 0.11/m^3. However, the EPI has delivered additional benefits such as an increase in farmers' income and an improvement in their water availability. Golf courses have also improved their water availability – mainly in the summer. With respect to cost reductions, the implementation of the EPI has involved savings in groundwater pumping cost, in fertilizer costs, and has avoided the need to build a freshwater pipeline. At the same time, it has enabled an increase in irrigated land.

Because the area is characterised by the use of groundwater with no surface supply for irrigation purposes, golf courses, the course managers and farmers consider that the most important benefit of the use of reclaimed water is that the guaranteed availability of water all year, even in summer when the yield of the wells decreases significantly. Hence, the EPI has an important risk-reduction role in the area.

The cost of the additional treatment for regenerating the treated water (tertiary treatment) and its distribution is paid by farmers and golf courses applying full cost recovery. Being a negotiation process between them, the percentage paid by each is different. The EPI has been designed to recover all the costs to implement water reuse projects. The revenue is collected by the Consortium of the Costa Brava (CCB) which is the institution that manages the water cycle in the TRB. The revenues are earmarked to pay for the regeneration and distribution of the water.

27.3.1.3 Distributional Effects and Social Equity

The use of reclaimed water for golf courses is generally accepted by the public. In fact, in the Tordera Delta area, the administration has not received any complaints regarding the project. However, the irrigation of crops with reclaimed water is still debatable.

Although the use of reclaimed water may lead to biological and chemical risks, we can consider that these are minimal since the Spanish Royal Decree 1620/2007 sets out strict parameters to be met by the reclaimed water according to its use. Education and monetary incentives are two key issues to encourage the use of reclaimed water. In the study area, important information campaigns addressed to both farmers and the general public have aimed to raise awareness about the positive effects and restrictions on the use of reclaimed water. The use of reclaimed water has enabled an increase in the irrigation area and consequently, new jobs have been generated. Although there is no statistical information relative to how the use of reclaimed water has altered employment, the farmers in the area estimated that there has been an increase of approximately 25 %.

Farmers are very satisfied with the quality of the reclaimed water. Because the area is characterized by the use of groundwater with no surface supply for irrigation purposes, farmers consider that the most important benefit from using reclaimed water is that the year-round availability. Farmers have been a very important part of the promotion of the use of regenerated water in the Tordera Delta. From the beginning, the administration was aware that the project would be successful if it had the support of farmers. For this reason, farmers have actively participated in the decision-making process.

For the golf courses, the use of regenerated water, rather than an increase in the benefits, has meant ensuring the availability of water for irrigation and, consequently the maintenance of its activity, especially in the summer when demand is higher. No health problems have been detected in golf courses workers associated with the use of reclaimed water. In some cases, the fact that the golf course is irrigated with regenerated water is visibly announced in order to show the commitment of the company to the environment, and so contributing to the environmental awareness of clients.

The third group of stakeholders is the inhabitants of the municipalities affected. Because a proportion of the reclaimed water is used for the irrigation of municipal gardens and to improve the water flow of Ridaura River, the opinion of wider public is of paramount importance. Before the use of regenerated water for the irrigation of gardens, the administration launched a major information campaign that included publication of brochures, lectures at schools, institutes, councils, associations of neighbors, cultural associations, etc. Education is perceived as a key issue to ensure the acceptance of the use of reclaimed water. Local people are aware of the fact that part of the improvement of the environmental quality of the River Ridaura is a direct consequence of the considerable efforts made in the treatment and regeneration of wastewater. In this case, transparency in decision-making process and the access to information have played key roles. Given that the water regeneration involves high costs, it was considered vital that the local population knew the causes for the development of water reuse projects and their associated benefits.

To avoid contact with regenerated water, information boards have been placed in the gardens irrigated with reclaimed water. The local environment and the appearance of the garden have not declined after irrigation with regenerated water. The environmental organizations of the area have publicly shown their satisfaction with the use of regenerated water for garden irrigation and for the maintenance of the flow of the Ridaura River.

27.3.2 The EPI Setting Up

27.3.2.1 Institutions

The culture of water reuse in the TRB has positively influenced the design and implementation of the EPI. The planned reuse of water in Catalonia began in 1985 when the Costa Brava used a disinfected secondary effluent for golf course irrigation. On April 7th in 2005, the council of administration of the Catalonian Water Agency published an edict concerning the criteria for the process and administrative procedures for using treated water for the irrigation of golf courses and similar facilities. This agreement stated that in general, the irrigation of golf courses and similar facilities must be done by reclaimed water from private or public WWTPs.

Due to water scarcity problems, water reuse has become an important resource in Spain. Therefore, in 2007 the Royal Decree 1620/2007 of December 7th established the legal framework for the reuse of regenerated water. The norm defines the concept of water reuse, introduces the concept of reclaimed water and determines the requirements for reusing reclaimed water, and the procedures to obtain legal authorization. Moreover, it includes provisions related to the acceptable uses and precise quality requirements for each case. The Royal Decree was promoted by the Environment Ministry but has many links to Health Ministry. The requirements about the quality of the reclaimed water established by the Royal Decree 1620/2007 are a key point to implement the EPI in other areas in Spain.

Given that the TRB lies entirely in the region of Catalonia, the Administration responsible for developing and monitoring the entire process is the Catalan Water Agency (ACA). Moreover, within the territory of the TRB there is another water administration that is responsible for the direct management of the water resources of the area: the Consortium of the Costa Brava (CCB). It is an autonomous organisation created in 1971 and composed by 27 municipalities along the Girona coast. It is worth noting that the CCB is one of the pioneering institutions in Spain in the development of water reuse projects. The participation of the CCB in the negotiating process has been essential for promoting the use of reclaimed water. This administration has not only provided the legal and institutional framework to develop the projects but it has actively participated in making water reuse a reality.

The existence of a wide legislation in the field of wastewater reuse has affected both the design and implementation of EPI. The fact that Royal Decree 1620/2007 does not allow the use of regenerated water to supply households, determined the design of the EPI. The fact that golf courses cannot irrigate with freshwater has also favored the implementation of EPI.

27.3.2.2 Transaction Costs and Design

Because the instrument is based on the negotiation process to promote the use of regenerated water, the main participants involved were farmers and golf course managers. Nevertheless, the water administration, mainly the CCB, also has a representative role in the sense that by the use of monetary incentives it also promotes the use of reclaimed water and all the legal authorizations required are approved by it.

The tradition of water reuse in the area was a key factor in selecting the EPI – which was implemented in several phases. One of the first negotiations to promote the use of reclaimed water was established in 2003 between the Mas Pijoan Ranch and the nearby golf course. In subsequent years, other farmers and golf courses have negotiated the use of reclaimed water by sharing infrastructure and costs. The administration procedures are mainly associated with obtaining the legal authorization for water reusing (Sala 2009).

There were no problems with regard to asymmetric information since both farmers and golf courses shared information relative to the costs of implementing water reuse projects.

One of the premises to develop the water reuse project was that the users paid the total cost of the regeneration and conveyance of the water (tertiary treatment since secondary treatment is paid by citizens in accordance with polluter pays principle). The fact that much of the initiative to develop the project has been by water users has facilitated the work. While it is true that there are additional costs for the regeneration and distribution of water, such as administrative costs, these can be considered as negligible.

The voluntary agreement to encourage the use of reclaimed water has played a vital role for the success of the projects developed subsequently. However, it is true that the negotiation between the ACA-CCB and users was not overly complex since the need to increase the supply of water was very clear. In this sense, the ex-ante transaction costs were minimal.

Ex-post transaction costs are basically associated with the monitoring of the quality of the reclaimed water. In this sense, an analysis is needed every month in order to check that the reclaimed water meets the quality criteria required by Spanish law. It is estimated that the annual cost of these analyses is approximately EUR 1,000/year.

27.3.2.3 Implementation

The voluntary agreement and the negotiation process as a mechanism to promote the use of reclaimed water is a very flexible instrument that can be adapted to local conditions ex-ante and ex-post implementation. The conditions of the negotiation are adjusted depending on the actors involved in the process and their water needs.

In the field of water reuse, public perception and participation play key roles. It is necessary that society understands the benefits and the risks associated with the use of reclaimed water. In this context, farmers were very satisfied with the quality of the products obtained by using reclaimed water and the perception of the use of regenerated water for golf course irrigation has had very good acceptance by local population. The good acceptance of the use of regenerated water for different purposes has influenced positively the implementation of the EPI.

Public participation did not play an important part in the design of the EPI but it was essential in the choice and implementation of the instruments. Before choosing the instrument, the water administration conducted environmental awareness campaigns and held informative meetings with farmer associations and managers of golf courses. We can say that the administration acted as a catalyst for the negotiating process.

Cooperation between stakeholders, namely farmers, golf courses managers and residents, is a key aspect for the success of a water reuse project. None of them had a dominant position which influenced the implementation of the EPI since the strategy applied was win-win. The legislation that governs the entire process is at the national level and therefore, the quality requirements of the reclaimed water and the administrative procedure must conform to this legislation. The next hierarchical level is determined by the regional legislation which develops and applies the state legislation. However, we must not forget that given the special feature of this type of projects adaptation must be made to local conditions. For this reason, cooperation between the regional and local administration was key.

The fact that the Spanish Royal Decree 1620/2007 describes the administrative procedure for obtaining the authorization for water reuse might facilitate the implementation of this EPI in other areas with water scarcity problems. In this context, the authorization procedure requires that the petitioner submit a water reuse project, with the River Basin Authority being the responsible to examine the documentation presented and report on the compatibility of the application with the Basin Hydrological Plan. The existence of a clear norm both related to the administrative procedure and the water quality criteria would facilitate the implementation of this EPI in other areas.

27.4 Conclusions

The area of Tordera Delta is characterized by high level of competition for water. Hence, the water administration considered that it was necessary to promote a closer cooperation among stakeholders in order to increase the use of regenerated water and therefore, decrease the pressure on the local aquifers.

The EPI has succeeded since after its implementation a significant volume of regenerated water has been reused. The use of voluntary agreement for promoting water reuse has taught some lessons: (i) crop production and golf course irrigation is now independent from variable rainfall patterns and groundwater availability; (ii) a mutual win-win strategy reliability and flexibility can be offered to the parties involved; (iii) the social perception relative to water reuse has been improved by the EPI implementation; (iv) the culture of water reuse has conditioned the choice and the implementation of the instrument; (v) sharing the information relative to previous experiences was essential to develop new water reuse projects; (vi) overall, transaction costs are very low and are associated with monitoring the quality of the regenerated water; and (vii) the objectives of the instrument were vaguely defined and in qualitative terms.

There are three main enabling factors that have contributed to the success of the instrument which are as follows:

– Win-win strategy: parties involved in the negotiation process must obtain benefits as a result of the cooperation. The agreement should be approved directly or indirectly by the administration since it has to authorize the water reuse project.
– The social acceptance of the water reuse: all the stakeholders should be aware of water scarcity problems and the challenge that they face. It is vital that local people knew the causes for the development of water reuse projects and their associated costs and benefits.
– Institutional framework: if the parties reach agreements but the institutional framework is not well defined, or is not conductive to water reuse, then the project implementation will be difficult.

Regarding disabling factors, if the economic assessment of the water reuse is negative for one or two parties involved there is no room for the negotiation. The stakeholder who takes the initiative in the negotiating process must previously ensure that the other party will also obtain benefits.

References

ACA. (2007). (Catalonian Water Agency). Implementation of the water framework directive in Catalonia. Pressure and impacts analysis. Barcelona. Available from: http://aca-web.gencat.cat. Accessed 22 Aug 2014 (in Catalonian).

ACA. (2009). (Catalonian Water Agency). Catalonian water reuse program. Available from: http://aca-web.gencat.cat/aca/appmanage…eLocale=es. Accessed 22 Aug 2014 (in Catalonian).

Borràs, G. (2002). Water reuse in Catalonia, Spain: The case of the Tordera Delta. Proceedings of the IWA regional symposium on recycling in Mediterranean region.Datzira.Irarlio, 26–29 Sept 2002.

Borràs, G., Soler, M., & Sala, L. (2007). Summary of data concerning the quality of the reclaimed water produced at the Blanes Reclamation Plant (Costa Brava, Gironda, Catalonia). Proceedings of the 6th IWA Specialist conference on wastewater reclamation and reuse for sustainability: Guiding the growth of water reuse. Antwerp, 9–12 Oct 2007.

CCB. (2011). Consorci de la Costa Brava. Available from: www.ccbgi.org. Accessed 22 Aug 2014 (in Catalonian).

FAO. (2010). From waste to resource: The economics of wastewater reuse in agriculture, by Winpenny, J., Heinz, I. and Koo-Oshima, S. FAO Water Reports No. 35. Rome.

Muñoz, J. & Sala. L. (2007). Environmental criteria for alternative removal in treated wastewater. Proceedings of the 6th IWA Specialist conference on wastewater reclamation and reuse for sustainability: Guiding the growth of water reuse. Antwerp, 9–12 Oct 2007.

Sala, L. (2009). Water reclamation and reuse in Costa Brava (1989–2009): Lessons learned and practical contributions. Workshop IINOVA-MED CONFERENCE: Innovative processes and practices for wastewater treatment and re-use in the Mediterranean region. Girona, 8–9 Oct 2009.

Sala, L. (2010). Evolution of water reuse practices in Tossa de Mar: From environmental reuse to urban non-potable reuse. Workshop of water reuse at National Taiwan University. Taipei, 17 Nov 2010.

Sala, L., & Serra, M. (2004). Towards sustainability in water recycling. *Water Science and Technology, 50*(2), 1–7.

Spanish Ministry of Industry, Tourism and Trace, Spain Energy White Book. (2009). www.mityc.es/energia/balances/Balances/LibrosEnergia/Energia2009.pdf. Accessed 22 Aug 2014 (in Spanish).

Water Reuse Association. (2012). Seawater desalination costs. White paper: Alexandria, VA 22314 USA.

Chapter 28
Key Conclusions and Methodological Lessons from Application of EPIs in Addressing Water Policy Challenges

Carlos M. Gómez, Gonzalo Delacámara, Alexandros Maziotis, Jaroslav Mysiak, and Manuel Lago

Abstract This final chapter presents the overall balance of collection of cases presented in the whole book. In line with the structure of the book, rather than assessing the EPIs themselves, this analysis focuses on their potential and actual contribution to the goals of water policy as the main criterion to discuss the screening, design and implementation of the EPIs. Furthermore, the discussion focuses on two critical aspects that may determine EPIs' success or failure: the first one is the need to deal with the multiple goals that are distinctive of water policy; in the water policy arena any instrument is expected to serve to development, financial, environmental and other social goals at the same time and any instrument must consider the trade-offs implied. The second relates with how the EPIs chosen match within the institutional set up which, at least, is essential to define property rights and to reduce transaction costs, and then to make a given EPIs a viable option to improve

C.M. Gómez (✉)
Departamento de Economía, Universidad de Alcalá and IMDEA-Water Foundation,
c/ Plaza de la Victoria s/n, 28803 Alcalá de Henares (Madrid), Spain
e-mail: mario.gomez@uah.es

G. Delacámara
Madrid Institute for Advanced Studies in Water (IMDEA-Water Foundation),
Punto Net 4, floor 1, Edificio ZYE, Parque Científico-Tecnológico de la Universidad
de Alcalá, 28805 Alcalá de Henares (Madrid), Spain
e-mail: gonzalo.delacamara@imdea.org

A. Maziotis
School of Mechanical, Aerospace and Civil Engineering (MACE), University of Manchester,
Manchester M13 9PL, UK
e-mail: alexandros.maziotis@gmail.com

J. Mysiak
Fondazione Eni Enrico Mattei (FEEM), Isola di San Giorgio Maggiore, 30124 Venice, Italy
e-mail: jaroslav.mysiak@feem.it

M. Lago
Ecologic Institute, Pfalzburger Strasse. 43/44, 10717 Berlin, Germany
e-mail: Manuel.lago@ecologic.eu

M. Lago et al. (eds.), *Use of Economic Instruments in Water Policy*,
Global Issues in Water Policy 14, DOI 10.1007/978-3-319-18287-2_28

water governance. Through the revision of the cases presented in the book this chapter stresses the difficulties as well as the importance of building fact based evidence about the virtues of EPIs. In the absence of that, EPIs selection and design is mostly guided by its presumed rather than its actual contribution to water policy. The chapter also contains a balance of the contribution of EPIs to help improving water quality, reduce scarcity and manage drought risks and to protect and restore river ecosystems and concludes with the main lessons learnt from the wide set of experiences covered all along the book.

Keywords Economic instruments • Water resources management • Water policy • Water scarcity • Water quality • River restoration

28.1 Introduction: EPIs a Means for a Purpose

The experiences revised in this book put the main focus in the ends rather than in the instruments themselves. Any assessment of the convenience of a more extended use of incentives, rather than prescriptions, needs to be based upon the potential of EPIs to make a real contribution to the actual goals of water policy. For that same reason, for example, price levels cannot be claimed to be right or wrong; rather, pricing schemes can be said to be adequate or inadequate to help achieving policy goals (i.e., reducing water scarcity, increasing resilience to extreme events or restoring and protecting the status of water resources).

Although arguments in favour of using EPIs to make water decisions more flexible and adaptable have been put forward, it is expected that such arguments in favour or against an extended adoption of EPIs would have to be based on proven facts and testable empirical evidence. The search for those experience-based judgements is built on the *ex-post* assessment of a significant number of EPIs in Europe and beyond discussed in the different sections of the book.

Nevertheless, reaching a set of strong, precise and easily transferable conclusions is an elusive task. Conclusions depend on many framework conditions, such as the institutions in place and the driving factors behind the EPI adoption. Furthermore, once these conclusions are widened to a more general framework they become contingent and less accurate.

One must also be aware that established instruments were assessed through criteria stemming from a new water policy approach. The set of principles used in the book to judge the outcomes of the instruments analysed were not in place when the majority of the EPIs were implemented and the approach of water policy has changed so far in many essential features.

But the interest of past experiences does not lie in the assessment of existing EPIs for the sake of it but rather on lessons that can be drawn towards a better response to current and future water policy challenges. Many EPIs considered so far were originally conceived to maximize water service flows available to an economic use (such as hydropower or household demand). Even in these cases, as below, there are important lessons to be drawn from past experiences. Therefore, the main concern is not only to sort out the real contribution of prevailing EPIs to the objectives they were designed for but on their potential to serve the goals of contemporary water policy.

Water policy is never defined by just one objective. In fact it is a mix of at least three main objectives: some environmental targets, some development goals, and the guarantee of financial sufficiency. Ideally, any water policy must result in a simultaneous contribution to the three objectives and help improve water resources, foster economic performance, and be financially feasible when not profitable. Yet, the relative importance of each one of these objectives has changed through time. While financial objectives are instrumental, not so long ago developmental objectives were called to play the leading role (and success in water policy was dependent on the success in coping with the increasing water demand from growing urban areas, irrigated agriculture and other water using economic activities). Today, the importance of environmental objectives has been upgraded and success in water policy is measured by the ability to coordinate all the demands of water services in the economy with the improvement and adequate conservation of water sources.

The actual purpose of each particular EPI and its potential to contribute to the specific objectives of water policy – being it environmental, developmental or financial – needs to be recognised. The experiences revised in this book show important contributions of water policy to promote and sustain particular economic activities (such as agriculture, hydropower or tourism development) as well as to make the provision of water and sanitation services financially sustainable. There is a relevant number of what one could call successful EPIs but just a few were linked to significant improvements in the water environment. However, it was identified that there is still room for innovative EPIs designed *ad hoc* to serve environmental objectives and, in particular, to manage the challenge of coping with increasing water scarcity, droughts and flood risk, poor water quality and degraded water ecosystems.

EPIs are never implemented in isolation. As any other policy instrument they are part of a policy mix jointly with command-and-control instruments and, although not commonly, with other EPIs. That is to say that the intended and effective role played by any EPI in the policy mix needs to be considered. This is a good reminder that EPIs should not be seen as a replacement for existing institutions but as a way for them to adapt and as a step forward towards better policy responses to existing water challenges. The real question becomes what contribution EPIs can really make to improve water institutions and current water policy mix. Part of the answer needs to be found in the particular goal to which existing EPIs were designed and implemented. Evidence shows that some EPIs, such as tariffs and charges, have been successful as financial cost-recovery mechanisms and that water markets have resulted in effective means to foster agricultural and hydropower generation. Moreover, no equivalent advances have been experienced so far in the effective contribution of existing EPIs to guarantee the protection of water resources and to deliver positive environmental outcomes. Experience gained in financing and sectoral development can probably be used to improve EPIs purposely designed to cope with managing water resources sustainably.

The remaining part of this concluding chapter illustrates a comparative revision of the *ex-post* assessment case studies presented in this book.

28.2 Why Evidence of Success Is Elusive?

In the evolving and uncertain scenarios where water policy needs to be assessed it may be hard to ascribe observed environmental outcomes to specific policy instruments in place, which is also the case for traditional command-and-control instruments. In practice, though, searching for answers has actually proved to be a real challenge hardly leading to accurate and robust responses and, very often, only to approximate (when not vague) answers. This may include a wide-ranging set of fairly common combinations of lack of data, ill-defined objectives, poorly designed instruments, lack of transparency and many other actual barriers. Nevertheless, the lack of information should not be seen as the only hurdle to effective water policy (and EPI) design and analysis. It is not a valid argument to show the supremacy of command and control over EPIs since both kinds of instruments rely, *a priori*, on the same information basis.

Within this context, some other reasons, identified in the literature (OECD 2011), may be found to be more significant, as they compromise the potential of the emerging interest in water policies buoyed on innovative EPIs. For instance, the fact that policy approaches relying more on individuals' freedom and decisions as a result of rational choice may lead to more uncertain outcomes if compared to legally prescribed and properly enforced actions.

A necessary condition for an EPI to have a direct effect over the status of water resources depends on its effectiveness to change the demand for water services: reducing water use or wastewater loads, installing more effective water use devices, improving water use practices, engaging in water restoration measures, etc. There are two main reasons why no relevant effect over water resources might be captured through the use of the methodological approach applied in the revision of the case study chapters presented in this book: either the outcome was not intended (i.e., the environment is a good pretext to make taxes acceptable and even for rent seeking and regulatory capture) or the outcome was actually intended but the EPI failed because of a wrong design of its delivery mechanism (a flat rate instead of a marginal price, too much moral hazard, no monitoring and enforcement in place, too low prices and too inelastic demand, the one who pays is not the one who cares for pollution or for water use…). This would be the case of a wrong (ill-defined) EPI, but an EPI after all.

A number of results presented help illustrate the rationale to raise doubts about many positive environmental effects that are usually taken for granted. One of them is that water markets always allow for more efficient allocation of water resources without any additional detrimental effect over the environment. Others are that investing in environmental protection is all that is needed to improve the status of watercourses, or that saving water at one point will always reduce water pressures and improve the status of all water bodies.

In what follows we discuss how far we have been able to progress in looking for a precise answer to what contribution to the sustainable management of water resources is delivered by the EPIs analysed in the book and, although the overall balance is basically positive, we place more emphasis on the most arguable issues; at the end of the day, these are also those offering a higher potential for learning.

28.3 What Water Policy Goals Were the EPIs Expected to Serve at the End?

The contemporary perception of what water policy is all about is still recent and not fully adopted by existing institutions either in Europe or abroad. Contemporary water policy consists in a mix of two targets: improving and protecting water resources, on one side, and finding the way to progress in the production of goods and services in the economy without generating additional damaging effects over the environment, on the other. More recently, a looking forward objective of increasing resilience and adapting to a more water uncertain future has been added to the picture. The main emphasis may be placed on one target or the other but in any case failure or success needs to be judged in terms of the real contribution to the sustainable management of water resources.

However, this approach to water policy has not always been in place. Evidence collected in this book shows that the new approach of water policy has not yet been completely assumed by all real-world institutions and stakeholders involved in water policy. In fact, many of the assessed EPIs had been in place long before Good Ecological Status was set in 2001 as the overarching aim of European Water Policy, or the prevalent importance of water ecosystems and the services they provide was realised by the UN Millennium Development Goals, or the most fundamental need to consider water as an economic good was mainstreamed for all different water policy facets in the Declaration of Dublin in 1992. In many of the reviewed case studies these objectives were not recognized as the central criteria for their initial design and objective setting.

The fact that we are assessing "old" material with new (or even emerging) approaches became more than evident in many case studies. For example, even in some recently implemented water policy EPIs (both within and outside the European Union), the intended and actual environmental outcomes had not even been considered as something relevant for the design or the implementation of those instruments (this is evident, for example, in the Chilean water market, see Chap. 19).

For the same reason, intended environmental outcomes are imprecise in many case studies, if existing at all, and information systems originally designed for their assessment were not supposed to provide any relevant information in that respect.

Practical examples do not always fit nicely with the shared perception that EPIs are definitely means to an end. Very often, ends cannot be easily identified (not to mention if it is in terms of the collectively agreed status of water bodies). Clear instruments without any identifiable purpose (at least in what concerns water policy) are nothing more than a rarity. Some EPIs, for example, have been able to survive long after the obsolescence of the original objectives for which they were conceived. See, for example, the water load and the water resource fee in Hungary (Chap. 4), which were already in place before Hungary's accession to the EU, and even to the economic downturn that came along the evolution from a centrally-planned towards a "free" market economy. The survival of these instruments owes more to their convenience to raise public revenue rather than to the social and political commitment to improve water governance and preserve the environment.

The role of these EPIs for the environment is not completely irrelevant (as water prices in Hungary are higher than in other water-abundant countries), but the main lessons to be learnt are mostly related with how a probably well-meant instrument has been gradually transformed to serve purposes that are now drastically different from initial ones. As a matter of fact, these objectives may not even be linked to cost-recovery, since revenues are not earmarked anymore to water works or water conservation measures. The perception, in Hungary but also in the rest of the EU, that the maintenance of such charges will still do some contribution to the environment may be one of the relevant factors explaining its political acceptability.

Although the recovery, preservation and effective protection of water resources are aimed at playing an increasing role in water policy, real-world EPIs are better characterized by a mix of both: the conventional (developmental) and the still emerging (environmental sustainability) objectives.

The former tends to consider water management as an instrument of development policy. In accordance such perception water policy goals are subordinated to development objectives to which water management is expected to contribute to such as energy development in Germany (Chap. 13), and the Po Basin in Italy (Chap. 12); irrigation expansion, as in the water markets in Colorado (Chap. 21), Murray-Darling (Chap. 20) and Chile (Chap. 19); tourism services (or Chap. 24) and land settlement, as in some of the above-mentioned non-EU studies on water allocation mechanisms.

The real difference lies in whether the EPI has resulted in more water to be used in the economy (a legitimate economic development objective) or rather in more water available for environmental purposes (which can reduce scarcity and drought vulnerabilities in the future: a sustainable development objective).

Otherwise, the modern perception of water management upgrades the importance of water policy and is focused on coordinating and accommodating all these sectoral policies into a collective strategy aimed at making sustainable the use and conservation of the available resources.

This distinction is still essential to understand the environmental outcomes intended and actually delivered by any particular EPI. This is clear in the promotion of hydropower in Italy (Chap. 12), where the EPIs largely rely on subsidies that are expected to deliver a better environmental status without jeopardizing the hydropower sector performance. In spite of pursuing the same goal, the design of each one of these EPIs widely differs, and so does the outcome delivered. In Germany (Chap. 13) environmental outcomes depend on the EPI's performance while in the Italian case study (Chap. 12) it hangs on the performance of other command-and-control alternatives (and the EPI's aim is to foster investment in hydropower generation). The first example is closer to modern water policy while the second still gives priority to economic development objectives.

Following the same line of argument, subsidies for drinking water conservation in Cyprus (Chap. 7) are concerned with solving a drinking water supply problem with minimal financial costs, rather than with the restoration or conservation of water resources. Policy communication is also a concern and, in this case, the appropriate meaning of the self-declared goal ("making the provision of water services sustainable") does actually refer to solving the financial challenges of the water utility rather than environmental challenges of the Cypriot economy. The same happens with subsidies to promote the use of recycled water in southern Spain (Chap. 27), which have increased the amount of available resources but have not showed any improvement in the status of freshwater sources.

A number of EPIs have proven the potential of pricing schemes, markets and voluntary cooperation to promote economic development in many areas, but there is still room for improvement to enhance the effective contribution of EPIs to protect the environment and to manage water resources in a sustainable way.

28.3.1 Trade-offs Between Financial and Environmental Objectives

Existing EPIs make evident that, rather than environmental concerns, the potential for revenue raising needs to be recognized as an (if not the most) important motivation to include prices in the water policy mix.

To assess past experiences and also to design workable EPIs it becomes crucial to distinguish between financial objectives (such as cost-recovery and revenue collection), on one side, and economic objectives (inducing socially desirable behavioural changes in order to improve efficiency and sustainability of water use), on the other.

As above, the distinction is not always clear, as most instruments are a combination of financial and economic instruments, but some examples in the extreme may help clarify not only the distinction but also its practical significance. For example, a water-trading scheme is a pure economic instrument (as it changes behaviour in a presumably efficient way), but does not help to fund the public budget. On the contrary, a flat-rate tariff for water is a pure financial instrument (as it collects money) but does not change current water demand. Moreover, public auctions of water use rights and volumetric tariffs are a blend of both financial and economic instruments. The distinction is of utmost importance for obvious reasons: financial instruments that leave behaviour unchanged cannot deliver any environmental outcome and, although they may contribute to make the provision of water and sanitation financially viable, they do not necessarily result in a real contribution to make water resources management more efficient and really sustainable.[1]

In addition, the conditions for an effective price instrument are precisely the opposite of those for a revenue raising tax: the purpose of a price instrument is to change behaviour and thus it should be an ineffective means of raising revenue. Conversely, the objective of a revenue raising tax is to maximise revenue and this requires that the effect of the tax on behaviour is minimal.

Some case studies show interesting trends aiming at transforming financial instruments into real incentives to change water users behaviour. The practical question may be formulated as follows: given a possible choice between changing behaviour and raising revenue, what is the policy preferred option? The answer to this question has also been changing over time. Traditionally, financial instruments were clearly the favoured ones: flat-rate tariffs for water (on a head count or surface basis) were considered as appropriate since they did not require special monitoring equipment.

Precisely because flat rates do not discourage water demand, they are associated with a more stable flow of revenues, which makes them suitable as a reliable cost-recovery mechanism. Irrigation prices in Southern Europe and household water tariffs (Chaps. 27, 7 and 8 in Spain, Cyprus and Italy).

Yet, flat rates may not be efficient (as they may foster squandering), neither fair (as income levels or actual water use may not be taken into account to set prices), but they have been socially acceptable as far as water was not scarce and its costs were not too high as compared to household budget or business turnover.

Yet, things can rapidly change when water becomes scarce and unpredictable or as a result of the implementation cost of more stringent environmental standards. In other words, water scarcity and social preferences may be important drivers in the transition from financial to economic (and environmentally relevant) policy instruments (as stated in the Tagus River example in Spain; see Chap. 18).

At least three of the case studies chapters show, for example, how increased scarcity and higher marginal provision costs can bring to surface the inefficiency of flat rates. If water expenditure becomes relevant in household and farm budgets, responsible users may have the incentive to highlight it through, for example, the installation of a metering device or by accepting to pay a higher unit price in exchange of being charged for its real consumption rather than by the average consumption of all water users. Hence, driven by equity concerns and by individual incentives, the previous financial instrument cannot only become fairer, but also a real EPI with the ability to reduce water demand and improve its allocation in the economy. This story can be illustrated by case studies on water tariffs for irrigation in Emilia Romagna (Chap. 8), the move towards water metering and the progress towards a water budget rate structure increasingly applied by water utilities in California (Chap. 11) or even Israel (Chap. 10).

[1] For the very same reason that pure economic instruments might not be financially sustainable in the long term.

28.3.2 One Instrument One Purpose. Does It Makes Sense as a Sensible Policy Recommendation?

As introduced above the question remains, are financial and economic objectives of water policy clearly different to each other? If aware of the difference, one would not fall in the common mistake of ascribing the effective outcome delivered by one instrument to another one.

For example, since the Council Directive 91/271/EEC of 21 May 1991 concerning urban wastewater treatment was passed, EU Member states have the obligation to control wastewater and to treat point effluents. The environmental outcome delivered is then attributable to the installation of these plants and not to the cost-recovery mechanisms chosen by each country to guarantee the operation and the renewal of these plants (most of them, especially in Southern Europe, built with Cohesion and Structural funds not recovered by prevailing effluent charges).

It seems irrelevant, though, whether fees, taxes or other financial instrument are in place since the building and operation of treatment plants is not a voluntary decision (thus, it does not depend on any specific financial instrument).

Should effluent charges have any environmental outcome, one would need to search for it in its effect over the demand for water services. As a matter of fact, because of stringent environmental standards as a result of European Directives and financial instruments implemented to support them, this is the main reason why water prices have increased all across Europe (i.e., someone needed to pay for the required upgrade in WWTPs, monitoring schemes, etc.). Paradoxically, an alleged quality instrument is demonstrated to have actually been the most powerful quantity instrument (for example, more than 90 % of price increases and of the associated water demand reductions in Spain are due to the internalization of new wastewater treatment costs).[2]

If financial and environmental objectives are different to each other the more reasonable policy option is a mix of two instruments, each one conceived to serve one of both purposes.[3] This book provides an example of this kind of innovative instruments; in this case, an operational mix of financial and economic incentives. The former is intended for funding the real objective of the instrument mix; the latter to induce changes in behaviour in order to promote the environmental objective of water policy. A financial instrument (a water tariff) is intended to collect the money required to induce the improvement in water quality (through a set of subsidies to foster given practices).

[2]Although operation and maintenance cost-recovery levels differ from place to place those investment costs covered by cohesion funds have not been recovered at all (as they are actually allowed by the Spanish law: legally that would be the equivalent to a double levy for the same service – one to the EU and the other to water users – of the same costs; Maestu and Villar 2007).

[3]In the previous example there is also a policy mix of one instrument to improve water quality (the compulsory setting of wastewater treatment capacities) and another for cost-recovery (a kind of effluent charge).

It is the case of the water abstraction charge combined with compensation payments in Baden-Württemberg (see Chap. 5).[4] In this case, the genuine environmental outcome to be assessed was that of the good farming practices inducing subsidies while that associated with water pricing (the supporting financial policy instrument) is of course very relevant but for a different reason: as a support to make the whole policy acceptable and financially feasible.

28.3.3 *From Prescription to Actual Choice: The Critical Importance of a Sound Design*

The real difference between command and control and EPIs, as alternative or complementary instruments for water policy, is that the latter relies on chosen rather than on legally prescribed individual decisions. Hence, the main purpose of any EPI must consist in adapting the diverse individual decisions of households, firms and farms (driven by their own knowledge, budgets, tastes and, basically, by their individual interests) to the courses of action that may be considered as the most appropriate from a social or collective welfare perspective. Water authorities, presumably representing the common interest and other stakeholders have the ability to decide on the rules of the game and then to direct decisions of all the individual agents.

Any incentive scheme has two essential requirements to be a practical one:

- The first one consists in widening the array of decisions available for each water user involved (i.e., buy different amounts of water, sell and buy water use rights, deliver a higher or lower effort to prevent water degradation, etc.), which are attractive enough so that agents are interested in taking part in the game (in the abundant mechanism-design literature this is called the participation or the rationality condition; see Börgers 2010).[5]
- The second condition, and the really important one, is that the action chosen by the agents must result in a real contribution to the policy goals (e.g., the efficient and sustainable use of water); this is the so-called incentive compatibility condition.

Experience shows that many poorly designed EPIs might comply with the first but not the second condition.

[4] This is a good example of the "one instrument for each purpose" golden rule recommended for the optimal design of incentive schemes. In this case, the desired behaviour is furthered by subsidies and financing is pursued through water prices. It would be a real mistake to understand the water price as a kind of a quantity instrument (which would not make any sense in a water-abundant country like Germany: water prices in excess for water provision costs are not exclusive to water scarce countries).

[5] The participation condition means that people must, for example, buy water, accept a subsidy and be willing to engage in water trade. But also that people prefer the alternatives proposed by the EPI rather than maintain the status quo. All that means that water users have something to gain by participating in the game proposed by the EPI.

In dry areas where close sources have been already exhausted anyone is glad enough to accept non-conventional recycled or desalinated water for free, not even with the need of a subsidy (so that the participation condition of the incentive scheme is fulfilled). This is not necessarily a step forward towards reducing water scarcity or to recover freshwater sources (and the incentive compatibility condition fails). See, for example, Chap. 27 on Spain. Along this line, experience shows that firms might have rather obvious incentives to voluntarily accept the installation of water saving devices specially if they are financed by the water authority (participation), but it does not automatically lead to lower water consumption as the water saved can be used for more water-intensive crops or to increase the irrigated area (as in the case of many subsidized programs to modernize the irrigation infrastructure in Spain or in Chile).

The ecologically friendly electricity programme in Germany (Chap. 13) provides incentives for energy companies to install costly infrastructure (e.g., fishing passability), especially as that is compensated by a 20-year flow of guaranteed revenues. However, their proper maintenance and operation is not ensured (as current behaviour is not monitored and the only enforcement criterion is the installation of the infrastructure).

Likewise, water trading is supposed to be a means to increase the overall allocation of water amongst places and economic activities. Provided transaction costs are not exorbitant,[6] the participation condition is more likely fulfilled when there are important differences in the marginal value of water giving place among potential buyers and sellers and mutually beneficial agreements are feasible (so that the participation condition is met).[7]

Nevertheless, in many water right trading schemes, incentive compatibility is not guaranteed. Representative examples show that the option to trade water may put into use a substantial amount of resources that in the absence of trading opportunities would have remained in Nature. In this case, water markets can paradoxically contribute to increased water scarcity and to spread water scarcity along the territory. This is already shown in the water transfers in the Middle Tagus in Spain (and it is even more evident in the Henares irrigation district as shown in Chap. 18), but it has also been proven, at a much higher scale, in the Murray-Darling basin in Australia (Chap. 20).

[6] Both the Chilean (Chap. 19) and Australian (Chap. 20) markets have a similar system of pro-rata share of water stocks, intended to reduce transaction costs and to eliminate opposition to transfers.

[7] In all case studies on water markets, one may have expected major differences in water prices across uses and that these differences might persist beyond what can be explained by asymmetries in conveyance costs and water quality, suggesting that water markets may have not developed fully enough to optimize efficiency gains. Yet, the comparison of Chaps. 19, 20 and 21 is not straightforward.

28.3.4 Are There Ways to Compare Intended and Observed Environmental Outcomes and Reach a Robust Conclusion on the Benefits of Any EPI?

Many water policy EPIs and command-and-control instruments are implemented and advocated on the basis of presumed rather than real outcomes. As proven in the chapters presented in this book, the environmental effectiveness of even the most popular and better-accepted examples is subject to serious shortcomings.

In many cases, even when the desired outcomes were observed, changes are hard to link to the EPI in place. Water demand has been severely reduced indeed in Hungary but the best candidates to explain such a trend are, first, the economic downturn and, second, the more stringent water regulations implemented for Hungary's accession to the European Union. The water load fees only played a marginal role (Chap. 4).

In a different context, how much of the recent expansion of hydropower in Italy is a response to a combination of peaking oil and coal prices and the implementation of the Kyoto protocol and how much to the substantial subsidies paid for renewable energy? The environmental outcome delivered is in the answer to this tricky question (Chap. 12).

Even when EPIs apparently fail, things might have been worse in its absence: without the Danish pesticide tax (Chap. 6) diffuse pollution would have been worse. The failure to reduce the Water Treatment Index only shows that the pesticide tax was only able to partially compensate for the powerful incentives to extend agricultural surfaces and yields resulting from high commodity prices. Additional difficulties can be found in Chap. 24, where a particular geological configuration generates a 20-year lag between implemented actions and the assessment of environmental outcomes.

In addition to that, EPIs are applied in combination with other instruments and the observed outcome is the result of a policy mix. Design analysis tends to fall in the embedding mistake when considering that all the benefits of improving ecological status can be attributed to the EPI (Chap. 8), which may be as fallacious as considering the EPI's outcome as irrelevant. A better option consists in recognizing the individual changes in behaviour that were induced by the EPI, and the associated changes in pressures and environmental impacts.

In that case there would also be some scope for contradictory results: some initiatives may be failing because of the success rather than the flaw of the overall water policy. The voluntary agreement to restore the river regime in the lower Ebro is currently being revised once the effectiveness of controlled floods to remove the invasive algae (and other microorganisms) disturbing the operation of power plants is lower than only ten years ago (Chap. 26). As a result of that, the power company is now less interested than before in the agreement. A plausible reason might be the rapid improvement experienced by water quality as a result of the installation of sewers and water treatment plans all along the river (despite flowing water being still low, macrophytes can now grow stronger).

In many cases, changes in behaviour guarantee the reduction of water pressures. The case is more evident when these pressures are directly observed as in the certification of hydropower plants or the observed water quality before and after the installation of a water treatment plant. Water fees in Germany (Chap. 3) have effectively serve as a financial instrument to fund the capital operation and maintenance of water treatment plants but have also acted as an economic instrument given its potential to increase water prices and to reduce water demand (and water loads). The same situation can be found in most EU countries as a result of the implementation of the urban wastewater treatment Directive (91/271/EEC) and progress in cost-recovery.

Another important advantage over command and control relies on its capacity to manage social conflicts[8] while opening the option for mutually beneficial agreements amongst stakeholders (such as in Chaps. 26, 23 and 17).

28.4 Are EPIs Suitable Instruments to Cope with Current Water Policy Challenges?

Above all, the real question is whether EPIs, when properly designed and implemented, can make a real contribution to improve water policy decisions. In particular, to what extent they are able to cope with the real challenges of water governance. Some of them are of a global scale, like coping with climate change and the severe water uncertainty linked to it; others are just local (site-specific), such as the degradation of water sources nearby. Given the variety in the nature and scale of water challenges, the still preliminary answer to how EPIs can contribute to their solution is organised in three particular categories: preventing the degradation of water quality, tackling increasing water scarcity and improving impoverished river ecosystems. Some categories for which there are not still EPIs in place, such as global warming, and some others, for which no particular EPI was considered within this book's choice (such as flood and drought risk), are not discussed.

28.4.1 What Is the Potential of EPIs to Reverse the Degrading Trends in Water Quality?

The chemical quality of water in surface and groundwater sources depends on both, the natural conditions of the river basin and the pressures exerted by humans and their economic activity. In an integrated river basin framework one needs to recognize that the measures able to improve that quality are not only those *end-of-pipe*

---

[8] In Chap. 17, offsetting to solve salinity problems in Australia is assessed as a cost-effective alternative in comparison to the conventional regulatory approaches (i.e. standards), as it allows environmental improvements to be achieved at a seemingly significant cost reduction.

alternatives designed to reduce pollution loads. Measures primarily designed to save water and to reduce abstractions have a real effect over water quality as they facilitate dilution, oxygenation and transport of pollution loads. Similarly, water quality is also the result of the self-treatment potential (i.e., natural assimilation capacity) of water bodies. There are obvious links between quantity, quality and system restoration measures that ought to be considered in an integrated approach.

Nevertheless, for the purpose of extracting some broad lessons from the case studies considered in this book, the potential of improving water quality by managing point and non-point pollution sources is discussed. Each one of them poses a distinct challenge for water policy: *point-source pollution* is a single and identifiable kind of harmful activities that need to be controlled by focused regulations and precisely defined licenses concerning the volume and content of the effluents discharged into the water environment. Individual decisions with respect to environmental outcomes are only possible once certain safe minimum standards are in place. Pricing schemes are almost exclusive to countries which have already implemented water policies and markets over water point pollution loads are not an option as significant scale economies of wastewater collectors and treatment plants convert them into a natural monopoly without a possible choice for households and other water users. Not surprisingly, normative prescriptions play the dominant role. These ideas are not in contradiction with the existence of instruments such as load fees (in Hungary, Chap. 4), point pollution charges, effluent taxes (in Germany, Chap. 3) and trading mechanisms in Ohio, USA (see Chap. 15).

On the other hand, *diffuse pollution* represents a particular challenge for water policy. The main reason is that the resulting quality of concerned surface and groundwater bodies is the consequence of many individual actions scattered throughout different places. In addition, decisions in a particular place might affect distant water bodies in a way that is not completely understood by available land use and impact assessment models and tools. Individual actions are in general unobservable and in practice it is almost impossible to determine how much any farmer or other water user contributed to the observed degradation in a water body. This is why tariffs (as in the Danish Pesticide Tax, Chap. 6) or use rights, for example, cannot possibly be defined on the effective contribution to nitrate concentration in a river stream. EPIs, when feasible, are mostly addressed to change behaviour patterns which are far from but meant to be closely linked to actual pressures exerted by water users over the environment (as in Chaps. 5, 23, 24, and 25). The case of salinity offsets in Australia (Chap. 17) can be considered in a different group.

EPIs aimed at reducing *point-source pollution* must be understood in the context of water policy development in each country. In Hungary (Chap. 4), effluent loads basically respond to the adaptation of environmental standards required in the accession process to the EU. The water load fee, implemented in 2004, may have played a relevant role in reducing water demand after it was transferred to consumers in higher water tariffs, and also indirectly it might have reduced pollution loads. Even the financial contribution to water policy of this instrument is disputable as proceedings make part of general public budget, and the improvement in water quality is a proven outcome of the installation of wastewater treatment plants mostly funded by EU cohesion funds.

In the EU, wastewater charges have been introduced as a cost-recovery mechanism along with the implementation of the Council Directive 91/271/EEC concerning urban wastewater treatment. The technical requirements, the volume and the composition of effluents permitted for any place, depending on the status of receiving water sources, were already defined. Little scope, if any, was left to individual decisions. Both the contribution of these prices to improve the ecological status of water bodies and to the success of water management plans to implement the WFD (some of them still to be approved) are uncertain. In Spain (and also in Hungary), water pollution is priced by volume and according to the effluent allowances are granted to each wastewater plant, there is no way to reward (through lower charges or fees) improvements in the quality of the effluent beyond what is legally prescribed. In both cases the instrument might have contributed to increase water prices and to reduce water demand (paradoxically performing better as a quantity rather than a quality instrument), and concerning water quality they may become an important element for the financial sustainability of sanitation services provided by water utilities in both countries.

In a similar way the German effluent tax (Chap. 3) is one piece of a policy mix, which also consists of discharge permits, pollution limits and mandatory technological standards. The policy mix, as in the previous two examples, has been mostly successful in obtaining its objectives but the real contribution of the effluent tax is impossible to single out. The tax is also based on permitted effluents both in volume and composition in such a way that incentives for further pollution reduction without technological change are missing. However, at least three complementary instruments may have played a significant role in reducing pollution and increasing the dependability of water quality targets. First, monitoring systems help verify that pollution limits are not surpassed and to set non-compliance fines that provide an incentive to stay within limit values. Second, along the implementation process three-quarters of private enterprises and two-thirds of municipalities had increased, accelerated, or modified their abatement measures for water pollution in anticipation of the charge. Finally, although the role of the effluent charge to reduce pollution substantially faded once the prescribed limits were obtained, firms still have the option to prove they are below these limits and are subsequently eligible for a tax rebate. The incentive has worked better for private than public utilities.

A trading mechanism is only feasible in exceptional circumstances for point pollution. The basic requirement consists in having many pollution sources within a common water body, so that a unit of pollution in one point can be exchanged for a given amount of pollution in another. As shown in the water quality trading (WQT) programme in Ohio (Chap. 15), this exceptional trading schemes may be of use to share water treatment burdens amongst the different sources reducing compliance costs whilst guaranteeing desired pollution limits. Trading can also facilitate a smooth transition to more stringent pollution limits. The pilot scheme allows for improving the design of the instrument and further results are expected to assess the transferability of these results.

The main problem of *diffuse pollution* is that it is almost impossible to ascertain how much any farmer or other user contributed to the observed degradation in a water body. The consequence is that tariffs are regarded as an appropriate EPI, as in the Pesticide Tax in Denmark (Chap. 6). This tax does not distinguish among locations and is homogeneous for the whole area of reference. The effectiveness of the tax is assessed according to the Treatment Frequency Index (TFI), a simple but limited indicator that measures the quotient between the fertilizer applied and the amount required by existing crops at a national level. The TFI shows that water policy has clearly failed to deliver the intended outcome of stabilizing the TFI at 1.7, but things may have been worse in the absence of the effluent tax (the highest pesticide tax in Europe). In such a context it is impossible to know whether the reason for this failure is the low price-elasticity of fertilizer demand or that despite being elastic its positive effect has been compensated by scaling commodity prices, high biofuel demand or any other factor explaining agricultural growth or other. Even if a TFI lower than 1.7 had been reached, this could not have been interpreted as any successful indicator at all. It is only an average indicator (compatible with water bodies in poor conditions) and it is still not clear what effective environmental outcome a 1.7 TFI would deliver. The main lesson is that tackling diffuse pollution by taxing proxies for pollution and using far but practical indicators to assess its success is associated with high uncertainties about its effectiveness.

An alternative lies in approaching diffuse pollution from the perspective of managing land and water ecosystems as economic assets and finding the way to reconcile the diverging pressures exerted by their users. Rather than taxing the use of an observable input with unobservable consequences over the environment, this alternative is about adapting the observable practices of water users in order to maintain or protect a desired status of a river basin. Improvements in the status of water bodies are economically feasible as far as the willingness to pay of potential beneficiaries of such improvements is higher than the compensation required for those in charge of delivering them. Four chapters follow this logic: Chaps. 5, 23, 24 and 25.

In the Dorset case study (Chap. 23), 52 out of 74 farms made voluntary cooperative agreements (with an initial economic compensation) with the regional water utility regarding implementation of measures to abate nutrient pollution, reducing water drinking provision costs and increasing water security. In Baden-Württemberg (Chap. 5), compensatory payments are financed with water abstraction charges. In the Evian case (Chap. 24) the private company helps farmers complying with standards and adopting sustainable practices. Additionally, The New York City Watershed Agricultural Program (WAP) (Chap. 25) has been able to define individual Whole Farm Plans (WFP) of 416 farms and to find the financial agreements to guarantee their adoption.

A number of logical arguments make the environmental outcomes delivered by these alternatives disputable. Effectiveness is still to be proven in Dorset, alternative explanations do exist for reduced pollution in Baden-Württemberg, command-and-control constraints might have played a dominant role in ensuring the quality of the protected Evian ranges, and there is not a plausible counterfactual to demonstrate that the reduction on the phosphorus pollution experience in New York could not have been obtained anyway. Nonetheless, these are all success stories and, in spite of the lack of robust empirical evidence, it is more likely that reasons rest in some important advantages over alternative EPIs (as the above-mentioned product tax) and command-and-control instruments.

EPIs can help enhancing the economic value of on-site environmental services provided by water resources. For example, in Dorset, adoption of good practices has cleaned out peak pollution events. Likewise, individual farmers do not have the skills or financial resources to identify best practices (especially when they depend on local circumstances – like soil types, moisture content or other agronomic factors) and the collaborative scheme can reduce information costs facilitating the coordination (as in Chaps. 23, 24, and 25). All this might not have a discernible impact in the short term but definitively it is a step forward to reduce uncertainty over the long-term status of conservation of water bodies (degradation risks have been severely reduced in the cases considered).

Even if the environmental status remains stable, the transition in farm production allows for a higher welfare level making the financial compensation redundant (in Chap. 5 the collaborative scheme proceeded after farmers stopped receiving side payments). In Evian, Dorset and the Cat-Del basin in New York cooperation is a means to empower local users with the conservation of a natural and economic asset, which outsiders depend on but, thanks to the cooperative agreement, that is also critical to the sustainability of their economic activities. All these reasons are difficult to experimentally link with data but are powerful arguments, however, in favour of long-term positive environmental effects and contribute to reduce uncertainty over the conservation of natural assets.

Finally, a special mention needs to be made to salinity offsets in Australia (Chap. 17), where reducing salinity in different points can compensate for excess in salinity in one point. Although the scheme allows to maintain and eventually reduce salinity overall, command and control is still required to locally monitor excess salinity. The EPI is intended to provide water users with an alternative to adapt decisions to increased salt loads and more stringent regulations and has also served to finance restoration projects with the potential to reduce salinity loads. In short, salinity rate threats in Australia have been abated over the period, and various salinity mitigation initiatives, including offsets, may probably claim at least some credit for it.

28.4.2 What Is the Potential Contribution of EPIs to Cope with Increasing Water Scarcity?

Managing water quantity means coping with the challenge of combining welfare increases and the production of those goods and services provided by the economy with the limited ability of water ecosystems to provide those activities with a continuous and dependable amount of required water.

The true question in this respect seems to be whether EPIs can make a real contribution to deal with excess demand of water services (water scarcity) and with the uncertainty in water provision (drought risk). The strategy adopted to handle these demanding tasks includes one (or a combination) of the following intermediate targets to which EPIs are expected to make a significant contribution:

- Improving water resource allocation everywhere and among economic uses in order to increase the potential of the economy to improve the provision of goods and services within the limits of available water resources (such as in water markets in Chile, Australia and Colorado (USA), assessed in Chaps. 19, 20 and 21, respectively).
- Making water allocation to alternative uses contingent to available resources every time in order to reduce welfare losses and provide a better response to droughts (Chap. 18).
- Increasing the technical efficiency in the production of water services so that they can be obtained with lower withdrawal rates from freshwater sources (by improving irrigation techniques, reducing leakages in water distribution networks, etc.). This can be the result of EPIs especially aimed at this goal (as in Chap. 10 in Israel and 8 in Italy) or an indirect effect of other EPIs (in Ohio, USA, New York, USA and California, USA; see Chaps. 15, 25 and 11).
- Replacing water provided by the natural environment by alternative resources intensive in human-made capital or non-conventional water sources such as reused or desalinated water (See Chap. 27 in Spain).
- Reducing water demand from households, agriculture and manufacturing. This is the case of water metering in Italy (Chap. 8), the tailoring of rate structures in California (USA) (Chap. 11) and water taxes in Italy (Chap. 8).
- Some additional instruments are mainly aimed at subsidizing desired behaviour, such as the subsidies for drinking water conservation (Chap. 7 on Cyprus) and the incentives to promote the use of recycled water (Chap. 27 in Spain).

Normative instruments have traditionally pursued these intermediate objectives of water policy but, as this book's case studies make clear, incentives are playing an emerging role.

Experience with water markets shows their significant role in finding mutually beneficial agreements between buyers and sellers, thus increasing the production of goods and services and making water trades a convenient instrument to promote different economic activities. These development objectives were the main driver in the original adoption of current water-trading schemes and concerns on their environmental outcomes is still an emerging issue.

Evidence shows that trading schemes may have increased pressures over water resources (by putting into use water that might not have been used in the absence of markets). This has been the case of Chile and the Murray-Darling basin in Australia (Chaps. 19 and 20), where available resources are said to be over-allocated (although there is no empirical evidence on this for Chile, where this statement would accept a number of non-minor nuances). On the other side, physical interactions between water bodies along a river basin and externalities that may arise still make it difficult to find a set of property rights that can be efficiently traded. For instance, in Chile increased activity in consumptive water use markets has generated increased conflicts with downstream users due the effects of water use rights over return flows.

Voluntary trading can play a critical role in stabilizing the economy and in providing an effective drought management alternative, provided all stakeholders are involved and provisions are made to compensate for third-party effects (Chap. 18).

Water scarcity on its own is a driving factor to increase water efficiency. Scaling up marginal costs makes the reduction of leakages in urban distribution networks more profitable and better irrigation devices are more advantageous when they avoid paying for more expensive and less dependable amounts of water (Chap. 10).

Modern technologies allow replacing freshwater for alternative sources opening up the opportunity to recover overexploited sources. However, it gives priority to increasing available resources rather than reducing pressures over the environment (Chap. 27). Experiences in Spain show that farmers are willing to accept alternative resources as buffer stocks to cope with droughts but reluctant to give up freshwater use rights in exchange.

Water demand management alternatives become more attractive when scarcity and more stringent environmental requirements increase the provision cost of water. In all these cases EPIs can be built upon the willingness of water users to adapt behaviour to the new circumstances (Chap. 8). It is difficult to say if lower consumption levels happen because of the EPI or just because people with meters already used less water before meters had been installed.

Water taxes are also useful to reduce water demand (Chap. 8), as well as a parallel improvement in household access to water supply and sanitation. However, there is a lack of sufficient and reliable data and further evidence is needed to confirm their actual effectiveness.

Abstraction fees have also been common, although their outcome has been by far less successful. More innovative approaches for water demand reduction such as the rate structure tailoring in California (USA) (Chap. 11) have been applied. Although it is generally regarded as a success, its applicability is heavily burdened by information availability and monitoring costs.

Subsidies for drinking water conservation (Chap. 7) were implemented to adapt drinking water demand to production capacity, rather than with the ability of the environment to provide the required resources in the long term. The EPI is compatible with subsidizing the construction of boreholes, which may be a success in avoiding financially costly alternatives for drinking water in a water-stressed country, but it is certainly a disputed instrument for promoting the sustainable use of surface and groundwater. The same can be said, for example, of incentives to promote the use of recycled water (Chap. 27).

28.4.3 Restoring River Ecosystems

There are EPIs that use voluntary agreements between parties at stake that can play a relevant role in river restoration programmes (to target specific environmental problems and specific changes in operation to improve environmental status of water bodies), as long as cooperation is designed in such a way that all parts can derive mutual benefits from it (Chap. 26).

Other EPIs use subsidies and aim at improving local river conditions by setting the necessary incentives to develop environmentally friendlier hydropower generation (Chap. 13). The impact of these EPIs on rivers remains unclear. In spite of the several measures and actions that have been taken to improve the water status in both case studies, there are no comprehensive studies showing the overall change in the ecological status of the water bodies. However it can be concluded that, at least for the German case (Chap. 13), there was an improvement of water bodies next to hydropower plants fulfilling environmental conditions, although again the magnitude cannot be exactly determined.

Finally, there is scope for subsidies whose objective is not necessarily targeting the mitigation of negative environmental effects from hydropower installed capacity, but basically the extended use of the technology, supported by command-and-control measures (Chap. 12). Although not the very instrument, but the policy mix it belongs to, can be considered a real contribution to the ecological status of improvement of water bodies as required by the WFD.

28.5 Some Lessons Learnt

EPIs are still part of a new approach to water policy. Stavins (2001) described "market-based instruments" (just a type of EPIs), as a "relatively new set of policies". More than 10 years later, they can still be seen as new to a large extent. This remains fundamentally true despite their recent upsurge. Although the evidence presented in this book is extensive, this should not leave the reader with the impression that EPIs have replaced, or are close to replacing, the dominant command-and-control approach to water management. Furthermore, even in those places where these "new" approaches have been used in a very genuine form and somewhat successfully (such as water quality trading systems in the USA or water use right markets in Chile, Australia or again the USA, for instance), they have not always performed as anticipated.

Information Quality, a Critical Factor but Not an Alibi There remains a great deal of uncertainty especially over the potential role of pricing-based EPIs, and water use right trading systems, for water demand management and allocation. EPI-WATER is aimed at shading light on this 'twilight'. To date, it is clear that reducing uncertainty would be highly contingent on the improvement of information systems and the availability of proven facts and testable empirical evidence. However, one should not conclude that nothing relevant might be said because of the lack of information, since this is also an essential characteristic of the assessment of command-and-control instruments. Decision-making on water management will definitely be improved with better information but cannot be dependent just on that. Information, after all, is not the only (scarce) element of decision-making.

Neither Generalization nor Relativism Conclusions hereby presented cannot be generalized to all EPIs and situations. This is but the synthesis of conclusions after the ex-post assessment of a few case studies. On one hand, though, it must be recognized that such comprehensive assessments are not very recurrent in the literature; on the other, further research could be done to draw some conclusions on the transferability of some of these experiences.

Failure of an EPI Does Not Necessarily Mean a Flawed EPI The review of experiences based on pricing (including taxes and fees), reveals that while they can have some effect in reducing water use, it is still not clear, that they are always more effective in doing so than other instruments. This does not preclude anything about their soundness but rather points out the need to emphasise on the delivery mechanism (that is on instrument-design issues). The failure of an EPI to meet its pre-determined objectives is not necessarily equivalent to a flawed EPI but the symptom of a bad design (not to mention other institutional variables).

Different Objectives of Water Policy EPIs are argued to be able to fulfil one or more social objectives: financial sufficiency of water policies, economic development, and environmental sustainability, amongst others (i.e., equity concerns). This implies that they may play different roles: an incentive function, a fiscal or financial one (not necessarily the EPI itself but a linked financial instrument), part of a liability regime, etc. Thus, the choice and design of the EPI should depend on which functions the instrument is desired to address. In the restricted conditions of a perfectly competitive market the price that falls out of the market, for instance, is argued to fulfil all three objectives. But in reality it may be preferable to address the three different functions separately and not to assume the best approach for one is the best approach for all.

One Goal, One Instrument: A Sensible Approach Cost-recovery (i.e., revenue raising) concerns have traditionally been the primary driver of reforms to water pricing. As the reader may have seen in the above analysis in this concluding chapter, though, despite being a legitimate social objective, cost-recovery is not an economic goal but a financial (thus instrumental) one. Financial goals should be clearly distinguished from economic incentives, aimed at inducing chosen behavioural changes. Cost-recovery mechanisms do emphasise on revenue collection (e.g., who covers fixed costs, what tariff structure is more convenient to maximize income, etc.). Hence, the way these questions are addressed does not necessarily have anything to do with efficient pricing, whose motivation should be to optimise water use and social welfare.

High Potential for EPIs Aimed at Environmental Objectives A relative success can be claimed for on the grounds of cost-recovery and economic development (i.e., hydropower expansion); however, results are definitely more uneven as to their environmental outcomes. This poses a challenge for future research, since there is room for innovative ad-hoc EPIs to meet specific environmental objectives: tackling water scarcity and droughts, managing flood risks, improving water quality, restoring damaged water ecosystems, etc.

The Divergent Role of Information in Instrument Comparison Transaction costs have precluded some actions which might otherwise be desirable from an efficiency perspective e.g., charging domestic consumers the actual cost of wastewater collection and treatment; that is, according to volume and load of pollutants. Conversely, some charging systems (e.g., charging surface water runoff by the volume produced) have only become possible with the reduction in transaction costs e.g., the availability of GIS databases of land use. In water management, information has typically been expensive and can be considered as part of transaction costs, EPIs typically require more differentiation (and hence more information than command-and-control systems). In a complementary sense, EPIs save information as well (i.e., setting a price and observing behaviour is not that demanding, markets might be a way of revealing preferences, etc.).

A Critical Question: The Definition of Water Rights A critical issue in the implementation of markets is a clear but nonetheless full definition of water rights or entitlements and of the associated risks. It is also important to account for the interactions between surface and groundwater resources (no specific provisions can be found in many of the assessed systems). Setting a trading scheme can be an answer to managing competing water demands, especially in scarcity-prone areas. Main concerns, though, remains on third-party effects (for instance, linked to the definition of rights on water return flows) and environmental externalities, as well as transaction costs (which should be minimized but not neglected, since they play no minor roles in some occasions).

The Paramount Importance of the Policy Mix EPIs are usually only one element of a larger policy mix. They are often combined with other policy instruments (being EPIs or not), into a water policy or management strategy. EPIs are therefore never implemented in isolation and should be assessed as a part of larger policy packages. Innovative EPIs do not need to be 'new' EPIs but rather better designed (but well-known) instruments or the combination of a number of them.

Economic Incentives for Behavioural Change Pricing and trading schemes are not always easy to implement (due to high transaction costs, equity concerns, social acceptability, institutional complex demands, etc.). The same could be said of payments for environmental services, which are also difficult to implement in societies with advanced water regulations and institutions, especially in EU countries where water resources are public-domain assets and where private (use) rights can only be issued under certain conditions. Side payments for good practices are not easy to accommodate within existing regulations in most EU countries and will require important legal amendments besides other transaction costs. All these considerations may lead the reader to think of a reduced scope for EPI implementation. However, this assessment shows that the potential for voluntary agreements based on economic incentives is high. At the end of the day, what defines an EPI is not an explicit monetary payment (although most of them will imply one), but the economic incentive to modify behavioural patterns regarding water use.

References

Börgers, T. (2010). Rational decisions. *Journal of Economic Literature, 48*(1), 153–156.

Maestu, J., & Villar, A. (2007). Precios y Costes de los Servicios del Agua en España. Informe integrado de recuperación de costes de los servicios de agua en España, artículo 5 y Anejo III de la Directiva Marco de Agua. Madrid: Ministerio de Medio Ambiente.

OECD. (2011). *Water governance in OECD countries: A multi-level approach* (OECD studies on water). Paris: OECD Publishing.

Stavins, R. N. (2001). Experience with market-based environmental policy instruments. Resources for the Future, Discussion paper 01–58. Washington, DC: Resources for the Future.

Index

M. Lago et al. (eds.), *Use of Economic Instruments in Water Policy*, Global Issues in Water Policy 14, DOI 10.1007/978-3-319-18287-2

R

S

T

U

V

W